MECHANISMS of HARM

Mechanisms of Harm

Lori Weintz

ISBN (digital): 9781630692612
ISBN (print): 9781630692629

Brownstone Institute

MECHANISMS of HARM

Medicine in the Time of Covid-19

LORI WEINTZ

BROWNSTONE
INSTITUTE

CONTENTS

Introduction 1

1. Wait! There's a pandemic? 3

"We can make radical changes to our lifestyles:" 4
Bill Gates – Anticipates vaccinating the whole world: 5
Pandemic wargames: 5
The World in 2019 - Poverty and Hunger at their Lowest Ever: 6
Moderna and worldwide vaccine production factories: 7
WHO wants top-down control in future pandemics worldwide: 7
Fauci calls lockdowns "inconvenient," says a vaccine is coming soon: 8

2. Remdesivir 11

Remdesivir was removed from the Ebola trial because of toxic effects: 11
National Institutes of Health employees profit from drug patents: 12
The Rise of Remdesivir despite Poor Performance: 13
Dr. Fauci chooses remdesivir as the "standard of care" in the U.S.: 15
May 1, 2020 FDA grants Remdesivir Emergency Use Authorization: 15
Remdesivir costs less than $6 to manufacture, but costs $3k for treatment: 16
Remdesivir is nicknamed "Run-death-is-near!" 17
Patients on remdesivir frequently experience serious adverse events: 18
World Health Organization advises against remdesivir to treat Covid-19: 18
Riches from Remdesivir: 19
April 25, 2022 FDA approves remdesivir for infants and children: 19
Covid-19 does not pose a serious risk to children and young people: 20
Unprecedented government interference in the doctor/patient relationship: 21
Remdesivir, known to damage kidneys, approved for patients with kidney disease: 22
Lawsuits against Gilead for downplaying clinical dangers of remdesivir: 22

3. Hydroxychloroquine (HCQ) 23

NIH "study" says HCQ not effective in Covid-19, increases mortality: 23
Dr. Vladimir Zelenko uses HCQ to successfully treat over 2,000 Covid patients: 24
Dr. Didier Raoult successfully treats over 1,000 patients with HCQ/AZ combo: 25
HCQ suddenly classified as a "poisonous substance" in France: 25
Doctors who question the official narrative are persecuted: 26
Dr. Meryl Nass reviews toxic use of HCQ in Recovery and Solidarity trials: 27
Corruption of the medical journals: 27
Toxic doses of HCQ were administered to trial participants: 28
Killing Hydroxychloroquine: 30
June 15, 2020 – FDA revokes EUA approval of HCQ to treat Covid-19: 30
FDA's withdrawal of EUA for HCQ leads to patient deaths: 31
Profit motives behind killing HCQ: 31

4. Stages of Covid-19 Infection 33
Severe Covid and the cytokine storms of Wuhan and Delta in 2020-2021: 34
Covid-19 is primarily a risk to the elderly and those with chronic illnesses: 34
Covid Wuhan and Delta variants were a serious disease for the vulnerable: 35
Standard Protocol: "Go home, but come back to the ER if your lips turn blue:" 36
Neglect was often part of the "Covid treatment protocol" in hospitals: 37
Discarding the Hippocratic Oath during Covid: 38

5. Real Doctors, Not Bureaucrats, Find HCQ Successfully Treats Covid-19 39
July 2020 - Doctors state HCQ and other off-label drugs could end the pandemic: 39
Character assassination of dissenting doctors: 40

6. Ivermectin is "Powerfully Effective" Against Covid-19 43
Dr. Pierre Kory testifies before the U.S Senate about ivermectin: 44
Early treatment of Covid-19 critical to relieving the healthcare system: 44

7. The War on Ivermectin and HCQ 47
The powers that be didn't *want* an already approved effective Covid treatment: 47
Official propaganda ramps up to suppress over-the-counter Covid treatments: 48
False rumors of hospitals filling with people who took veterinary doses of ivermectin: 50
Mainstream Media and Social Media censor speech to "prevent misinformation": 50
Ivermectin manufacturer, Merck ditches ivermectin for expensive Molnupiravir: 51
Mainstream media fails to report objectively: 51
Ivermectin: "Entire countries or regions have relied on it." 52
In 2023 a government lawyer says FDA never prohibited ivermectin: 53
Walking the tightrope, FDA "spokesperson" wobbles on ivermectin: 54
Emotional abuse perpetrated by the FDA: 55

8. Paxlovid 57
Effective monoclonal antibodies nixed by the government in favor of Paxlovid: 57
Paxlovid is ineffective and expensive: 59
FDA Chief breaks FDA rules and promotes Pfizer's Paxlovid to the public: 59
CDC promotes expensive antivirals, vilifies off-label inexpensive antivirals: 61
I'll rub your back if you'll rub mine: FDA commissioners work for Big Pharma 62

9. Power of the purse – NIH Distributes Billions for Research and Studies 65
Don't cross Fauci if you want funding from NIH for your research: 65
The silencing of dissident voices harms science and medicine: 67

10. These Are Not your Father's Vaccines 69
How traditional vaccines work: 69
New-technology vaccines are the winners in the Covid vaccine race: 70
A vaccine inserts a pathogen, a gene therapy inserts instructions to your cells: 70
October 2021: CDC changes the definition of vaccine 71
Novel lipid nanoparticle technology made the mRNA shots possible: 71
The difference between the mRNA and adenoviral vector Covid shots: 72
Novavax protein subunit vaccine: 73

11. mRNA shots are a fundamental shift in vaccinology 75

Known in April 2020: the spike protein should not be the basis for mRNA shots 76
Autoimmunity: 77
Antibody dependent enhancement (ADE): 77
Immune Escape, also known as viral escape or antigen escape: 79

12. mRNA has Never Been Successfully Used in Vaccination 81

Fauci "Sometimes a vaccine looks good in trials, but actually makes people worse:" 83
The problem is not just the Spike Protein, it's the mRNA platform: 84
Vaccine misunderstanding...or intentional deception? 85

13. Clinical Trials Did NOT Show Covid Shots Were "Safe and Effective" 87

"95% effective" was based on mathematical deception: 88
How Relative Risk is calculated: 89
How Absolute Risk is calculated: 90
The actual relevant number was "only 4% effective," not "95% effective: 90
The vaccinated fared worse than the placebo group in Pfizer clinical trials: 91
This was not an honest mistake – it was a narrative driven by profits and power: 92

14. The More Covid Shots You Take, the More Likely You Are to Get Covid 95

Informed consent must be based on an accurate risk-benefit ratio: 95
Cleveland Clinic study - more Covid shots equal more Covid infections: 96
IgG4 antibodies – when the immune system is trained NOT to fight Covid: 97
2023: Emergency's over, but $1.4 billion is allocated to creating Covid vaccines: 99
CDC: Vaccinated are susceptible to the new variant, so get vaccinated: 100

15. Early Warning Voices about the Dangers of the Covid-19 Shots 103

Dr. Sucharit Bhakdi - Germany: 103
Dr. Peter McCullough - U.S.A: 105
Dr. Geert Vanden Bossche - Belgium: 106
Dr. Luc Montagnier - France: 107
Dr. Byram Bridle – Canada: 108
Naomi Wolf, PhD – U.S.A: 109

16. There has Never Been an Effective Vaccine for a Respiratory Virus 111

Diseases have not been eradicated through vaccination: 112
"Wrong shot, wrong protein, wrong virus" 112
Dr. Ryan Cole becomes another target for the censorship complex: 114
2016 study outlined some of the problems with lipid nanoparticle technology: 114

17. Israel: Pfizer's guinea pig, and the World's canary in the coal mine 117

Myocarditis in young men post-vaccination: 117
The vaccinated become the bulk of Covid hospitalizations: 117
Israeli Ministry of Health suppresses adverse vaccine events: 119

18. Myocarditis Harms Show Up Early in the U.S. Vaccination Campaign 121

Myocarditis cases rise in 16 to 17 years olds in U.S.: 121
VAERS shows a safety signal, so the government questions VAERS data: 122
Majority of VAERS-reported deaths occurred within 2 days of Covid vaccination: 123
VAERS is thought to represent only 2% of total adverse events: 123
Increased Adverse Event reports to VAERS are NOT because More Shots were given: 125
D-Med, the Military medical record system, shows a rise in myocarditis: 127
Before FDA Emergency Use Authorization, they already knew about myocarditis: 128
FDA adds myocarditis warning to Covid shots – recommends all receive them: 130

19. Why Children, especially, Should Not Be Given the Covid-19 Shots 131

European countries discontinue the Covid mRNA shots for young people: 132
The mRNA Covid-19 shots may permanently harm infants and toddlers: 132
Loss of Bifidobacteria – another possible risk to infants due to Covid-19 shots: 134
Dr. Robert Malone issues a warning against Covid shots for children: 135
Evidence of suppressed immune systems in children emerges in fall 2023: 137

20. Pharmaceutical Companies Have No Liability for Vaccine Harms 139

Establishment of the National Childhood Vaccine Injury Act - 1986 139
The number of vaccines on the childhood schedule exploded after 1986: 139
Dr. Anthony Fauci has been the architect of diminishing childhood health: 140
Putting the Covid-19 shot on the childhood schedule shelters Big Pharma: 141
Corruption – the collusion of Big Pharma, the Government, and the Media: 142
PREP Act 2005: additional liability protections for Big Pharma and others 143
Vaccine injury compensations through the PREP Act are almost nonexistent: 143

21. Vaccine Mandates, and Hatred Toward the Unvaccinated 145

Pres. Biden's infamous and inflammatory September 2021 address to the Nation: 146
The private sector pushes back. Daily Wire refuses to enforce vaccine on employees 148

22. Cardiac Events, Deaths Rise in Athletes and Young People 151

The "fact-checkers" and legacy media try to normalize abnormal medical events: 152
Fatalities after Covid shots "many folds higher" than after flu shots: 154
Life insurance reports highest excess mortality ever recorded: 155
Analyst finds excess deaths correspond with vaccine mandates in Q3 of 2021: 155
The fact-checkers are also compromised by ties to Big Pharma: 157
Insidious lie: "Long Covid caused by natural infection, not by the vaccines" 158

23. Covid Vaccines killed 114,134 Americans within 3 Months in 2021 161

Scatterplot charts that show correlation between Covid vaccination and Covid deaths: 161
Quarter 3/2021 Vaccine Deaths Attributed to Covid-19: 164
Vaccination makes people more susceptible to Covid infection: 166

24. Worldwide Data post-Covid Vaccination Shows Reduced Live Births 169

Reproductive system adverse events in women post-Covid vaccination: 170
Maternal and fetal complications post-vaccination: 170
Trillions of dollars kept the vaccine rollout in motion: 172

25. Dangerous Variations in Covid-19 vaccine batches 175
Bait and Switch: 176
The mRNA vaccines erroneously contain DNA fragments: 178
"Why weren't we told? Where is the informed consent?" 179
Questions for Pfizer, regulators, and politicians: 181
Censorship of Science: 182

26. V-Safe: CDC's (skewed) Covid-19 vaccine injury report system 185
CDC knew early studies reported that the mRNA shots caused serious harm: 185
Legal action against the CDC was required before it would release V-safe data: 188
The CDC ended V-safe enrollment on May 19, 2023: 188

27. Lawsuits Against the Federal Government 191
Missouri v Biden: government suppression of free speech through social media: 192

28. Covid Authoritarians use Fear, Impose Mandates in fall 2023 195
A few charts that show the futility of mandates to prevent Covid spread: 195

29. Domestic Terrorism and Questioning the Official Covid Narrative 199
Homeland Security labels Covid- response questioners "domestic terrorists:" 199
Federal Judge Terry Doughty does not agree with Homeland Security: 201
"Freedom of speech – not reach:" 202
Orwell's 'Ministry of Truth," referred to by Judge Doughty, is from the book *1984*: 203

30. Covid-19: Handled as a Biosecurity Threat, not a Health Emergency 205
Pres. Eisenhower's 1961 warning of a too powerful military-industrial complex: 205
Covid-19 vaccines were a military countermeasure: 206
Dr. Kory: "VAERS exploded...like seeing the battlefield strewn with soldiers:" 210
Lockdowns weren't a form of "going medieval," they were brand new in 2020: 211
Dictionaries change the definition of lockdown: 212
Cybersecurity Infrastructure Security Agency determined if you were "essential:" 213

31. Every War Involves Propaganda 215
The Propaganda of Policy Decisions based on Modeling: 218
The Propaganda of Lockdowns and 15 Days to Slow the Spread: 221
The Propaganda of Face Masks: 227
The Propaganda of Asymptomatic Spread: 237
The Propaganda of PCR Tests, Tests, Tests, Tests.... 238
Propaganda about Hospitals Being Overrun with Covid Patients: 242
Propaganda through Inflated Covid-19 Death Counts: 243
Propaganda of Blaming the Unvaccinated for the Continued Pandemic: 245
Propaganda of The Science™ 248
Propaganda of "The Experts:" 250
The Propaganda of Endless Sanitizing to "Stop the Spread" 254
The Propaganda of Super-Spreader Events: 257
Propaganda of Equating Covid-19 with the Spanish Flu: 259
Messaging about Covid-19 – a Master Class in Propaganda: 260

32. The Propaganda of Prizes and Awards for Pandemic Players 265
Nobel Prize for technology that led to "two highly effective" mRNA Covid vaccines: 265
Anthony Fauci Announced as the 2024 Inamori Ethics Prize Winner 268
Peter Hotez receives Inaugural IDSA Anthony Fauci Courage in Leadership Award 269
A Russian immigrant who could enlighten Hotez, if the doctor would listen: 270
The White Hats versus the Black Hats: 271
33. The Covid-19 Vaccine Merry-Go-Round 275
30-year Director of FDA vaccine research resigned in protest of Covid boosters: 275
The Illusion of Consensus for Covid-19 Vaccines: 277
Covid shots post-National Health Emergency: Team Reality vs The Narrative 277
Our World In Data – Covid-19 in 2023: 282
34. Pandemic Response: A Success or Failure? It Depends... 283
35. How Do You Decide What Is True? 289
Journey of a "lefty liberal" from being Asleep to being Awake: 290
36. The Culpable Are Attempting to Rewrite History 293
Former NYC Governor Andrew Cuomo Lies About His Pandemic Role: 293
California Governor Newsom tries to reframe his pandemic totalitarianism: 294
AFT President Randi Weingarten Lies: "I fought to reopen schools in 2020": 295
The moment calls for truth, not for self-justification: 297
Don't Let Them Get Away With It: 300
Medical Terms Glossary (Layperson's Version) 305
Acknowledgments 319
About the Author 321
Endnotes 323
Index 361

Because the ability of future generations to pursue life, liberty, and happiness depends on the protection of these unalienable rights by courageous people, who understand today.

For we wrestle not against flesh and blood, but against principalities, against powers, against the rulers of the darkness of this world, against spiritual wickedness in high places.

EPHESIANS 6:12

INTRODUCTION

No one has the full story or complete knowledge about the Covid-19 pandemic and our response, but we should be able to examine what happened and ask questions. This compilation is an attempt to analyze our pandemic response with a critical eye.

This report is not a medical study or conclusive document. It presents various ideas and concerns shared by so-called "Covid dissidents" who, from the beginning, were concerned that our pandemic response was causing more harm than Covid-19 itself. No doubt there are other relevant points, and points of view, that are missing from this collection, but it's a start. My hope is that this book will lead to thoughtful consideration, fruitful conversations, and the pursuit of additional knowledge.

Perhaps the most important thing to note in a discussion of medicine in the time of Covid-19 is this: If there had been effective treatments for Covid-19 using already FDA-approved medicines (which there were), there would not have been a need, or a legal basis, for the approval of emergency use medicines and vaccines.

If someone had hurt your family and friends in an underhanded manner, and had the capacity and intent of doing it again, would you want to know? Maybe you could prevent future harm if you were aware of how and why they did it in the first place.

On May 11, 2023, the Covid pandemic emergency officially ended in the U.S. My concern, and the impetus for this report, is the belief that the Covid-19 pandemic response was a dress rehearsal. So much was accomplished during the pandemic by those who embrace authoritarianism under the guise of safety and protecting society. Emboldened, they will try to force us all into a repeat, as soon as they can stir up the next fear

– another variant or a new pathogen, climate change, compromised food supply, energy shortage, international unrest – any "emergency" will do.

You may, or may not, believe that the Covid-19 pandemic merited the response we gave. Either way, consideration should be given to what we were forced to do, and what was lost over the past four years – all under the banner of fighting a virus.

The following compilation of topics is an effort to explain the How and Why of Medical Harms during the Covid-19 pandemic, focusing on four main aspects:

1. The suppression of early effective treatments for Covid-19
2. Remdesivir and other profitable, but ineffective, Covid-19 treatments
3. The development and implementation of the Covid-19 vaccines
4. How Covid-19 was used to impose stringent, anti-democratic measures

These four aspects will be addressed, not in any certain order, but throughout this discussion. Details relating to medical and scientific expertise are linked to original source material such as studies, reports, and the writings and statements of licensed professionals. A Medical Terms Glossary (Layperson's Version) is located at the end of the book.

For more, please see On Topic With Lori.[1] My work can also be found at Brownstone Institute.[2] There are so many voices - I sincerely thank you for giving some of your time and attention to mine!

Lori Weintz

1. WAIT! THERE'S A PANDEMIC?

It is sad that the shutdown will be harder for poorer countries than richer countries.

BILL GATES

TED interview | March 24, 2020

Public health can never be about just one illness; it has to be about the health of the population as a whole.

DR. AARON KHERIATY[3]

Scholar on Medical Ethics

An epidemic is an outbreak of disease that spreads quickly and affects many individuals at the same time. A pandemic is an epidemic that was once localized but is now appearing in other countries and even on other continents. The difference between an epidemic and pandemic isn't in the severity of the disease, but the degree to which it spreads.

Before Covid-19 there had been other pandemics. But in the past 100 years, with the exception of the Spanish flu in 1918, the other pandemics came and went without much notice for the majority of the world's population. For example, much press coverage of the first SARS in 2003[4] neglected to report there were only 774 total deaths worldwide. Likewise, the heightened reporting of the 2012 MERS[5] pandemic failed to summarize that there were only 858 total deaths. In contrast, recurring influenza [6] strains kill an average of 400,000 people worldwide each year.

In January 2020, when we started to hear reports of a new respiratory virus in China, something was different. Those images of people dropping dead in the street, and the modeling from the Imperial College of London forecasting millions of deaths if we didn't take drastic measures, set us on an unprecedented pandemic-response course.

But while disoriented citizens worldwide were coming to terms in March 2020 with the statement, "There is a pandemic," enforced by the dramatic shutdown for "two weeks to slow the spread," others did not appear to be so taken aback. If nothing else, what most of us saw as an unprecedented calamity, they saw as an anticipated opportunity.

"We can make radical changes to our lifestyles:"

The **World Economic Forum (WEF),** long having watched for a global crisis to further its agenda of a "Great Reset" [7] of capitalism declared, "One silver lining of the pandemic is that it has shown how quickly we can make radical changes to our lifestyles."

WEF Founder Klaus Schwab stated, **"The pandemic represents a rare but narrow window of opportunity to reflect, reimagine, and reset our world."** Schwab lectured that the Great Reset "will require stronger and more effective governments...and it will demand private-sector engagement every step of the way." All voluntary and for the greater good, of course. (Later in the pandemic, the Great Reset was rebranded Build Back Better, but the phrases are basically interchangeable.) Interestingly, **while many in leadership seemed to be using the same words and talking points, anyone who pointed this out was called a conspiracy theorist,** as shown in the following clip:

Build Back Better (2 min)[8]

The WEF's stance could be considered the quirky views of a few wealthy elites, if it were not for the fact that so many government and corporate leaders are affiliated with the organization. At the beginning of the pandemic, the WEF's stated intent was that "15 days to slow the spread" would mean "radical changes to our lifestyles" – forever.

Bill Gates – Anticipates vaccinating the whole world:

At the same time the WEF was busy reimagining our world, philanthrocapitalist Bill Gates enthusiastically shared his certainty in both a March 2020[9] interview and another in April[10] that everyone would need to be vaccinated before the world could reopen and return to normal. This idea fit conveniently into Gates' worldview, where vaccines and pandemics are an investment opportunity[11], and public health is mostly about anticipating the next deadly virus.

The Bill & Melinda Gates Foundation, along with the WEF, and other partners, is one of the founders of the Coalition for Epidemic Preparedness Innovations (CEPI[12]). CEPI's website states it is "an innovative global partnership working to accelerate the development of vaccines against epidemic and pandemic threats." One of its goals is the **100 Days Mission to deliver a vaccine in 100 days against the next disease**, "to give the world a fighting chance to extinguish the existential threat of a future pandemic virus."

CEPI's mission aside, viruses and bacteria have not been considered an "existential threat" to humanity for the most part, but simply a part of the microbial planet[13] we inhabit. Covid-19 was known early on to impact the elderly and those with chronic health issues. Healthy people and the young were at very little risk for serious outcomes, but somehow the narrative of Covid-danger took off like wildfire. A certain pandemic playbook was being implemented.

Pandemic wargames:

The Bill & Melinda Gates Foundation has been a big player in pandemic wargames [14] during the past 20 years. In fact, the Gates Foundation, World Economic Forum, and Johns Hopkins University, among other players, had just completed Event 201[15] in October 2019, in which they simulated a severe coronavirus pandemic. Participants in Event 201 were from governments (including the Chinese government), public health, Big Pharma, media, finance, public relations, and academia.

As with previous disease wargames, a large focus of the **Event 201** simulation involved control of citizens during the pandemic. The simulation covered "how to use **police powers** to detain and quarantine citizens, how to impose **martial law**, how to **control messaging** by **deploying propaganda**, how to employ **censorship to silence dissent**, and how to **mandate masks, lockdowns**, and **coercive vaccinations** and conduct track-and-trace **surveillance** among potentially reluctant populations." (*The Real Anthony Fauci*, by Robert F. Kennedy, Jr, Chapter 12 Germ Games, p. 382, and Event 201 p.426-435)

Each of these pandemic simulations ends with coerced mass vaccination to conquer the virus. The human immune system, which has effectively interacted with the microbial world from the beginning of time, is not considered a significant player in the virus-to-vaccine worldview. Many of the people who participated in Event 201 pivoted to being key players in the Covid-19 official pandemic response, mirroring what they had practiced.

The World in 2019 - Poverty and Hunger at their Lowest Ever:

While those conducting 20 years' of virtual pandemic wargames were focused on vaccines, the real world made great strides in handling disease and poverty in more organic ways. **Improvements in sanitation, clean water, nutrition, education, and increasing access to safe and inexpensive energy sources all contributed to better health worldwide.** During the past century, the advent of antibiotics, and advances in diagnostics and surgery skills also increased life expectancy and quality of life.

Prior to the Covid-19 pandemic, poverty and hunger were decreasing throughout the world, due in part to all these advances, and also because of the interconnected world economies. **During the pandemic**, government and public health leaders imposed **sanctions and mandates**, at the expense of all other aspects of human health and well-being, that have **slowed and even reversed much of that progress.**[16]

As explained by Dr. Jay Bhattacharya[17] of Stanford University:

> Over the past 40 years, the world's economies globalized, becoming more interdependent. **At a stroke, the lockdowns broke the promise the world's rich nations had implicitly made to poor**

nations. The rich nations had told the poor: Reorganize your economies, connect yourself to the world, and you will become more prosperous. This worked, with 1 billion people lifted out of dire poverty over the last half-century.

But the lockdowns violated that promise. The supply chain disruptions that predictably followed them meant millions of poor people in sub-Saharan Africa, Bangladesh, and elsewhere lost their jobs and could no longer feed their families.

Far from recognizing the tremendous harms inflicted by their dystopian and totalitarian views and policies, **the Schwabs and Gateses of the world are pushing for increased top-down control of people in all areas of their lives.**

Moderna and worldwide vaccine production factories:

Moderna is ramping up the construction of mRNA vaccine factories all over the world[18], while fear-mongers and profiteers assure us there will be another pandemic soon[19]. They insist we must continue to fund gain-of-function research[20]. Gain-of-function is a term referring to the modification of a virus, through laboratory experimentation, to make it more contagious and/or more virulent. The theory is that we manufacture dangerous viruses in the lab, so we can develop vaccines that counter them – just in case an enemy has the same idea. Gain-of-function is controversial research and those who continue to promote it are apparently ignoring the now widely acknowledged possibility that the SARS-CoV-2 virus escaped from a lab. (see here [21], here [22], and here [23])

WHO wants top-down control in future pandemics worldwide:

The World Health Organization (WHO) has presented changes to the International Health Regulations (IHR) treaty that will allow it to impose a response[24] on any country where it identifies a health threat, or *potential* threat, without consulting the people or their government. On September 20, 2023, the U.N. issued a Declaration[25] to back the WHO whenever it declares a viral variant to be a "public health emergency of international concern." (It should be noted that because Bill Gates is the second largest donor [26] to the World Health Organization, just behind the U.S.; he has an outsized influence[27] on the WHO's focus[28] and resource allocation.)

Those who imposed lockdowns, mask mandates, and segregation of the unvaccinated have doubled down on their insistence that they handled everything just right – except for maybe they should have done it all sooner and harder. The **European Commission and the WHO have announced a "digital health" partnership** to establish a Digital Vaccine Passport[29] system as a way of tracking every citizen in the world.

> What we are up against is not a misunderstanding or a rational argument over scientific facts. It is a fanatical ideological movement. A global totalitarian movement... the first of its kind in human history.
>
> **CJ HOPKINS** | Author & Playwright | October 13, 2020[30]

Fauci calls lockdowns "inconvenient," says a vaccine is coming soon: While Schwab and Gates shared their enthusiasm for harsh pandemic measures, Dr. Anthony Fauci, then director of the National Institute of Allergies and Infectious Diseases (NIAID) participated in a Facebook interview with Mark Zuckerberg[31] on March 19, 2020, three days into "flatten the curve." In talking about the lockdown, Fauci, **with almost psychopathic understatement, termed the unprecedented physical separations of people, and closures of everything, "inconvenient," and "disruptive of society."** Then he dropped a bombshell, "At the end of 15 days we'll reevaluate and see if what we've done has any noticeable impact…I project we will go longer than two weeks. It doesn't just turn around over a week and a half or two."

Fauci also explained to Zuckerberg that not only was **a vaccine being developed** (he expected one ready to use within 18 months)**, but there were therapeutics that would treat Covid-19 disease in the meantime**. He specifically mentioned two by name: Hydroxychloroquine (HCQ), and Remdesivir. With regard to HCQ Fauci said, "There's a lot of buzz on the internet about…hydroxychloroquine [a drug that] has been approved for decades, very cheap, used in malaria and certain autoimmune diseases."

Dr. Fauci failed to mention that the National Institutes of Health (NIH) of which NIAID is a part, found in a 2005 study [32] during the first SARS pandemic, that **chloroquine** (a precursor to HCQ) [33] was a **"potent inhibitor of SARS coronavirus infection and spread"** in cell culture studies. (emphasis added) Chloroquine had also shown promise against MERS in vitro (test tube).

Prior to Covid-19, HCQ was a well-established drug known to have few side effects, and almost no interactions with other drugs. Available over-the-counter in many countries including France, India, Mexico, and many African nations, HCQ had been on the World Health Organization's list of Essential Medicines (meaning it is required to be inexpensive and readily available) for decades (see WHO list p. 55),[34] and had been taken safely by billions of people.

Perhaps Fauci didn't mention the great potential of chloroquine because prophylactics (preventative drugs) and therapeutics (drugs that treat disease) were not really on his radar. **The real focus was soon to be President Trump's Operation Warp Speed -** the rapid development of a Covid-19 vaccine - which was already well under way. **But making some money off of expensive therapeutics, such as remdesivir, also appears to have been appealing.**

2. REMDESIVIR

Tragically, the government-backed mechanical (ventilators) and pharmaceutical (remdesivir, mRNA shots, etc) interventions didn't work to remedy the respiratory illness problem. Instead, they added an additional layer of chaos on top of the virus mania that had captured the world.

JORDAN SCHACHTEL

Investigative journalist | December 13, 2023 [35]

In the interview with Zuckerberg, in addition to mentioning HCQ, Fauci said, "There's a drug called remdesivir, which is...developed by Gilead as an antiviral. We tried it in Ebola. It didn't work as well as some of the other drugs, but it's there." It was really, really there; remdesivir manufacturer Gilead spent $2.45 million lobbying Congress[36] in the first quarter of 2020, which was also when Congress drafted and passed the Coronavirus Aid, Relief and Economic Security (CARES) Act that included funding for vaccines and treatments in response to the pandemic.

Remdesivir was removed from the Ebola trial because of toxic effects: "Didn't work as well" was a gross understatement. During the 2018 Ebola drug trials [37] funded by the National Institutes of Health (NIH), remdesivir (brand name Veklury) [38] was one of four different drugs used to treat Ebola. Those in the remdesivir group had the highest overall deaths, with a mortality rate of over 50 percent in the first 28 days. Trial participants who received remdesivir also had significantly elevated markers for liver and kidney damage, leading the safety board to terminate its use mid-trial.

A drug with remdesivir's profile was not a good candidate for much of anything. It's not that it just wasn't effective against Ebola; it's toxic. Why would Fauci even consider it?

Robert F. Kennedy, Jr explains one reason in his book *The Real Anthony Fauci* (*TRAF*). It has to do with perception. The NIH had to appear to be doing *something* about the pandemic in the short term, even though the long term goal was the vaccine. RFK, Jr states:

"Optics required that NIH devote some resources to antiviral therapeutic drugs; critics would complain if [Fauci] spent billions on vaccines and nothing on therapeutics. However **any licensed, repurposed antiviral that was effective against Covid for prevention or early treatment... could kill his entire vaccine program because FDA wouldn't be able to grant his jabs Emergency Use Authorization.** Remdesivir, however, was an IV remedy, appropriate only for use on hospitalized patients in the late stages of illness. **It would therefore not compete with vaccines.**" (*TRAF* p. 64 emphasis added)

National Institutes of Health employees profit from drug patents:

> *The fact that the FDA's advisory committee, the Vaccines and Related Biological Products Advisory Committee (VRBPAC), the one that met today and voted to recommend to the FDA to grant the EUA (for the Covid-19 vaccines) that's really important...because what it shows is that the process that we have here in the United States is, decisions and recommendations are made by independent bodies... we want to make sure that we impress the American public that decisions that involve their health and safety are made outside of the realm of politics, outside of the realm of self-aggrandizement, and are made, in essence, by independent groups.*
>
> **DR. ANTHONY FAUCI** | Director NIAID | December 11, 2020

Dr. Fauci states that decisions involving American's health and safety are made "outside of the realm of politics," but nothing could be further from the truth. The other reason RFK, Jr. gives for Fauci's interest in remdesivir is money. National Institutes of Health (NIH) employees are allowed to put their name on patents[39], and thereby profit from product approval. For example, NIH is listed on the patent for Moderna. Tellingly, Anthony Fauci's[40] household net worth increased during the pandemic from $7.5 million to $12.6 million.

In addition to the conflict of interest at NIH, with employees profiting[41] from products they approve, a large portion of the operating budget for NIH, which includes agencies such as the Food and Drug Administration (FDA), is provided by pharmaceutical companies[42] – the very companies whose products the FDA regulates. In short, **there was money to be had in remdesivir, and none to be had by using already approved medicines that were no longer under patent restrictions, and therefore inexpensive.**

Investigative journalist Sharyl Attkisson found that at the time of the remdesivir review, eleven members of the NIH's Covid-19 Treatment Guidelines panel had financial ties to Gilead[43]. **A review of the panel members during the past year finds myriad relationships[44] between NIH panel members and various drug companies.** Gilead, flush with the prospect of remdesivir becoming the treatment for Covid, predicted remdesivir would bring in $3.5 billion[45] in 2020 alone.

In April 2020, knowing remdesivir's dismal and dangerous profile, Fauci's NIAID began the clinical portion of a 29-day remdesivir trial[46] on hospitalized U.S. volunteers. The Protocol Details Summary [47] states that, "The drug has been tested before in people with other diseases," but neglects to mention its complete failure in the Ebola trial.

The Rise of Remdesivir despite Poor Performance:

Before Fauci's remdesivir study was completed, a study conducted in China[48] was released that should have put remdesivir out of the running.

The Chinese study found that **intravenous remdesivir**:

- did not provide significant clinical or antiviral effects in seriously ill Covid-19 patients,
- was not associated with a difference in time to clinical improvement,
- caused adverse events in 66% of remdesivir recipients

In short, patients were worse off taking remdesivir than if they received the placebo.

RFK, Jr. points out that, "The Chinese study was a randomized, double-blind, placebo-controlled multi-center, peer-reviewed study, published in the world's premier scientific journal, *The Lancet*." "In contrast, Dr. Fauci's NIAID study was, at that point, still unpublished, not peer-reviewed, its details undisclosed. It employed a dirty placebo and had suffered a sketchy mid-course protocol change," which is taboo in clinical studies. (When the NIAID study couldn't show a reduction in disease or mortality, the end goal was changed to a reduced hospital stay.)

A **dirty placebo, also known as "spiked," or "phony" placebo,** refers to the use in a drug trial of an "active comparator" rather than a pure placebo such as a saline injection. In this case, patients in the treatment arm were given remdesivir. Patients in the "placebo" arm were given remdesivir that contained the same ingredients, but in an inactivated state. As explained by RFK, Jr., **Dr. Fauci's NIAID has developed the use of "spiked" placebos over 40 years to "conceal adverse side effects of toxic drugs for which they seek approval.**" (*TRAF* p. 65)

Following is a chart from the Arms and Interventions section of the study[49] showing that the NIH study did indeed employ a dirty placebo. As shown on the left, 200 mg of Remdesivir, with some inactive ingredients, was used as the Placebo Comparator, and 200 mg of Remdesivir was used as the Experimental drug. Logic would indicate that determining the effect of a new, unapproved drug would be better served by placing it opposite an actual placebo, instead of opposite a modified version of the same drug.

Arm	Intervention/treatment
Placebo Comparator: Placebo 200 mg of Remdesivir placebo administered intravenously on Day 1, followed by a 100 mg once-daily maintenance dose of Remdesivir placebo while hospitalized for up to a 10 days total course. n=286.	Other: Placebo The supplied placebo lyophilized formulation is identical in physical appearance to the active lyophilized formulation and contains the same inactive ingredients. Alternatively, a placebo of normal saline of equal volume may be given if there are limitations on matching placebo supplies.
Experimental: Remdesivir 200 mg of Remdesivir administered intravenously on Day 1, followed by a 100 mg once-daily maintenance dose of Remdesivir while hospitalized for up to a 10 days total course. n=286.	Drug: Remdesivir Drug Remdesivir is a single diastereomer monophosphoramidate prodrug designed for the intracellular delivery of a modified adenine nucleoside analog GS-441524. In addition to the active ingredient, the lyophilized formulation of Remdesivir contains the following inactive ingredients: water for injection, sulfobutylether beta-cyclodextrin sodium (SBECD), and hydrochloric acid and/or sodium hydroxide.

Dr. Fauci chooses remdesivir as the "standard of care" in the U.S.:

> *The NIAID spent upwards of $70 million in taxpayer[50] money in connection with the development of remdesivir to treat Covid-19. It's a legitimate question to ask why remdesivir was even considered in 2020. It had been pulled from the Ebola trial in 2018, due to elevated markers for liver and kidney damage and a trial participant mortality rate over 50 percent.*

Fauci, knowing the Chinese study was about to be published, which would likely impede the FDA's approval of remdesivir to treat Covid-19, headed off the bad news in an Oval Office press conference on April 21, 2020 declaring "quite good news." Fauci said the NIAID's clinical trial showed remdesivir was reducing the hospital stay of infected Covid patients by about four days. He neglected to mention the significant number of adverse events in both arms of the study[51], and that **twice as many remdesivir subjects as placebo subjects had to be readmitted to the hospital after discharge, raising the question of early release to make the numbers look better.** Fauci said he was ending the study, would give remdesivir to the placebo group, and that remdesivir would be America's new "standard of care" for Covid. (*TRAF* p.66)

Holocaust survivor Vera Sharav, who has dedicated her life to exposing corruption in the clinical trial industry, states:

> Dr. Fauci had a vested interest in remdesivir. He sponsored the clinical trial whose detailed results were not subject to the peer review he demanded for the drugs he regarded as rivals, like hydroxychloroquine and ivermectin. Instead of showing transparent data and convincing results, he did 'science' by fiat. (*TRAF* p. 67)

May 1, 2020 FDA grants Remdesivir Emergency Use Authorization:

On April 29, eight days after Fauci's above referenced Oval Office press conference, the **Chinese study [52] that showed serious adverse reactions to remdesivir in 66% of trial recipients,** was published in the *Lancet*. Undeterred, the NIH issued a News Release [53] stating that the "NIH clinical trial shows Remdesivir accelerates recovery from advanced Covid-19."

Two days later on May 1, 2020 the FDA [54] made remdesivir the official new Emergency Use Authorized (EUA) Covid drug.

Remdesivir costs less than $6 to manufacture, but costs $3k for treatment:

The FDA's recognition of remdesivir as the standard of care for Covid-19 meant that Medicaid and insurance companies could not legally deny it to patients, and that doctors and hospitals that failed to use it could even be sued for malpractice. (*TRAF*, p. 67) Gilead, which manufactures remdesivir at less than $6 for a course of treatment[55], charges around $3,000 (depending on insurance) for a 5-day treatment. In contrast, a 5-day course of HCQ costs less than $20. Somehow, the chairman of Gilead thought this was a bargain:

> As the world continues to reel from the human, social and economic impact of this pandemic, we believe that pricing remdesivir well below value is the right and responsible thing to do.
>
> **DANIEL O'DAY** | June 29, 2020[56]
>
> Chairman & CEO, Gilead Sciences Manufacturer of Remdesivir

It's true that research and design, and manufacturing costs, are all part of drug development, so perhaps the huge disparity in price manufacture vs price to market is somewhat justified. But costs aside, the safety profile on remdesivir versus HCQ is sobering. When remdesivir was fully approved in October 2020[57], the FDA stated, "Drug interaction trials of Veklury (remdesivir) and other concomitant medications have not been conducted in humans," meaning it was not known what type of adverse interactions it might have with other drugs.

Remdesivir was known to cause kidney problems, and increase levels of liver enzymes, a sign of liver damage. Remdesivir had not been tested for safety[58] in women who were pregnant or breastfeeding. HCQ, on the other hand, has been in use for almost 70 years and is safe to prescribe to children and adults of all ages, including pregnant women and nursing mothers.

Remdesivir is nicknamed "Run-death-is-near!"

> *All of these protocols, the fear mongering, the isolation, the toxic medications – I walked away feeling like I had participated in medical murder.*
>
> **GAIL MACRAE** | Nurse in Los Angeles

Nurse Gail Macrae[59] worked at Kaiser Permanente Santa Rosa Medical Center in California during the pandemic. As **media hype insisted that hospitals were overflowing with Covid patients,** Macrae observed the opposite. "They were never full of patients," said Macrae. In fact, from the onset of Covid, for the whole first year of the pandemic, not only was **the hospital mostly under capacity**, but Macrae's shifts as a contract nurse were often canceled for lack of patients. A compilation of total patients in Los Angeles County hospitals from March 2020 to March 2021 confirms Macrae's experience:

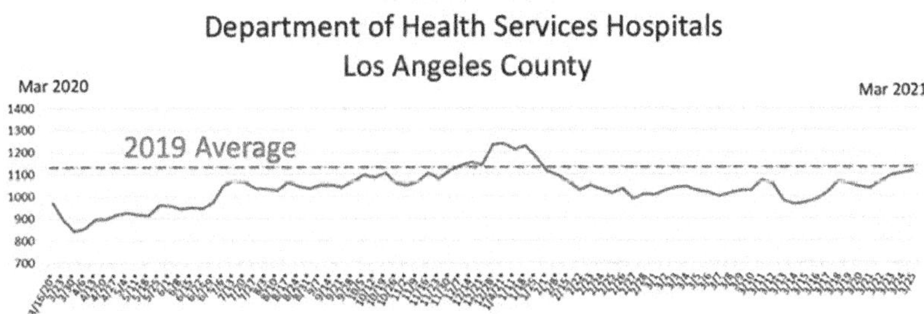

In the above chart, compiled from Los Angeles County records, the dotted red line shows the 2019 average for patients in Los Angeles County hospitals was around 1,150. The blue line shows that from March of 2020 to March of 2021, hospitals had fewer patients than in 2019, with the exception of one blip above-average in Dec 2020 to Jan 2021, during the Delta wave.

Patients on remdesivir frequently experience serious adverse events[60]: In addition to being troubled by the dishonest media coverage, Macrae also found herself dismayed by the Covid-19 treatment protocols. She and the other nurses would talk amongst themselves and question why the experimental EUA antiviral remdesivir was being administered to patients in later stages of disease, when an antiviral would not be very effective. **They were seeing no improvement in patients receiving remdesivir, and in fact, upon administration many patients were going into multi-organ failure.** (During the pandemic, some nurses renamed remdesivir[61] "run-death-is-near.") Macrae brought up her concerns with supervisors and was told, "This is protocol. This is all we have. There is nothing else we can give."

World Health Organization advises against remdesivir to treat Covid-19: In October 2020 the World Health Organization (WHO) advised against[62] the use of remdesivir in hospitalized Covid-19 patients. The WHO, echoing the China study of April 2020, said there was **no evidence[63] that Remdesivir had positive impact on "mortality, need for mechanical ventilation, time to clinical improvement, and other patient-important outcomes."** This finding was based on the results from in the WHO's Solidarity trial[64] involving 11,330 adults in 30 countries.

Dr. Andre Kalil, with the National Institutes of Health, dismissed the WHO's Solidarity trial[65], claiming "poor quality study design," and pointed to the results from the NIH trial that showed remdesivir led to reduced hospital stays in adults with Covid-19. Dr. Kalil was the principal investigator on the NIH trial to which he was referring, which was the same trial that was ended by Fauci[66] after his April 2020 White House press conference. The same trial that showed serious adverse events in "131 of the 532 patients who received remdesivir (24.6%) and in 163 of the 516 patients who received modified remdesivir (31.6%)."

Although trial data and **actual field experience** were **showing that remdesivir did not improve outcomes** for hospitalized Covid patients, **remdesivir was given full FDA approval in October 2020[67].** In granting approval, the FDA cited Fauci's NIH study, and also approved remdesivir for use in pediatric patients 12 years and older, based on a small study sponsored by remdesivir manufacturer Gilead Sciences. The fox guarding the henhouse comes to mind.

Riches from Remdesivir:

While HCQ and ivermectin were being sabotaged by the FDA, NIH, and mainstream media, the **financial incentives offered by the federal government for using remdesivir, and other official protocols in the hospital setting were phenomenal.**

An analysis[68] by the Association of American Physicians and Surgeons stated, "The CARES Act provides incentives[69] for hospitals to use treatments dictated solely by the federal government under the auspices of the NIH." Financial incentives for using federally approved protocols included:

- A "free" required PCR test in the Emergency Room or upon admission for every patient, with government-paid fee to hospital.
- Added bonus payment for each positive COVID-19 diagnosis.
- Another bonus for a COVID-19 admission to the hospital.
- **A 20 percent "boost" bonus payment from Medicare on the entire hospital bill for use of remdesivir.** (emphasis added)
- Another and larger bonus payment to the hospital if a COVID-19 patient is mechanically ventilated.
- More money to the hospital if cause of death is listed as COVID-19, even if patient did not die directly of COVID-19.
- A COVID-19 diagnosis also provides extra payments to coroners.

The federal government also offered a $10k funeral reimbursement[70] to families of patients whose death certificates said they died of Covid.

Attorney Thomas Renz and CMS (Centers for Medicaid Services) whistleblowers[71] have calculated a total increased payment of at least $100,000 per patient. These financial incentives led to treatments that often were not in the patients' best interests. Certainly, hospital boards that lost millions during the hospital lockdowns – no patients allowed unless you had Covid-19 or a heart attack, so to speak – were eager to refill their coffers.

April 25, 2022 FDA approves remdesivir for infants and children:

Despite its dangerous profile, the FDA approved[72] remdesivir "in certain high risk situations, such as hospitalization, on April 25, 2022. At that time FDA also approved emergency use authorization of remdesivir in infants.

Patricia Cavazzoni, former Pfizer employee and current director of the FDA's Center for Drug Evaluation and Research, stated, "As Covid-19 can cause severe illness in children, some of whom do not currently have a vaccination option, there continues to be a need for safe and effective COVID-19 treatment options for this population."

A doctor at a North Carolina children's hospital, upon learning of the FDA approval stated, "We need proven antiviral treatment options, like remdesivir, that can help treat some of the most vulnerable in our society: children."

These doctors apparently were unaware, either through incompetence or purposeful ignorance, of the hundreds of studies showing success with less toxic medications.

Covid-19 does not pose a serious risk to children and young people: Let's be very clear. Infants, children, and young people are not at risk due to Covid-19 disease, something that we knew early on (see here[73] and here[74]). During the entire pandemic, Covid-19 infection in children was mild. CDC records (as of 29 Nov 2023) show that in the 0-17 age group, through the now almost four pandemic years, there were 1,696 total deaths attributed to Covid, which accounts for 1.29% of all deaths in that cohort during that time period. In other words, 98.71% of deaths in those aged 0-17 were due to other causes.

Yet so much harm was inflicted in the form of face masks, isolation from friends and extended family, closed schools, the closing of playgrounds and skate parks, the cancelation of extracurricular activities and once in a lifetime events such as Homecomings and performing tours, and the overall disruption of normal life. The same can be said for the 18-29 cohort. Covid accounted for only 3.04% of all deaths in that age group during the almost four years of the pandemic. While deaths due to Covid-19 were not high, the *response* to Covid-19 led to a marked increase in substance abuse[75], as depression and anxiety[76] soared, and suicide attempts[77] and suicides increased.

From the beginning of the pandemic, it was known that disabilities and deaths due to Covid-19, in people under 50, almost without exception occurred in those who had underlying health problems[78] to which Covid was added.

In an analysis of death certificates, the National Center of Health Statistics, which compiles the record of all deaths in the U.S, reported[79] in December 2020 that:

> For 6% of deaths, Covid-19 was the only cause mentioned. For deaths with conditions or causes in addition to COVID-19, on average, there were 2.9 additional conditions or causes per death."

In other words, only 6% of Covid deaths were actually primarily caused by Covid – the other 94% were deaths "with Covid," alongside a host of other health conditions. This startling 6% figure included all age groups, even the most elderly and infirm.

During Covid-19, society broke the unwritten contract to protect the next generation. The mistreatment of our children and young people during the Covid-19 pandemic, for a disease that posed no risk to the young, must surely be identified as one of the most immoral moments in our Nation's history.

Unprecedented government interference in the doctor/patient relationship: Government interference in the practice of medicine was one of the issues discussed in a May 18, 2023 session of the House Select Subcommittee on the Coronavirus Pandemic. As stated by one of the panel members, Rep. Rich McCormick, who is a practicing doctor in Georgia:

> As [the pandemic] continued, I became very aware that the government was the biggest problem of all...They interjected themselves between the professionals and patients. They kept families apart. They didn't let people even die with dignity, or with any choice in their own health care.

This interference in the doctor/patient relationship is evident in the continued used of remdesivir, despite renal failure and/or other adverse events[80] in the majority of Covid patients. Also, the financial incentive to ventilate patients continued, despite the known high mortality rate for those ventilated[81].

Federal reimbursement[82] for using remdesivir and mechanical ventilation in hospitalized Covid patients continued through September of 2023, although the official end of the national health emergency was on May 11, 2023.

Remdesivir, known to damage kidneys, approved for patients with kidney disease:

> *The combination of the terms "acute renal failure" and "remdesivir" yielded a statistically significant disproportionality signal with 138 observed cases instead of the 9 expected.*

> **Study published December 19, 2020**[83] | Clinical Pharmacology & Therapeutics

Further evidence of corruption in the FDA approval process is found in the July 14, 2023 approval[84] of remdesivir for treatment of Covid-19 in patients with kidney disease. Dr. Peter McCullough states[85], **"Remdesivir can cause both kidney injury and liver damage…The FDA approval action defies logic** and will be added to a long list of acts that will be considered malfeasance…"

Stories abound[86] of people who were subjected to remdesivir and ventilation, without proper informed consent, sometimes even against their will. Other accounts tell of judges having to issue orders for patients to be allowed to try IVM and HCQ in hospital, despite the safety profile and lack of contraindications with other drugs – meaning, it couldn't hurt to try and it might help.

Lawsuits against Gilead for downplaying clinical dangers of remdesivir: Lawsuits are beginning to be filed by people who feel their loved ones were killed[87], not by Covid-19, but by remdesivir, including a class-action lawsuit[88] filed in California on September 26, 2023. The class-action lawsuit claims remdesivir manufacturer Gilead Sciences "misrepresented and/or omitted the true content and nature of the drug." One of the attorneys pursuing the case states, "It's a terrible drug…you can see there's a big difference in the creatinine levels and the blood levels, kidney readings after they get the remdesivir."

> Lamentably, Dr. Fauci's failure to achieve public health goals during the Covid pandemic are not anomalous errors, but consistent with a recurrent pattern of sacrificing public health and safety on the altar of pharmaceutical profits and self-interest.

> **ROBERT F. KENNEDY**, Jr. Intro, *The Real Anthony Fauci*

3. HYDROXYCHLOROQUINE (HCQ)

There's a lot of buzz on the internet about...hydroxychloroquine [a drug that] has been approved for decades, very cheap, used in malaria and certain autoimmune diseases.

ANTHONY FAUCI

NIAID Director | White House Covid Task Force | March 19, 2020

Fauci, back in April of 2020 couldn't just ignore HCQ, despite his preference for the profitable remdesivir - not with all that "buzz on the internet." "The buzz" most likely had to do with doctors who were having success treating Covid patients with hydroxychloroquine. Plus on March 28, 2020 the FDA had given HCQ Emergency Use Authorization to treat Covid. **It was important to appear that HCQ was given a serious chance, but the study the NIH put together was a half-hearted attempt at best, and designed to fail at worst.**

NIH "study" says HCQ not effective in Covid-19, increases mortality:

In April 2020 the NIH did a short retrospective study[89] on HCQ, meaning they analyzed the records of a couple hundred men over age 65 in VA hospitals, and found "no evidence that use of hydroxychloroquine, either with or without azithromycin, reduced the risk of mechanical ventilation in patients hospitalized with Covid-19." The analysis also concluded that overall mortality increased in patients treated with HCQ. What the NIH Veterans Administration study failed to point out is that HCQ is most effective at onset of symptoms during the viral replication stage, not when a Covid patient is ill enough to be hospitalized and in the inflammation stage of the disease.

An Associated Press[90] article published when the VA report was released on April 21, 2020 stated, "The drug (HCQ) has long been known to have potentially serious side effects, including altering the heartbeat in a way that could lead to sudden death." Someone was feeding the media misleading information about HCQ. **It was well known that HCQ would only impact the heart in an adverse manner if given in too high doses, based on its decades' long safety profile.** Also, there were multiple real-world examples that countered the NIH "study."

> I became suspicious when the health authorities started an overt campaign against hydroxychloroquine, which I knew was a safe drug because I'm from the industry, and I had familiarity with it. I looked specifically into the issue that they were falsely assigning to it, which is QT prolongation (a heart-signaling disorder) and arrhythmias associated with drug-induced QT prolongation. It happened to be the area of focus of the last company that I had and worked with in pharma. I knew what they were saying about this drug was absolutely not true.
>
> **More importantly, the regulators knew perfectly well that they were saying things that were not true.** That immediately gave me pause. I started thinking, "They're professionals, they know this issue, they know this data, yet they're saying things that are not true." That led me to start questioning the whole thing. If you catch an official or a professional lying about something straight to the public, what else are they lying about? (emphasis added)
>
> **SASHA LATYPOVA** | Former Pharmaceutical Executive | June 17, 2023[91]

Dr. Vladimir Zelenko uses HCQ to successfully treat over 2,000 Covid patients:

Early in the pandemic, family physician Dr. Vladimir "Zev" Zelenko developed a successful protocol[92] for treating Covid-19. In the early months, Dr. Zelenko treated over 2,000 Covid patients. For those in need of medication he used a combination of HCQ, zinc, and azithromycin (AZ) or doxycycline, depending on the patient. His goal was to treat at-risk patients during Stage

1 of Covid, the viral infection stage, which kept it from developing to severe disease. Of the 2,000 patients, many of them elderly, all recovered without long-term effects, except for two patients who had other severe health problems and passed away. Dr. Zelenko's treatment kept 84% of his Covid patients out of the hospital. He observed long-term lung damage only in those who had been hospitalized and put on a ventilator.

Dr. Didier Raoult successfully treats over 1,000 patients with HCQ/ AZ combo:

In Marseilles, France, Dr. Didier Raoult's team conducted a study of 1,061 patients[93] treated for Covid with a combination of HCQ and azithromycin from March 3-April 9, 2020. The study reported, "A good clinical outcome and virological cure was obtained in 973 patients within 10 days (91.7%)."

> The HCQ-AZ combination, when started immediately after diagnosis, is a safe and efficient treatment for COVID-19, with a mortality rate of 0.5%, in elderly patients. It avoids worsening and clears virus persistence and contagiosity in most cases.
>
> **Interpretation of Marseilles, France study** | Conducted March 3- April 9, 2020

A microbiological and clinical scientist, Raoult is the most highly-published infectious diseases expert in Europe, and was founder and head of the IHN Mediterranee research hospital, the premier infectious disease facility in France. Raoult was familiar with earlier studies of HCQ as an infective inhibitor of coronavirus disease progression. His report was likely influential in the FDA's initial approval of HCQ for the treatment of Covid.

HCQ suddenly classified as a "poisonous substance" in France:

HCQ was over-the-counter in France for decades before some behind-the-scenes political[94] maneuvering led to its reclassification as a "poisonous substance[95]" in January 2020.

When Raoult released his findings in May 2020, prescriptions[96] for HCQ went from an average of 50 per day to several hundred, and then even thousands. The French government quickly acted[97] to recommend it not be prescribed for Covid except in clinical trials, in part based on the falsified Surgisphere study, which will be addressed in this section.

Raoult continued to have success using HCQ, combined with other drugs, as a treatment for Covid-19. From March 2020 through December 2021 Raoult conducted a retrospective cohort study[98] of 30,423 Covid-19 patients. A pre-print version of the study concluded that, "HCQ prescribed early or late protects in part from COVID-19-related death."

It would appear that Raoult poked a hornet's nest in conducting a routine study using routine drugs with decades-long safety profiles. After the pre-print of the study was published in March 2023, a group of French research bodies[99] called for Raoult to be disciplined for the "systematic prescription of medications as varied as hydroxychloroquine, zinc, ivermectin and azithromycin to patients suffering from Covid-19 without a solid pharmacological basis and lacking any proof of their effectiveness."

Just to review: Hydroxychloroquine and ivermectin are on the World Health Organization's list of essential meds and have almost no side effects or interactions with other medications. Zinc is an essential nutrient found in a variety of plant and animal foods and is available in pill form in any drug store. Azithromycin is an antibiotic that has been widely prescribed for decades and is also a WHO essential medicine. Multiple doctors and hundreds of studies[100] have found ample evidence that these and other off-label meds were effective in treating Covid patients. Just what are these French research bodies so worried about?

Doctors who question the official narrative are persecuted:

Before proceeding further, it's necessary to point out that **every doctor mentioned in this paper, who has questioned the official Covid-19 narrative, has experienced public and professional persecution.** Enter any of their names in a search engine and a list will appear of negative articles smearing their character and credentials. They-have lost jobs, been censured, suffered financial losses, and been subjected to threats and actions against their medical licenses and certifications.

> It's important to ask ourselves why these professionals, who prior to the Covid-19 pandemic were well-respected and had successful careers, would subject themselves to the kind of

ridicule and harm – both professional and financial, that they've experienced for questioning the official pandemic narrative. It would be much easier to go along and stay quiet.

This turning against freedom of speech and thought in medicine, and in other areas of our lives, should be of concern to all of us.

Dr. Meryl Nass reviews toxic use of HCQ in Recovery and Solidarity trials:

In June 2020 Dr. Meryl Nass was a practicing physician in Maine. In previous years Dr. Nass had testified before Congress multiple times with regard to the Anthrax scare and biomedical terrorism, among other issues. Having been contacted by the Indian Health Ministry with some concerns, Dr. Nass was led to analyze two large HCQ studies[101] – the Recovery Trial and the Solidarity Trial.

The Recovery Trial[102], a joint effort with the UK government, the Wellcome Trust, and the Bill & Melinda Gates Foundation, had ended in early June, concluding that HCQ did not mitigate Covid, and led to higher death rates in patients.

The WHO's multi-nation Solidarity Trial[103] had recently resumed the HCQ arm of the study which they had paused in May 2020 due to the reports from the Surgisphere study published in the *Lancet*, a premiere medical journal. The Surgisphere study[104], which claimed patients who received chloroquine or HCQ had 35% higher death rates, was retracted 13 days after publication, as its data was determined to be fabricated.

Corruption of the medical journals:

The high-impact medical journals played a massive role in the human toll of Covid by censoring positive studies of repurposed drugs like hydroxychloroquine and ivermectin. They published clearly fraudulent trials that were designed to fail; to show that ivermectin didn't work, and to show that hydroxychloroquine didn't work. They also manipulated trials showing the safety and efficacy of the vaccines.

DR. PIERRE KORY | Pulmonary & Critical Care Specialist | Co-founder FLCCC

The **Surgisphere** piece was a scandal[105] in the medical journal industry as to how such a **shoddy and falsified study** made it through peer review and into print. Rather than being an anomaly, the Surgisphere study became **emblematic of the corruption in the vetting and the peer review of articles in prestigious medical journals during Covid.**

One of the problems with **retracted studies** such as Surgisphere is they **continue to be cited as if they were legitimate.** Taros and colleagues found in an analysis[106] that "retracted articles were cited an average of 44.8 times" which was higher than average. They also found that the presence of the words "withdrawn" or "retracted" before an article title did not affect citation rates.

In the case of a falsified study, retraction is necessary. But many journals no longer provide clear explanations as to why articles are retracted. For example, Jessica Rose and Dr. Peter McCullough submitted a study on vaccine-related myocarditis to the journal *Current Problems in Cardiology*. The peer-reviewed study was accepted for publication, but then without explanation, it was withdrawn[107]. McCullough is certain the study was withdrawn because it did not support[108] the official narrative that Covid vaccinations are safe and effective. Rose and McCullough found that 3,569/3,594 (99.3%) cases of myocarditis requiring hospitalization were not co-associated with Covid-19 respiratory illness, but were temporally associated with Covid-19 vaccination. The study was eventually listed on Zenodo[109], a general-purpose open repository.

After the Surgisphere study was determined to be falsified, the Solidarity trial resumed. Dr. Nass noted that in Solidarity, Recovery, and REMAP[110] (another trial looking at possible Covid treatments), **hydroxychloroquine was only being administered to hospitalized patients.** HCQ is most effective during the early, viral replication, stage of Covid-19 disease, and is **not very effective for someone who is already ill enough to be hospitalized.**

Toxic doses of HCQ were administered to trial participants:

But aside from the fact that they were giving HCQ at the wrong point in Covid infection, Dr. Nass was alarmed to learn that **both the Solidarity and Recovery trials were administering toxic doses of HCQ to trial participants**. The health ministry of India, which had profound success using

HCQ to treat Covid-19, informed Nass that they had contacted the WHO with concerns that the Solidarity Trial was using **four times the standard dose.** Even worse, in the REMAP study, **patients who were targeted for HCQ administration were already on a ventilator or in shock.** That is, patients who were already near death were given toxic doses[111] of HCQ.

Nass states, "[HCQ] is very safe when used correctly, but not a lot more can potentially kill." The WHO's hired consultant in 1979, H. Weniger[112], had looked at 335 episodes of adult poisoning by chloroquine drugs, noting that a single dose of 1.5 to 2 grams of chloroquine base "may be fatal." The Recovery trial used 2.4 grams in the first 24 hours of treatment, and a cumulative dose of 9.2 grams over 10 days. The Solidarity trial used 1.55 grams of HCQ base in the first 24 hours. Nass concluded, "Each trial gave patients a cumulative dose during the first 24 hours that, when given as a single dose, has been documented to be lethal. (The drug's half-life is about a month, so the cumulative amount is important.)"

Nass confirmed that the massive dosage of HCQ in the trials was not recommended for *any* therapy condition according to the drug's U.S. label, and various pharmacology reference sources.

> The [trials were] not, in fact, testing the benefits of HCQ on Covid-19, but rather [were] testing whether patients survive toxic, non-therapeutic doses.
>
> MERYL NASS

On June 15, Dr. Nass contacted WHO Director General Tedros Ghebreyesus informing him of her findings, and pointing out that trial directors, and the WHO, would be liable for damages if trial subjects had not been informed of the known risks associated with high doses of HCQ. On June 17, the **WHO abruptly ended the Solidarity trial**, claiming the decision was based on the Recovery trial results, among other data.

Dr. Nass has continued to call out the medical ethics violations she observed during the pandemic. Specifically she has highlighted the **dangerous precedent of government interference in the patient-provider[113] relationship.** Nass states:

> [This battle is] about whether doctors and patients will be allowed in the future to decide on the care of the patient or whether there will be intrusions by the federal government, the insurance companies, the WHO [World Health Organization], the U.N. [United Nations], etc. who will be calling the shots and telling us what we can and can't do to treat patients.

In more than three decades of practice as a physician, there has not been one patient complaint against Nass. In fact the three patients the Maine Board of Licensure claims Dr. Nass harmed by treating them for Covid with off-label medications (including HCQ and ivermectin) were not even interviewed by the Board. They were, however, interviewed by Nass' attorney, and all three expressed appreciation for Nass' handling of their cases, and anger that Dr. Nass was being targeted by the Board. For her courageous efforts, Dr. Nass' medical license was put on probation[114] by the Maine Board on September 26, 2023. Dr. Nass has filed a countersuit[115] against the Board for retaliatory conduct against her, and for violation of her First Amendment rights.

Killing Hydroxychloroquine:

> *"Who or what is willing to maim and kill patients in order to kill hydroxychloroquine's use in Covid-19?"*
>
> **DR. MERYL NASS** | June 19, 2020[116]

June 15, 2020 – FDA revokes EUA approval of HCQ to treat Covid-19:

Relying on the skewed results from the Recovery Trial, on June 15, 2020, the FDA revoked its EUA approval of HCQ. The FDA alert[117] stated that "continued review of the scientific evidence available for hydroxychloroquine sulfate...to treat Covid-19" determined that "the potential benefits of...HCQ no longer outweigh the known and potential risks for the authorized use."

FDA's withdrawal of EUA for HCQ leads to patient deaths:

Contrary to policy-setting bureaucrats who were not treating patients, boots-on-the-ground doctors, such as Dr. Zelenko and Dr. Didier Raoult, were finding that HCQ and ivermectin (ivermectin is addressed later in this paper) were part of a regimen that prevented hospitalization for Covid, and reduced the length of time for illness. Dr. Paul Marik,[118] the second most published critical care physician in the world, recounted in a January 24, 2022 U.S. Senate panel discussion[119] his extreme distress when prohibited by the hospital board from using repurposed (off-label) drugs after the FDA withdrew EUA.

To clarify, off-label, refers to using a drug to treat a condition other than those listed on the product label. Off-label prescribing is a key component in the practice of medicine, as doctors use their knowledge and intuition to treat each patient. Dr. Marik noted in a later hearing[120] that 40 percent of drugs used in hospitals are off-label drugs, which use the FDA encourages, and "off-labeling is just a technical point about advertising." Specifically, a drug company cannot advertise a product for use in any capacity other than its FDA-approved use. Doctors, however, can prescribe off-label, and share the results with colleagues.

At the time of the FDA's withdrawal of EUA for repurposed drugs, mortality for Marik's patients with Covid had been 50 percent that of his colleagues, yet he was instructed to stop using the off-label protocol he'd developed, and to use remdesivir.

Marik reported emotionally, "For the first time in my entire [40-year] career, I could not be a doctor...I had to stand by idly watching these people die." For speaking out at his hospital, Marik lost his hospital privileges and was reported to the national practitioner database, potentially ending his medical career.

Profit motives behind killing HCQ:

With regard to the suppression of HCQ use, Dr. Nass concluded, **"WHO and other national health agencies, universities and charities have conducted large clinical trials that were designed so hydroxychloroquine would fail** to show benefit in the treatment of Covid-19, **perhaps to advantage much more expensive competitors and vaccines in development,** which have been heavily supported by Solidarity and Recovery trial sponsors and WHO sponsors."

Robert F. Kennedy, Jr explained[121] it this way in 2022:

> There's 400 studies that show benefits from hydroxychloroquine and almost 100 studies, I think 99, that show extreme benefit...of ivermectin. And there's a handful of studies that are government-produced, WHO-produced, financed by Bill Gates that say that there was no benefit, but those studies have a lot of problems.

The WHO's Solidarity trial examined remdesivir, hydroxychloroquine, and two other drug combos for effectiveness against Covid-19. The **WHO's statement that remdesivir should not be used was ignored by Fauci and the FDA.** The WHO's sabotage of HCQ was used to suppress the life-saving drug in favor of toxic remdesivir, and also to make way for the unnecessary Covid shots.

4. STAGES OF COVID-19 INFECTION

If you think about a virus, what's the purpose? What's the virus trying to do? It's trying to stay alive...And if the virus kills someone, if it kills the host, it dies with the host. So it totally defeats the purpose.

Because the goal of a virus is to survive, replicate, and spread, it tends to evolve toward being more infectious and less deadly. There are exceptions and other factors, but in general... that's what virologists expect to see occur with SARS-CoV-2, the coronavirus that causes COVID-19.

NORTHEASTERN (UNIVERSITY) GLOBAL NEWS[122]

December 13, 2021

In natural Covid infection, the virus enters the body through the nose and mouth. The spike protein of the virus binds with the ACE2 receptor in the nose and the virus replicates for a few days. The immune system either takes care of it there, or the virus continues into the lungs and a person becomes symptomatic while continuing to fight the infection.

For many people Covid has not been much more than a cold. Some have been completely asymptomatic. Others have had body aches and chills, fever, nasal stuffiness, nausea, cough, loss of taste and smell, exhaustion, and weakness, among other symptoms. Covid-19 can be a nasty disease, but from the beginning it had a 99.98% recovery rate, which means most people recover from Covid infection. The pattern for viruses is for them to become more transmissible, and less lethal. SARS-CoV-2 was no exception in this regard. With the advent of the Omicron variant, Covid became much milder than the original Wuhan and the Delta variants had been.

An important factor of natural Covid infection is the fact that **those who recover from natural infection have immune systems trained to mount a response to all parts of the virus – not just the spike protein.**

Severe Covid and the cytokine storms of Wuhan and Delta in 2020-2021:

The toxic part of the SARS-CoV-2 virus is the spike protein. For those unfortunate enough to experience severe Covid, the disease can progress to the point that the spike spreads through the bloodstream to the body's organs, causing a cytokine storm. Cytokines are molecules that promote inflammation. In a cytokine storm, too many cytokines are released, leading to an over-activation of other immune cells such as T cells, macrophages (a type of white blood cell that helps eliminate pathogens), and natural killer cells. **Essentially, in a cytokine storm, the immune system attacks the body it was meant to protect.**

As the excess activity of these immune cells continues, damage occurs to the endothelial lining of the circulatory system and the alveoli of the lungs, and eventually leads to thrombosis, which involves blood clotting, circulation disorders and multiple organ failure.

Covid-19 is primarily a risk to the elderly and those with chronic illnesses:

Most people who contract Covid recover, even those who have some days of being very ill. Stanford professor John Ioannidis, a meta-research specialist, determined from early data that Covid-19 had a case fatality rate lower than influenza. (Ioannidis' later meta-analysis[123], based on more data from around the world, placed the overall case fatality rate at 0.20 percent, but the figure was almost 0.0 percent for children and young people.)

Covid-19 disease has mostly impacted the elderly and those with other serious illnesses and conditions, such as diabetes and obesity. For example, in the U.S. over 80 percent of Covid deaths were in the 65 and older population[124], and almost all deaths, regardless of age, were in those with comorbidities. U.S. data in 2021, which held consistent in later evaluations, found that not one healthy child in the 0-12 age group died of Covid[125]. In Sweden, where schools remained open throughout 2020 for primary and middle school students, there was not one Covid death[126] among the 1.8 million children enrolled. (Sweden also did not mandate masks or social distancing in the schools, or other settings.)

Covid Wuhan and Delta variants were a serious disease for the vulnerable: Because the original Wuhan variant and subsequent Delta variants of SARS-CoV-2 were more toxic, the failure to treat in early stages of the disease was an especially egregious lapse in medical ethics. The following chart released by cardiologist Dr. Peter McCullough in October 2020 shows the "Untreated Mortality Risk" in the stages of Covid-19 disease. The bar across the bottom shows the "Ambulatory Phase," "Hospitalization Phase," and "Death." The black curved line shows the risk of mortality rising as the disease progresses without treatment.

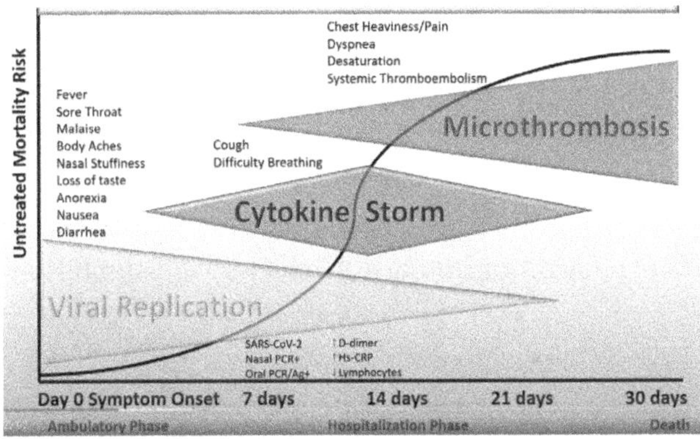

This chart is a screenshot from a video titled Ambulatory Treatment of Covid-19, published in an article by Dr. McCullough in Lyme Disease Association on December 7, 2020. Click on this link[127] and scroll down to the embedded video on the bottom left.

I personally know of two people who died early in the Covid-19 pandemic because of failure to treat during the beginning viral replication phase of their illness. One was a neighbor, age 60, who was in good health when she contracted Covid. The other was the husband of an acquaintance, in his late 50s, who was slightly overweight and struggled during cold and flu season each year.

Both were treated with the Standard Protocol, which is to say, they were *not* treated. Think about it. **Do you know of any other disease or condition in your lifetime in which the doctor sends you home without prescribing something to help with your symptoms?**

Standard Protocol: "Go home, but come back to the ER if your lips turn blue:"

When they became ill and **went to the doctor, or ER (I'm not sure which), both my** neighbor and the acquaintance mentioned above, were told there was no early treatment for Covid. They were **sent home**, told to take Tylenol for their discomfort, and to come back to the hospital if their breathing got so difficult that their lips turned blue.

Both dealt with worsening Covid-19 symptoms **until they were having difficulty breathing** and were very ill. **Both were subjected to the toxic protocol of remdesivir and ventilation in the hospital. Both died.** It is likely these individuals would have recovered from Covid had they been allowed early treatment, as would many thousands of others.

Dr. Peter McCullough explained [128] in early 2021, that the decision to do nothing led to the worst possible outcomes from Covid-19 – hospitalization and death. He noted that people were often home for 14 days, growing gradually more ill without treatment, until they were hospitalized. McCullough stated, "There's no immediate lethality to the virus. In fact we've got a long window of time to make a diagnosis, organize treatment, and prevent hospitalization and death." McCullough emphasized:

> The different unique aspect of the medical response to SARS-CoV-2 and Covid-19, was for the first time we had an infectious disease where the medical community settled into a group think - and this was supported by the NIH, the CDC, the FDA, the American Medical Association, all the medical societies - to tell doctors, "Don't touch this virus. Let patients stay home until they get as sick as humanly possible, and then when they can't breathe anymore, then go to the hospital"...**The federal agencies were enormously inept in terms of perceiving what this problem was, incredibly inept. It was just the opposite of what medicine had always been.** Medicine had always been early innovation by doctors, emperic treatment...It always started out with early empiricism (knowledge gained through observation and experience) then getting to guidelines and agency statements years later. (emphasis added)

Neglect was often part of the "Covid treatment protocol" in hospitals: Not only did the federal government financially incentivize a diagnosis of Covid, and interfere with the doctor/patient relationship, but the overall care of Covid patients was often callous. Certainly there were medical personnel who continued to give compassionate care, but there are also many accounts of neglect, and disregard for patients' treatment preferences and basic needs.

The experience[129] of many Covid patients was that of isolation[130], lack of food and water, and increasing disorientation. Loved ones often couldn't visit. Patients couldn't leave their room. Personnel were hidden behind masks, gloves, and often covered head to toe in protective garb. There was almost no human contact. Medical preferences were ignored[131] and needed medications for chronic conditions were withheld. Some patients were ventilated[132], not because they were having difficulty breathing, but because the medical staff didn't want an infected Covid patient openly breathing in the room. Being placed on a ventilator increases the mortality risk[133] for the patient.

> Last spring, doctors put patients on ventilators partly to limit contagion at a time when it was less clear how the virus spread, when protective masks and gowns were in short supply. Doctors could have employed other kinds of breathing support devices that don't require risky sedation, but early reports suggested patients using them could spray dangerous amounts of virus into the air, said Dr. Theodore Iwashyna, a critical-care physician at University of Michigan and Department of Veterans Affairs hospitals in Ann Arbor, Michigan.
>
> **Wall Street Journal** | December 20, 2020

There are many accounts of the elderly[134], and the disabled[135] being denied food, even being restrained, and being administered drugs[136] that suppressed them into unconsciousness and death.

A hospital stay is not someone's best moment, under the most ideal of circumstances. A hospital stay during Covid was the worst of circumstances for patient well-being. We have not yet come to terms with the shocking fact that malpractice and inhumane treatments were rampant[137] during the Covid-19 pandemic. Treatment was sometimes especially egregious if the patient was unvaccinated for Covid-19.

Patients who died under these circumstances may have had Covid-19 listed as the cause of death, but they were actually victims of medical abuse and neglect.

Discarding the Hippocratic Oath during Covid:

"First do no harm" applies as much to using medical knowledge to treat a patient's symptoms as it applies to not using harmful treatments. It doesn't mean to blindly follow orders, which is what so many in the medical profession did. For those who say, "But Covid was a new disease and we just didn't know how to treat it," comes this answer from Dr. Richard Urso, given during a panel discussion convened by U.S. Senator Ron Johnson in January 2022:

> That's not true, Senator. **We knew early on. We had treatment early on from the very first day in March.** That's a fabricated lie [to say we didn't know how to treat Covid]. It's scientific fraud to say that. **There was treatment for inflammation, there was treatment for blood clotting, there was even treatment that we could try for the virus, there's treatment for respiratory demise.** It was definitely an option...So it's been a fraud from the beginning. (*4:27:40*)[138]

But as seen in case after case the doctors, nurses, and scientists who did ask questions, did try to treat patients, and actually did engage in the practice of medicine, have paid an extremely high price.

5. REAL DOCTORS, NOT BUREAUCRATS, FIND HCQ SUCCESSFULLY TREATS COVID-19

We knew in March of 2020 there were a whole host of effective drugs which could have stopped this pandemic...The success is early treatment...If this was adopted in March and April of 2020, we would have saved hundreds of thousands of lives. We would have abolished this pandemic. It's a moral, ethical outrage that we were not allowed to treat patients with safe, effective, cheap repurposed drugs, in favor of big pharmaceutical control...

DR. PAUL MARIK

Chair of FLCCC | Pulmonary & Critical Care Specialist

July 2020 - Doctors state HCQ and other off-label drugs could end the pandemic:

In July 2020 a group of doctors from around the U.S., who had been successfully treating Covid patients, held a press conference they called the White Coat Summit[139], attempting to counter the maligning of HCQ. **Standing in front of the U.S. Supreme Court in Washington D.C., they announced that HCQ, combined with other** therapeutics**, was remarkably effective at treating Covid-19.**

The White Coat Summit doctors were finding in each of their practices that **HCQ administered at onset of Covid illness was preventing progression of the disease, almost eliminating hospitalizations and deaths. It was their professional opinion that properly administered HCQ could have a significant impact on bringing the pandemic to an end.** These doctors also spoke out against the fear mongering, the lockdowns, and especially the closure of the nation's schools, **citing the**

low transmission rate of Covid from children to adults, and the low impact Covid had on infected children.

One of the physicians, Dr. Richard Urso stated, "The whole political situation has driven the fear toward this drug." He explained that the safety profile for **HCQ is safer than aspirin, Motrin, and Tylenol,** but that the REMAP, Solidarity, and Recovery Trials had all used 2400 mg in the first day. Dr. Urso explained that as a prevention against Covid-19, only 200 mg twice a week of HCQ was needed. **But the clinical trials, Urso said, "used massive toxic doses and guess what they found out? When you use massive toxic doses, you get toxic results.**" He explained that HCQ concentrates in the lungs, and combined with zinc, is highly effective as both a prophylaxis and early treatment of Covid-19 disease.

Character assassination of dissenting doctors:

> *The media's COVID coverage has inarguably been an abomination. Instead of providing healthy skepticism, an adversarial relationship to authority and demanding accountability and acknowledgement of mistakes, the media chose to be establishment cheerleaders.*
>
> **IAN MILLER**, *Unmasked* | July 23, 2023[140]

The video received millions of views before being blocked by YouTube for "misinformation." Mainstream media, and social media platforms were not interested in these experts who were finding success in treating Covid. Being motivated by Big Pharma advertising dollars[141], and government pressure[142], they were **busy being an enforcement arm of the government against "misinformation."** (See the Twitter files[143] published in April 2023, this analysis in *The Hill* published September 13, 2023[144], and the ruling of the 5th Circuit Court on October 3, 2023[145] in Missouri v Biden).

Character assassination of each of the doctors in the White Coat Summit video began on that day and has continued throughout the pandemic and afterward, as has been the case for any professional who has shared an opinion counter to the official narrative.

Dr. Peter McCullough, an internationally known heart surgeon, was also successfully treating Covid-19 patients with HCQ. One of the most published doctors in the world in his field, Dr. **McCullough developed a successful protocol[146] for early treatment of Covid-19 that he and colleagues published in the American Journal of Medicine on August 7, 2020. It included the administration of the** antivirals **HCQ, doxycycline, and favipiravir, together with other** off-label **and over-the-counter meds.**

6. IVERMECTIN IS "POWERFULLY EFFECTIVE" AGAINST COVID-19

When Covid came along it was clear that there was no evidence base to the strategies being promoted by the authorities, by the WHO and governments. And so I became interested in how I could help and promote evidenced-based strategies. That opportunity came when I saw Dr. Pierre Kory's testimony in front of the U.S. Senate, asking that ivermectin be used. And I was fairly intrigued as to why a doctor should have to plead with politicians to use a safe old medicine that's been around for ages. So I did a rapid review of the available evidence [on ivermectin] and it was clear it really should be used and there wasn't much to lose by giving it a try.

DR. TESS LAWRIE, U.K.

Researcher, External consultant to WHO | Co-founder World Council for Health [147]

American Thought Leaders, Dec 6, 2022

Having successfully treated hundreds of Covid patients, **Dr. McCullough updated his protocol [148] in November 2020 to include ivermectin as an effective Covid treatment.** McCullough published his update in the Lyme Disease Association journal, because major journals would no longer print a Covid protocol that didn't follow the official government narrative.

Dr. McCullough stated **that early outpatient treatment of Covid "is the only hope in reducing the risk of hospitalization and death."** He noted that ivermectin, in addition to being an anti-parasitic, has antiviral properties and works by inhibiting the viral entry into the nucleus of cells. **McCullough emphasized that neither HCQ or ivermectin cause cardiac**

toxicities, when given in proper dosages. On November 19, 2020, Dr. McCullough gave testimony[149] of his findings in a hearing held by the Senate Committee on Homeland Security & Governmental Affairs.

Dr. Pierre Kory testifies before the U.S Senate about ivermectin:

On December 8, 2020 Dr. Pierre Kory, pulmonary and critical care specialist and president of Front Line Covid Critical Care (FLCCC), testified at the Senate Committee[150] on Homeland Security and Governmental Affairs in a hearing on "Early Outpatient Treatment: An Essential Part of a Covid-19 Solution."

Dr. Kory noted that the FLCCC[151] is comprised of some of the most highly published physicians in the world, who have all been working from the beginning on how to treat Covid with existing meds. **Kory said he was "severely troubled" that there was no government task force assigned to identify** repurposed drugs **to treat Covid-19. "Everything has been about novel and/or expensive pharmaceutically engineered drugs. Things like tocilizumab, and remdesivir, and monoclonal antibodies, and vaccines."**

"We have filled that void," said Dr. Kory. He cited "mountains of data from the past three months,...from many...countries around the world, showing the miraculous effectiveness of **ivermectin. It basically obliterates transmission of this virus."** Dr. Kory pointed out that ivermectin led to the Nobel Prize in Medicine in 2015 for the doctors who developed it against the parasite that causes river blindness, and is a WHO Essential Medicine (see WHO list p. 8)[152]. **Ivermectin was now showing amazing strength as an** antiviral **and** prophylactic **against Covid-19.**

Early treatment of Covid-19 critical to relieving the healthcare system:

Dr. Kory expressed the exhaustion that he and other doctors were experiencing in treating Covid patients in-hospital, and **emphasized that early treatment of Covid-19 was critical to relieve the pressure on the healthcare system**. Kory described 11 randomized controlled trials from around the world of nearly 4,000 patients where ivermectin, compared to the placebo group, led to lives saved, less need for hospitalization, less transmission (even within households where one member was ill with Covid), and less case counts. Kory declared, **"[Ivermectin] is**

a fundamentally and powerfully effective therapy against Covid-19."

Dr. Kory called on the NIH, CDC, and FDA to urgently review the latest data on ivermectin and issue guidelines for its use, reversing the recommendation against ivermectin that had been issued months before.

But they did not.

7. THE WAR ON IVERMECTIN AND HCQ

Earlier this year the [FDA] put out a special warning that "you should not use ivermectin to treat or prevent COVID-19." The FDA's statement included words and phrases such as "serious harm," "hospitalized," "dangerous," "very dangerous," "seizures," "coma and even death" and "highly toxic." Any reader would think the FDA was warning against poison pills. In fact, the drug is FDA-approved as a safe and effective antiparasitic.

HENDERSON & HOOPER

Wall Street Journal Opinion | July 28, 2021

Early in the pandemic, doctors such as McCullough and Kory believed they were working hand-in-hand with the government to promote public health and bring an end to the pandemic. Unfortunately, many of the heads of government agencies managing the pandemic had less altruistic motives. **Billions of dollars in taxpayer funding had already been invested in flawed treatments and the rushed development of Covid vaccines. Positions and reputations were on the line.**

The powers that be didn't *want* an already approved effective Covid treatment:

If Covid-19 could be treated effectively with already FDA-approved meds, such as hydroxychloroquine and ivermectin, there would be no legal justification to give Emergency Use Authorization to the Covid vaccines in development.

> As stated in the *FDA's own rules*[153], under section 564 of the FD&C Act, "The FDA can use its **Emergency Use Authorization** authority to allow the use of unapproved medical products...to diagnose, treat, or prevent serious or life-threatening diseases **when certain criteria are met, including that there are no adequate, approved, and available alternatives.**" (emphasis added)

The Covid-19 pandemic was a moneymaker for certain players for whom there was not much benefit in finding inexpensive, already FDA-approved treatments that would bring the pandemic to an end. The truncated clinical trials for Pfizer and Moderna's Covid-19 vaccines were finishing up, and the mass vaccination of the public was about to begin - a rollout that would be promoted by the National Institutes of Health's CDC, FDA, and NIAID including a pricey taxpayer-funded advertising campaign[154].

Official propaganda ramps up to suppress over-the-counter Covid treatments:

What had been a somewhat subtle undermining of the use of existing and effective meds such as HCQ and ivermectin became an **all-out battle**. All of a sudden **HCQ and ivermectin, both with a long history of safety, were treated as if they were ineffective, harmful, untested drugs.** HCQ was represented as dangerous to your heart. The NIH came out with recommendations against using HCQ due to its "lack of effectiveness shown in clinical trials and its many side effects," referring to the same Solidarity, Recovery, and REMAP trials that sabotaged HCQ by giving lethal doses to patients.

A *Wall Street Opinion* piece in July 2021 asked, "Why Is the FDA Attacking a Safe, Effective Drug?" The authors pointed out, "Ivermectin fights 21 viruses, including SARS-CoV-2, the cause of Covid-19...Despite the FDA's claims, ivermectin is safe at approved doses. **Out of four billion doses administered since 1998, there have been only 28 cases of serious neurological adverse events, according to an article published this year in the *American Journal of Therapeutics*. The same study found that ivermectin has been used safely in pregnant women, children and infants.**"

> If the FDA were driven by science and evidence, it would give an emergency-use authorization for ivermectin for Covid-19. Instead, the FDA asserts without evidence that ivermectin is dangerous.
>
> **HENDERSON & HOOPER** | *Wall Street Journal Opinion* | July 28, 2021[155]

Yet ivermectin, the Nobel Prize-winning medicine was suddenly described as primarily a veterinary medicine. Like many other drugs, ivermectin is used in veterinary medicine as well as in human treatments. Over decades, ivermectin has been safely taken by billions of people[156], and is available over-the-counter in many countries.

On August 1, 2021, the FDA tweeted a meme with the caption, "You are not a horse. You are not a cow. Seriously, y'all. Stop it," along with an article[157] about "Why you should not use ivermectin to treat or prevent Covid-19."

With a tweet, the FDA relegated ivermectin to the realm of "horse paste," and shame. How could anyone be so gullible as to use horse medicine to try to treat Covid-19? An online search in November 2023, found the August 2021 FDA article was still in the top four results, and the dozens of articles[158] and studies[159] in favor of ivermectin absent. In a proper scientific environment, opposing views on ivermectin would both be present in the public dialogue so that people could read the information and decide for themselves. Certainly doctors would not be threatened with revocation of their medical license for off-label prescribing of FDA approved medications (see here[160] and here[161]).

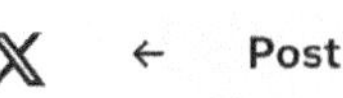

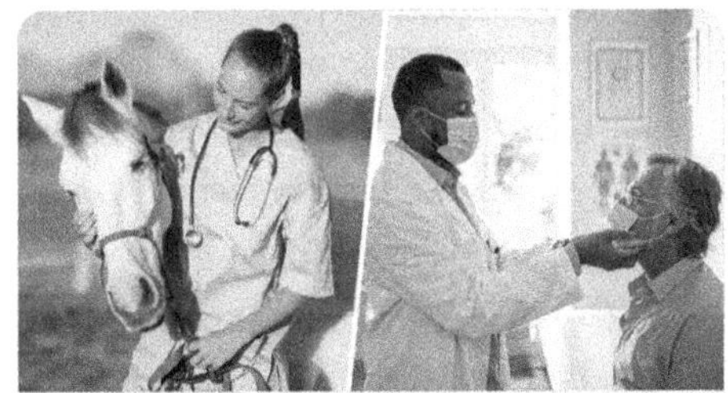

False rumors of hospitals filling with people who took veterinary doses of ivermectin:

With the FDA's pronouncement against ivermectin, allegations were made that people were taking veterinary doses of ivermectin and filling ERs[162], taking the space needed by others. Those claims were refuted by the ER departments[163] of the hospitals in question, but the rumors persisted. Exaggerated claims[164] were made, without comparison context, that poison control centers were being inundated with calls about ivermectin overdose. Pharmacists were instructed by the federal government[165], and also instructed by the American Medical Association and American Pharmacists Association, not to fill prescriptions for ivermectin, if they were for Covid-19 rather than for FDA-approved conditions. (see here[166] and here[167])

Mainstream Media and Social Media censor speech to "prevent misinformation":

> *Drug marketing is a big business, and companies are willing to spend a lot of money to offer you an easy solution to a health problem you may or may not have...* ***The United States and New Zealand are the only countries where drug makers are allowed to market prescription drugs directly to consumers****. The U.S. consumer drug advertising boom on television began in 1997, when the FDA relaxed its guidelines relating to broadcast media.*
>
> **Harvard Health[168] Publishing, 2017** | Harvard Medical School

Mainstream media, often compromised by Big Pharma[169] advertising dollars, **became a mouthpiece for the government, repeating what they were fed while deriding and suppressing dissenting voices who were questioning the narrative. Social media was pressured by the government to remove posts, and even the entire accounts,** of professional medical practitioners and everyday citizens, who were asking questions and discussing alternative pandemic treatments and responses. The **government's interference was so widespread and egregious that it is currently being sued** in *Missouri v Biden*[170], for violations[171] of First Amendment free speech rights.

Ivermectin manufacturer, Merck ditches ivermectin for expensive Molnupiravir:

Merck[172], the company that originated and still sells ivermectin, agreed with the FDA that ivermectin should not be used for Covid-19. That would seem like strong evidence against it, if not for two reasons. 1) Merck had filed to receive EUA approval[173] for molnupiravir[174], a new antiviral Covid-19 treatment, predicted to garner billions of dollars in sales, and 2) Merck would be liable for exorbitant fines if it promoted ivermectin for off-label use, contrary to FDA approval.

As explained by Henderson[175] and Hooper[176], "While off-label prescribing is widespread and completely legal, it is illegal for a pharmaceutical company to promote that use…The fines for doing so are outrageous. During a particularly vigorous two-year period, the Justice Department collected over $6 billion from drug companies for off-label promotion cases. Merck's lawyers haven't forgotten that lesson."

In an American Institute for Economic Research article[177], the **suppression of ivermectin** is explained this way: **"The reason is not that the drug is ineffective**; rather, the reason is that any expenditures used to secure approval for that new use will help other generic manufacturers that haven't invested a dime. Due to generic drug substitution rules at pharmacies, **Merck could spend millions of dollars to get a Covid-19 indication for ivermectin and then effectively get zero return. What company would ever make that investment?"**

Mainstream media fails to report objectively:

When two different studies on the effectiveness of ivermectin in treating Covid-19 were released in early 2022, mainstream media grabbed onto the TOGETHER trial[178], conducted among 3,515 patients in Brazil, that concluded ivermectin "did not result in a lower incidence of medical admission to a hospital due to progression of Covid-19." The **media entirely ignored** the observational study of 223,128 people in Brazil[179] as part of a "citywide prevention program using ivermectin for Covid-19." That trial found "ivermectin as a prophylactic agent was associated with significantly reduced Covid-19 infection, hospitalization, and mortality rates."

Dr. Kory states [180], "It is a tried-and-true tactic with effective and dastardly results. Big Pharma and other well-financed interests sponsor purportedly 'impartial' medical trials aimed at discrediting cheaper generic alternatives. Ignoring the flaws in the methodology, the media runs wild with the desired narrative, which is amplified by a well-orchestrated public relations effort. Social media shuts down alternative views and critiques. The result is fewer choices and higher prices for vaccines and anti-viral drugs—terrible for consumer health, but terrific for pharma companies' bottom lines." (emphasis added)

Ivermectin: "Entire countries or regions have relied on it."

> *[Ivermectin] has really carried us through the pandemic. Entire countries or regions have relied on it.*
>
> **DR. PETER MCCULLOUGH**

When Dr. McCullough made the above statement about ivermectin, he was likely referring to countries such as **Nigeria, which had a Covid death rate of 15 per million, one of the lowest in the world, compared to the U.S. with a Covid death rate of 3,492 per million – one of the highest in the world.** Nigeria has a much lower vaccination rate than the U.S., and a younger median age, but most of the population takes ivermectin and/or HCQ regularly to prevent river blindness and malaria. (To look at Covid numbers for each country, including Nigeria and the U.S., click here[181] and scroll down. Deaths per million in the population are shown in the 9th column to the right of each country.)

A more side-by-side comparison of using ivermectin to treat Covid, versus using the standard U.S. protocol, is that of Kerala and Uttar Pradesh in India. Kerala[182], population 35 million, used the U.S. Covid protocol, which specifies *not* using ivermectin. But in Uttar Pradesh, when cases soared[183], the government offered a free over-the-counter packet[184] that included ivermectin, to its 230 million citizens. Covid cases dropped dramatically[185] and rapidly in Uttar Pradesh in 2021 and remained low, with the exception of the initial Omicron surge. A comparison between the two states in June 2022[186] found Kerala with higher total cases and deaths than Uttar Pradesh, despite its much smaller population.

Others of the 33 Indian states followed Uttar Pradesh's lead,

including Delhi, Karnataka, and Uttarakhand, and also saw a dramatic drop[187] in Covid infections and deaths. To date, India overall has a significantly lower death rate of 378 Covid deaths per million[188], compared to the U.S.'s 3,492[189] Covid deaths per million, despite a lower vaccination rate, often more crowded living conditions, and hospital systems that struggle to meet the medical needs of a population of 1.4 billion.

A peer-reviewed study[190] published August 8, 2023, analyzed a "natural experiment" during the Covid-19 pandemic in which ivermectin was first authorized **in Peru**, then restricted. **"As determined in all 25 states... reductions in excess deaths correlated closely with the extent of IVM (ivermectin) use**... Several potential confounding factors, including effects of a social isolation mandate imposed in May 2020, variations in the genetic makeup of the SARS-CoV-2 virus, and differences in seropositivity rates and population densities across the 25 states, were considered but did not appear to have significantly influenced these outcomes."

Dr. Meryl Nass[191] and Dr. Harvey Risch[192] have both compiled excellent data, which they continue to update with current studies, on the effectiveness of ivermectin and other off-label treatments for Covid-19 disease. **Studies[193] showing the effectiveness of ivermectin to treat Covid-19 have continued to increase**, yet the FDA and CDC continue to ignore and demean this life-saving treatment.

In 2023 a government lawyer says FDA never prohibited ivermectin:

In a case[194] working its way through the courts, three doctors[195] claim that the U.S. Department of Health and Human Services "has illegally interfered with the practice of medicine," and caused them professional harm, including being fired for prescribing ivermectin to patients.

The plaintiff **doctors state that the FDA's communication during the pandemic regarding ivermectin, an already approved FDA med, was an overstep of authority.** "While the plaintiffs acknowledge the FDA's authority to regulate drugs...the FDA has no authority to 'prohibit, direct, or advise against off-label uses of drugs approved for human use."

> Nothing in the Federal Food, Drug, and Cosmetic Act shall be construed to limit or interfere with the authority of a health care practitioner to prescribe or administer any legally marketed

device for any condition or disease within a legitimate health care practitioner-patient relationship.

21 UNITED STATES CODE SECTION 396[196]

In the **August 10, 2023 hearings before the 5th Circuit Court, FDA lawyer Ashley Cheung Honold said the FDA made the statements about ivermectin** based on "reports" of consumers being hospitalized after self-medicating with ivermectin intended for horses. Attorney Honold said the FDA didn't require anyone, or prohibit[197] anyone, from doing anything. **She called FDA directives**, such as the horse tweet referred to above, **"merely quips."**

Honold continued, "FDA is clearly acknowledging that doctors have the authority to prescribe human ivermectin to treat Covid[198]. So they are not interfering with the authority of doctors to prescribe drugs or to practice medicine." (emphasis added)

We can ask **what makes the FDA think that supposed "quips" to suppress a potentially life-saving treatment during a pandemic can be considered appropriate.** In addition, how is it that these rumors about people being hospitalized for taking veterinary doses of ivermectin were found false[199] in 2021, but the FDA attorney still refers to them in a 2023 court hearing as though they're factual?

Walking the tightrope, FDA "spokesperson" wobbles on ivermectin: The FDA immediately backtracked[200], after Honold's statement in court led to various news organizations and public officials reporting that ivermectin is now authorized by the FDA to treat Covid.

The very next day, August 11, 2023, an unnamed FDA spokesperson made a statement to fact-checking service Politifact[201] that the FDA has *not* authorized or approved ivermectin for use in preventing or treating Covid-19. There was no official, attributable press release.

Another fact-checking service, Snopes[202], published on August 15, 2023, "Because the FDA never had — or even asserted that it had — the authority to block off-label use of FDA-approved drugs, and because the FDA had repeatedly reaffirmed that fact throughout the pandemic, claims that a new policy shift occurred in August 2023 regarding the prescribing of ivermectin for COVID-19 were false."

But the fact remains that because of the FDA's directives, doctors sent ill Covid patients home without treatments. Pharmacists were told not to fill prescriptions for ivermectin to treat Covid. Doctors were told by hospital boards and CEOs that they would be disciplined if they used ivermectin to treat Covid, and some were fired. Medical boards threatened, and took action against, doctors' licenses for treating Covid patients with ivermectin. And strong observational studies and randomized controlled trials that show ivermectin is effective against Covid have been consistently ignored[203] by the FDA.

Emotional abuse perpetrated by the FDA:

In a previous proceeding for this same case (Case 22-40802 5th Circuit Court) in November 2022, FDA attorney Isaac Belfer stated in court that the FDA was not issuing "directives" but was telling doctors what they "should do" when advising against ivermectin as a Covid treatment. **Anyone care to explain the difference between a directive and telling someone what to do? The FDA's claim that its statements about ivermectin were just guidance and "not mandatory" rings hollow.**

In the pattern of all emotional abusers[204], the FDA created the controversy around ivermectin, issued "shoulds" and employed shaming, and now is pretending it didn't do it. The FDA is essentially asking, "Why are these silly doctors suing us and saying we interfered with their practice of medicine? We know they lost their jobs, and patients died, but that's not our fault."

Honold and Belfer, the FDA, and the compromised fact-checking services all bring gaslighting to a whole new level. It would be almost amusing if so many people hadn't been harmed and even killed due to the suppression of inexpensive, off-label, effective life-saving treatments.

On September 1, 2023, the **5th Circuit Court of Appeals ruled that the three plaintiff doctors can continue their case against the government,** and remanded the case to the District Court. The 5th Circuit Court issued a finding that the FDA had overstepped its authority, stating, **"FDA is not a physician. It has authority to inform, announce, and apprise – but not to endorse, denounce, or advise."**

8. PAXLOVID

In approving Paxlovid, the Biden White House and the FDA also seemed to deliberately ignore hundreds of clinical trials conducted on hundreds of thousands of patients detailing the established safety and efficacy of IVM (ivermectin) and HCQ.

DR. DAVID GORTLER

Former FDA Senior Advisor

Aside from Gilead's molnupiravir, another example of a novel expensive drug approved by the FDA for treating Covid is the antiviral Paxlovid[205], brought to you by Pfizer. **Despite listing 30 drug interactions, and possible side effects including liver problems and severe allergic reactions, Paxlovid was granted Emergency Use Authorization by the FDA in December 2021.**

Former FDA advisor Dr. David Gortler[206] states, "In the case of Paxlovid, not only was evidence of failure deliberately ignored; prospective testing methodologies were altered mid-trial to favor a positive outcome when it became apparent that the Paxlovid trial results would not meet their original endpoints. **In fact, Pfizer had already opted to stop its Paxlovid trial, but then changed its mind after the FDA intervened via the White House."**

Effective monoclonal antibodies nixed by the government in favor of Paxlovid:

Worrisome mutations could also become more common if antibody therapies aren't used wisely — if people with COVID-19 receive one antibody, which could be thwarted by a single viral mutation, for

> *example. Cocktails of monoclonal antibodies, each of which can recognize multiple regions of the spike protein, might lessen the odds that such a mutation will be favoured through natural selection, researchers say.*
>
> **EWEN CALLAWAY** | Journalist at *Nature* | September 8, 2020[207]

When **Pres. Joe Biden gave his September 9, 2021 speech[208] mandating Covid vaccines, he also spoke of monoclonal antibodies**. This was likely a direct result of the decrease[209] in hospital admissions that Florida experienced by opening monoclonal antibody infusion clinics around the state. Essentially, the monoclonal antibody infusion provided immediate antibodies to the individual, boosting the person's immune system until the body began to make its own antibodies against Covid.

Biden acknowledged in his speech that Emergency Use Authorized monoclonal antibody treatments had been **shown to reduce the risk of hospitalization by up to 70%, and promised to make more treatments available for "free" across the country.** However**, the very next day**, *Forbes*[210] issued an article titled "Nation Short on Supply of Key Covid Treatment – Desperate States Told To Reduce Requests." The article reported that a **"new distribution strategy** from the HHS to ensure patients who need the treatments will be able to get it," involved **reducing deliveries to seven states**[211] along the Gulf Coast (so-called "red states"). The **rationale was that these states with "low vaccination rates," had been utilizing about 70% of the distributed supply.**

By December of 2021, **monoclonal antibodies were all but phased out[212] by the federal government, claiming that there wasn't clinical data to prove they were effective against the new Covid variants.** This conveniently drove doctors and medical facilities toward using paxlovid [213] and molnupiravir [214], both of which received EUA approval in December 2021. Paradoxically, although **the Covid-19 shots also were not supported by any clinical data showing they were effective against the new variants, the "shot in every arm" campaign continued.**

Paxlovid was fully approved on May 25, 2023, to treat "mild-to-moderate Covid-19." This despite the fact **that rebound[215] is a known possible reaction to Paxlovid, meaning the disease returns with worse**

symptoms upon completion of the course of treatment. Observationally, the rebound effect from Paxlovid is so prevalent that it notably occurred in President Joe Biden[216], CDC Director Rochelle Walensky[217], and then-NIAID Director Anthony Fauci[218].

Paxlovid is ineffective and expensive:

Dr. McCullough says of Paxlovid[219], "I don't think it's terribly effective; the CDC did put a warning out on it in May of 2022 that says it prolongs illness and causes rebound." McCullough has used Paxlovid in his practice, but never as a standalone medication. "People have learned that Covid's a serious illness, particularly in the elderly, and a single drug is not going to be enough," states McCullough. His chosen Covid treatment is an ivermectin-based approach combined with other meds.

Gortler summarizes, **"Paxlovid was a failure, but the White House had foolishly already paid Pfizer $5.3 Billion in advance**. Rather than admit failure and epic waste, the FDA then stepped in and with zero transparency, altered the established clinical trial parameters mid-trial to make Paxlovid's findings seem better than they were. Pfizer then completed the trial, declared Paxlovid a success and the White House doubled down on its $5.3 billion investment, spending a sickening total of $10.6 billion on Paxlovid."

FDA Chief breaks FDA rules and promotes Pfizer's Paxlovid to the public:

In what has become a continuous parade of non-contrite rule breakers, **FDA commissioner Robert Califf** admitted in an April 2023 interview[220] with Bob Wachter, Chair of the Department of Medicine at UCSF, that he **promoted Paxlovid as a treatment for Covid. "In normal times," Califf said, "the FDA should not be a cheerleader, the FDA is a referee...but in this case we were in the middle of a pandemic, people were dying at very high rates."**

Pfizer was prevented by FDA rules from advertising Paxlovid to the public, due to its Emergency Use Authorization, so Califf promoted it for them. Califf cited only one clinical trial as proof for Paxlovid's effectiveness, failing to note the poor parameters[221] of the trial that essentially made it irrelevant to treatment of the Omicron variant.

Investigative medical journalist Maryanne Demasi[222] says, "It's not clear why Califf felt the need to promote the use of Paxlovid given the U.S.

government had already committed to purchasing 10 million treatment courses at a cost of $5.29 billion."

Demasi quotes Jessica Adams, an FDA regulatory affairs expert who said, **"Something is really wrong with public health 'leadership' that thinks every norm can be thrown out the window in an emergency... The FDA has learned nothing during the pandemic and is setting terrible precedents for future emergencies."**

It appears that Adams is correct about the FDA having "learned nothing." In an April 24, 2023 interview[223], FDA Director Robert Califf expressed that the process of granting approval to drugs is a heavy responsibility:

> A well-intentioned person thinks, "I have a treatment that prevents this disease," but you don't see the benefits of that for years to come. But you will see the risk, and **to approve it with no demonstration that the benefit/risk equation is positive,** is a **very scary** thing to do, **particularly for big decisions that affect millions of people.**

An underlying sense, in reading those words, is that Director Califf is attempting to excuse the truncated approval process for vaccines and therapeutics during the pandemic. **The purpose of the FDA is to assure that the benefit/risk ratio is positive, and that safety is thoroughly evaluated *before* approving any new drug.** When millions of lives are impacted by your decisions, hoping you will see benefits in the future is not enough. Califf's attitude may explain something about why two 30-year veterans of the FDA resigned[224] in protest in September 2021.

Clayton J. Baker, MD, former Clinical Associate Professor of Medical Humanities and Bioethics at the University of Rochester, states that Califf is correct: "'To approve [a medical treatment] with no demonstration that the benefit/risk equation is positive, is a very scary thing to do.'" **Dr. Baker continues, "I would add that such an approval is more than scary. It is also thoroughly unethical,** as it violates all four of the pillars of medical ethics[225]. It is the very definition of medical adventurism. Furthermore, to coerce or mandate people to take such a treatment is outright monstrous."

Dr. Peter McCullough states[226] "The question on the table is why is the FDA so resistant to even inspecting or re-evaulating safety, let alone pulling these [Covid vaccine products] off the market. Why is Robert Califf, the FDA Commissioner so silent on vaccine safety?"

CDC promotes expensive antivirals, vilifies off-label inexpensive antivirals:

So, feeling unwell? Think you have Covid? Take an ineffective PCR test[227] and grab a useless mask – manufacturers will thank you. Go to CDC.gov[228] for your treatment options – Pfizer's Paxlovid $530 per treatment[229] course, Merck's Molnupiravir $712 per treatment[230] course, and Gilead's Remdesivir $3,120 per treatment[231] course.

Take note of the CDC warning "People have been seriously harmed and even died after taking products not approved for use to treat or prevent COVID-19, even products approved or prescribed for other uses. Talk to a healthcare provider about taking medications to treat COVID-19." Be sure and click on the hyperlink in the warning which takes you to an ancient March 28, 2020 CDC article[232] informing that "chloroquine...can cause serious health consequences, including death."

Note the most recently updated October 4, 2023[233] CDC advice for outpatient Covid-19 treatment:

> There is strong scientific evidence that antiviral treatment of outpatients at risk for severe COVID-19 reduces their risk of hospitalization and death. **The antiviral drugs nirmatrelvir with ritonavir (Paxlovid) and remdesivir (Veklury) are the preferred treatments for eligible adult and pediatric patients** who are at high risk for progression to severe COVID-19.

Ka-ching! And we haven't even looked yet at the money that incentivized companies vying for the contracts to develop Covid-19 vaccines. **One need only look at the revenues for some of the major drug companies in 2022 to get a feel for the financial largesse involved in Covid-19 treatments.**

In 2022, Pfizer's revenues[234] soared to a record $100 billion. Moderna's[235] full-year 2022 revenues were $19.3 billion. Gilead Sciences[236] annual gross profits for 2022 were $21.6 billion. Merck's[237] 2022 worldwide sales were $59.3 billion. In May 2021, CNN[238] Business ran an article titled, "Covid vaccine profits mint 9 new billionaires." It's to be assumed that the number has grown since 2021.

I'll rub your back if you'll rub mine: FDA commissioners work for Big Pharma

When Robert Califf was sworn in as head of the FDA on February 17, 2022, it was without the support of Senator Bernie Sanders[239], among others, who **viewed Califf's ties to Big Pharma as compromising his ability to fill the post.**

Califf, after leaving the FDA for a time in 2017, received consulting fees from Merck, Biogen, and Eli Lilly, before returning to government. Sanders pointed out that, according to Califf's financial disclosure form, he "owns up to $8 million in the stocks of major drug companies." Califf has also made millions of dollars as a consultant for more than a dozen pharmaceutical corporations.

Sanders stated, **"At a time when the American people pay the highest prices in the world for prescription drugs and as drug companies continue to be the most powerful special interest in Washington, we need leadership at the FDA** that is finally willing to stand up to the greed and power of the pharmaceutical industry." **With his ties to Big Pharma, Califf is certainly not that leader. Nor has anyone in a long time been that leader.**

> Not only have the drug companies spent over $4.5 billion on lobbying and hundreds of millions of dollars in campaign contributions over the past 20 years, they also have created a revolving door between the FDA and the industry. Shockingly, nine out of the last 10 FDA commissioners went on to work for the pharmaceutical industry or to serve on a prescription drug company's board of directors.
>
> **SENATOR BERNIE SANDERS (I-VT.)** | December 14, 2021[240]

Two FDA employees who were deeply involved in the development and approval of the Moderna vaccine for Emergency Use Authorization, have since left the FDA to take lucrative positions[241] with Moderna. The many conflicts of interest, and financial incentives influencing FDA regulators, make the whole organization suspect. The health and well-being of the people they're supposed to protect does not appear to be the main motives in the FDA's product approval process.

The big pharmaceutical companies have plenty of money to go around, despite having paid multi-billions of dollars in fines[242] for products that have harmed consumers.

> Instead of mobilizing an effective public-private response to the advertised problem, Operation Warp Speed and the Task Force served as a vehicle for further panic and the facilitating of taxpayer cash that ended up enriching the pharmaceutical industry. These taxpayer-funded, Covid-related slush funds ballooned to astronomical heights across two presidencies, delivering record profits to Pharma companies that took pains to bring themselves onsides with the people in charge in Washington, D.C...
>
> Operation Warp Speed and the resulting Task Force operation was, by all objective accounts, a catastrophic blunder, but that didn't stop many of its members from parlaying their roles on the high visibility government detail into successful post service gigs.
>
> **JORDAN SCHACHTEL** | Investigative journalist | December 13, 2023[243]

9. POWER OF THE PURSE – NIH DISTRIBUTES BILLIONS FOR RESEARCH AND STUDIES

"The practice of medicine has been corrupted."

DR. PIERRE KORY

Pulmonary and Critical Care Specialist | Founder FLCCC

There's that old saying, "Power corrupts, and absolute power corrupts absolutely." **The Covid-19 pandemic has laid bare unhealthy power structures that were already in place, but were operating largely unnoticed by most people just going about their daily lives.**

Every year, the National Institutes of Health distributes billions of dollars in grants and contracts for research and studies. In 2022 NIH research grants totaled $33.3 billion[244]. RFK,. Jr reports that "Between 2010 and 2016, every single drug that won approval from the FDA – 210 different pharmaceuticals – originated, at least in part, from research funded by the NIH." **The United States Department of Health and Human Services (HHS) is the named owner of at least 4,400 patents. Under an HHS policy, NIAID employees can earn up to $150k annually from drugs they help develop at taxpayer's expense.** (*TRAF*, p. 120- 121)

Don't cross Fauci if you want funding from NIH for your research:

> *The prospect of domination of the nation's scholars by Federal employment, project allocations, and the power of money is ever present and is gravely to be regarded...[I]n holding scientific research and discovery in respect, as we should, we must also be*

> *alert to the equal and opposite danger that public policy could itself become the captive of a scientific-technological elite.*
>
> **PRESIDENT DWIGHT D. EISENHOWER** | Farewell Address[245] | January 1961

At the same time Big Pharma is **granted taxpayer money** for product development, scientists and researchers who need **funding for research are largely dependent on being in the good graces of those who hold the purse strings at NIH.** In a 2006 *Harper's* article[246], Celia Farber exposed the **corruption and vendetta-driven system, headed by Dr. Anthony Fauci for 40 years, that has made NIAID an appendage of Big Pharma.**

RFK, Jr. explains that **with billions of dollars at his disposal, Dr. Fauci has had the "power to make and break careers, enrich – or punish – university research centers, manipulate scientific journals, and to dictate not just the subject matter and study protocols, but also the outcome of scientific research across the globe."**

"During his half-century as America's Health Czar," continues Kennedy, "Dr. Fauci has played a central role in crafting a world where **Americans pay the highest prices for medicine and suffer worse health outcomes compared to other wealthy countries."**

> Adverse drug reactions are among the nation's top four leading causes of death, after cancer and heart attacks.
>
> **ROBERT F. KENNEDY, JR.** | *The Real Anthony Fauci*, p. 119

Farber notes that **"genuine scientists"** are in the minority under the system Fauci nurtured and passed on to the next generation of regulators. These genuine scientists "look, sound, and behave like scientists. And to varying degrees, they **all live in a climate of both economic and reputational persecution...Fauci's vendetta system has many ways of crushing the natural scientific impulse – to question and to demand proof."** (*TRAF*, p. 118-119)

The silencing of dissident voices harms science and medicine:

> *There is perhaps nothing that opens the door to censorship wider than the fear of disease and the prospect of an early death. Indeed, there is nothing that matches a looming pandemic to generate fear. And there is nothing like fear to grease the skids of censorship.*
>
> **JAY BHATTACHARYA** & **STEVEN H HANKE** | September 7, 2023[247]

In **October 2020 three prominent epidemiologists,** one each from Stanford, Yale, and Oxford, **issued the Great Barrington Declaration[248] (GBD) calling for an end to the devastating Covid lockdowns.** The GBD highlighted the inhumane impact of lockdowns on the poor and vulnerable, and called for a return to traditional pandemic response of "**focused protection**." Namely, allowing society to open back up and normal life to continue, while taking actions to protect the elderly and immune-compromised, who were the only cohorts for which Covid-19 was a serious threat. In this manner herd immunity would be reached more quickly, which would in turn provide more protection for the at-risk.

Rather than opening a discussion to consider the legitimate concerns and proposals of these prominent professionals, **Dr. Fauci and then-FDA director Francis Collins labeled them "fringe epidemiologists" and called for a "quick and devastating takedown[249]" of them, and their ideas.** They were successful. The term "herd immunity" was redefined as a "strategy" of allowing the virus to spread without check ("let it rip") through the population. Something the GBD writers never promoted.

One of the GBD authors, Jay Bhattacharya[250], states "American government officials, working in concert with big tech companies, defamed and suppressed me and my colleagues for criticizing official pandemic policies—criticism that has been proven prescient. **While this may sound like a conspiracy theory, it is a documented fact, and one recently confirmed by a federal circuit court."**

Let's not forget that these compromised people at the National Institutes of Health, which is the umbrella organization for the CDC, FDA, and NIAID, are the ones who review, and often profit from, the drugs and vaccines that they approve. This unhealthy relationship between regulators and Big Pharma has never been on more spectacular display than with the Covid-19 vaccines.

10. THESE ARE NOT YOUR FATHER'S VACCINES

There have been two waves of injury to the world. The first has been the infection which preyed upon the frail and the elderly. And then the second wave of injury now has been the Covid-19 vaccines. The role of the WHO appears to be adverse in both of these.

DR. PETER MCCULLOUGH

Testimony, European Parliament | September 13, 2023[251]

As soon as the genetic sequence of SARS-CoV-2 became available in January 2020, the race was on. **Operation Warp Speed was an opportunity for advances in science, medicine, and pharmacology, along with some large profits to be made.** Various companies began developing a vaccine and there were initially more than 180 vaccine candidates, based on several different platforms[252], both traditional and innovative.

How traditional vaccines work:

Dr. Byram Bridle, Associate Professor of Viral Immunology in the Department of Pathobiology at the University of Guelph, summarizes vaccines this way:

> Generally speaking, there are many different kinds and types of vaccines, all with the same objective—tricking a healthy body into thinking it is under attack by a particular disease so the immune system will learn to create the cells and proteins necessary to swiftly destroy the pathogen if it becomes a threat.

The **basic method behind vaccination, until Covid-19,** is that a **small amount of the virus,** that has been killed (inactivated) or weakened (attenuated) so that it cannot cause disease, is **introduced to the body. The body perceives the pathogen and mounts an attack, creating an immune response that will be activated in case of future exposure,** protecting against disease.

New-technology vaccines are the winners in the Covid vaccine race:

> *[The vaccines] are not safe fundamentally, in their essence, in the way they are designed. If you look at how they are meant to work, they are not vaccines. And the reason why they have been approved so quickly is because they've been called vaccines.*
>
> **DR. TESS LAWRIE**[253] | Medical doctor and Researcher | Dec 6, 2022

But it was not traditional vaccines that ended up with emergency use authorization. It was the Pfizer and Moderna mRNA shots, and the Johnson & Johnson and AstraZeneca adenoviral vector shots, that were deployed in the U.S., and much of the Western world and developed nations.

A vaccine inserts a pathogen, a gene therapy inserts instructions to your cells:

A gene therapy and a vaccine are not the same thing. Dr. David Wiseman PhD in pharmacology explained in a December 2022 roundtable discussion[254] that the FDA has a vaccine checklist and a gene therapy checklist. The mRNA Covid shots meet the definition of gene therapy, which means they are a biopharmaceutical product, but the FDA used the vaccine checklist for the approval process.

In September 2019 the Securities and Exchange Commission (SEC) filing[255] for BioNTech states on page 21, "...in the United States, and in the European Union, **mRNA therapies have been classified as** *gene therapy* **medicinal products**...." Moderna's November 2018 (SEC) registration states, "**mRNA is considered a gene therapy** product by the FDA."

October 2021: CDC changes the definition of vaccine

In order to justify this lapse in regulatory processes, the CDC changed the definition of "vaccine." Until October 2021 the CDC defined vaccine as "a product that stimulates a person's immune system to produce immunity to a specific disease, protecting the person from that disease." Immunity is protection from an infectious disease. The new definition of vaccine is, "A preparation that is used to stimulate the body's immune response against diseases."

As explained[256] by Dr. Joseph Mercola, "So, a 'vaccine' went from being something that produces protective immunity, to simply stimulating an immune response." Meaning immunity to disease is no longer a required aspect of a vaccine – it just has to cause the body to produce antibodies. Also, it means that the gene therapy mRNA shots, which should have been evaluated as a biologic, now fit the definition of "vaccine."

> By avoiding the gene therapy checklist, the FDA did not need to check for cancer-causing indicators, or provide a pharmacokinetics study, which shows how the drug is distributed, absorbed, and cleared from the body.

Dr. Robert Malone, who pioneered mRNA technology in the 1980s and has been involved in various aspects of vaccine development throughout his career, has been **an early and consistent voice of warning with regard to the Covid shots.** In a December 2022 U.S. Senate panel[257] he confirmed that they are not a vaccine: "These mRNA and adenoviral vectors are gene therapy applied for the purpose of creating an immune response." (1:15:02)

Novel lipid nanoparticle technology made the mRNA shots possible:

The mRNA gene therapy shots (Pfizer/Moderna) contain a code that is carried to the cells via a lipid nanoparticle envelope. **If the mRNA were not enclosed in the lipid nanoparticle, it would be broken down by the body before delivering the message. The lipid nanoparticle, a** nanotechnology, **had been used in research for some time, but not in clinical use for humans.** Nanotechnology has to do with manipulating atoms and molecules at nanoscale. **There are many nanostructures in**

nature, but nanotechnology has to do with designing and producing such structures intentionally.

> As explained in a *Nature Reviews Materials article*[258], "Lipid nanoparticles are going into billions of arms in the form of Covid-19 mRNA vaccines, delivering, at last, on the promise of **nanotechnology to revolutionize drug delivery...Without lipid nanoparticles, Covid-19 mRNA vaccines would not exist."**

Dr. Malone and Dr. David Gortler explain[259] that lipid nanoparticles are "aggregates of both positively charged chemicals and negatively charged RNA. The result of the interaction between the oppositely charged lipids and RNA is "a 'self-assembling particle'...which forms a sort of blob of mixed fats (the "lipid nanoparticle envelope") coating one or more RNA molecules. Malone and Gortler state, "It is possible that these positively charged, synthetically-manufactured lipids, which do not appear to exist naturally, should be considered a novel biotechnology all by themselves..." Which means they should have received separate evaluation for approval, but they didn't.

The difference between the mRNA and adenoviral vector Covid shots: As just explained, the Pfizer and Moderna shots use a revolutionary drug delivery system, the lipid nanoparticle envelope, to carry the message to the cells telling them to produce the spike protein. The Johnson & Johnson (J&J) and AstraZeneca Covid shots are also gene therapy, instructing cells to make the spike protein using an adenoviral vector platform . University of Missouri Health Care explains that the adenoviral vector vaccine[260] also known as a DNA vaccine, is created by adding the gene for the coronavirus spike protein to an adenovirus, a common virus that causes colds or flu-like symptoms.

As explained[261] by Dr. Joseph J. Eron, of UNC School of Medicine, "The Johnson & Johnson shot uses a weakened common cold virus, called an adenovirus, as a vehicle to deliver a single coronavirus gene into human cells. That gene then provides the instructions for making the SARS-CoV-2 spike protein."

A May 2020 article[262] in *Chemical and Engineering News* (C&EN) explained, "Compared with some of the newer, experimental technologies—such as

Moderna's mRNA vaccine, which was the first to enter human trials in the US—adenoviral vectors are touted as a more tried-and-true approach. J&J calls its adenoviral vector platform a 'proven' technology." The article explained, however, that the only commercial adenoviral vaccine that had ever been available before the Covid shots was a rabies vaccine[263] used to immunize wild animals. C&EN stated, **"So far, no adenoviral vaccines have demonstrated they could prevent disease in humans."**

In May 2022, the FDA limited[264] the J&J Covid-19 vaccine to "people 18 and older for whom other vaccines aren't appropriate or accessible and those who opt for J&J because they wouldn't otherwise get vaccinated." The FDA stated that the change was made "because of the risk of a rare and dangerous clotting condition[265] called thrombosis with thrombocytopenia syndrome (TTS) after receiving the vaccine." TTS was defined by Healthline[266] as a new, "very rare" condition that "happens when a person has both blood clots and low platelet counts after receiving certain COVID-19 vaccines."

The article stated, "The mechanism that causes TTS after Covid-19 vaccination is not yet understood." However, Dr. Sucharit Bhakdi had been warning since March 2021, that the mRNA and adenoviral vector shots would cause blood clots, due to creation of spike protein in "forbidden" areas of the body, such as the circulatory system and the brain. It's just that the European Medical Association and the FDA weren't listening. (See Section 15, Early Warning Voices about the Dangers of the Covid-19 Shots.)

The Johnson & Johnson Covid-19 vaccine is no longer available[267] in the United States as of May 15, 2023.

Novavax protein subunit vaccine:

Less frequently used for Covid-19 vaccination in the U.S. is the Novavax adjuvant vaccine, given Emergency Use Authorization on October 19, 2022. Novavax contains the SARS-CoV-2 spike protein (but not mRNA), and uses a Matrix-M adjuvant. An adjuvant is a substance incorporated into the vaccine that enhances the immune response. The idea behind the protein subunit vaccine is that it trains the immune system to recognize the spike protein and mount an attack. The EUA for Novavax was renewed on March 28, 2023.

11. mRNA SHOTS ARE A FUNDAMENTAL SHIFT IN VACCINOLOGY

You're intervening in a series of nested complex systems. That is not a winning formula. One should be extremely cautious intervening in a complex system, let alone a complex system where you have the immune system inside of a population experiencing a pandemic. The chances that we know enough to do that well, are pretty low.

BRET WEINSTEIN, EVOLUTIONARY BIOLOGIST

Unherd interview[268] | June 6, 2022

Unlike **traditional vaccines**, which essentially stay at the injection site in the arm, the **lipid nanoparticles and adenoviral vectors carry the mRNA message throughout the whole body, causing the cells to make spike protein in unlimited amounts for undetermined amounts of time.** In addition, the Covid shot mRNA is messaging cells, such as those in the brain and the circulatory system, to make spike **protein in places that would not normally encounter the SARS virus during natural infection.**

Studies have shown that the amount of **spike protein in those who have been vaccinated is much higher than in those who have natural infection.** We were told that the mRNA would break down after a few days, and that the injection would stay in the arm. Neither statement was true.

An April 2023 study[269] from Italy found **vaccine-induced spike protein in 50% of the biological samples analyzed for up to 187 days after vaccination.** As the Italian study stated, **vaccine-induced spike**

protein can be distinguished from wild-type spike protein (natural infection) due to specific amino acid variations.

This means the National Institutes of Health, with its $45 billion budget could be tracking the duration of vaccine-induced spike protein, as part of its efforts to determine the impact of Covid-19 shots on recipients. But the NIH is not doing that. In fact, the NIH has stopped tracking the impact of the Emergency Use Authorized Covid shots on recipients altogether. (see Section 26 V-Safe: CDC's (skewed) Covid-19 vaccine injury report system)

In addition to their many other failings, the **Covid shots train the immune system to focus on the spike protein,** leaving the recipient **less able to fight subsequent variants that have a mutated spike protein**.

Known in April 2020: the spike protein should not be the basis for mRNA shots

> *The messenger RNA vaccine "is the genetic code for the potentially lethal spike protein part of the virus. It was the worst idea ever to install the genetic code by injection and allow unbridled production of a potentially lethal protein in the human body for uncontrolled duration of time. Everything we've learned about the vaccines since they've come out is horrifying.*
>
> **Dr. Peter McCullough** | Testimony to European Parliament September 13, 2023[270]

In the above quote, Dr. McCullough was testifying before a subcommittee of the European Parliament in September 2023, but it was known in 2020 that the spike protein should not be the basis of the vaccines.

The SARS-CoV-2 virus has 29 proteins[271], one of which is the spike protein. On April 9, 2020[272], James Lyons-Weiler, PhD in biology, published an article explaining something crucial about the SARS virus: "Only one immunogenic epitope in SARS-CoV-2 [has] no homology to human proteins." Homology is "Similarity often attributable to common origin."

To review: the theory behind the mRNA shots is that they instruct the body's cells to make the SARS-CoV-2 spike protein. The cells create the foreign spike protein, the immune system recognizes it as an invader, and mounts an immune response, without ever having been exposed to the virus. In the future, when the body is actually exposed to the virus, it will already know how to fight it, preventing illness. That was the theory. That was the hope. But science and medicine are not based on theories and hope. Scientific knowledge had shown previously that vaccinating to a single protein could create problems such as antibody dependent enhancement and autoimmunity. Instructing the body to *make* the single foreign protein takes the problem even a step further.

Autoimmunity:

Lyons-Weiler explained, "Homology between human and viral proteins is an established factor in viral – or vaccine-induced autoimmunity." **Autoimmunity** is a condition where the body is perceived as a threat and **the immune system attacks itself**. Lyons-Weiler said the parts of the virus that were homologous to human proteins, including the spike protein, should not be considered for vaccine development. Homologous proteins could lead to immunological priming, also known as antibody dependent enhancement (ADE). Immunological priming, or ADE, was observed in animal trials of mRNA vaccination during SARS-1 and MERS.

Antibody dependent enhancement (ADE):

Antibody dependent enhancement, also known as pathogenic priming, occurs when, instead of leading to strong immune response, exposure to the pathogen **acts like a Trojan horse[273], allowing the pathogen to get into cells more easily**, helping the viral proteins to invade the immune system and further spread. Lyons-Weiler pointed out that "immunogenic peptides in viruses or bacteria that match human proteins are good candidates for pathogenic priming...," which is not a positive.

He noted that if non-homologous parts of the epitopes were used, they might be "viable candidates for vaccine development." **Lyons-Weiler warned against basing a vaccine on proteins homologous to the human body, such as the spike protein.** Others also warned about focusing on a narrow part of the virus, such as just the spike, for an immune response.

Professor Angus Dalgleish, MD, is a prominent U.K. oncologist whose work has been cited more than 25,000 times. As the race began for vaccine development, **Dr. Dalgleish and others warned that the SARS-CoV-2**

spike protein was 79% homologous to human proteins and should not be used as the basis for a vaccine. As John Campbell, PhD explains[274]:

> If 79% of the spike protein is the same as human proteins, and you make antibodies to that spike protein, that means that 79% of those antibodies are going to cross-react with human proteins.

Dr. Dalgleish confirms Campbell's assessment and points out that **some of the cross-reactions will be weak,** some will be more troublesome, such as skin disorders, **but the most troublesome will present as cardial dysfunctions, neurological disorders, and explosive cancers.**

> Why did the 150 other people designing vaccines all go with the whole Spike protein? And we shouted and we said, 'Don't do it, don't do it!' And because all the Lemmings decided it was a good idea to jump with the Spike protein, they just ignored us and every vaccine candidate used the whole spike protein.
>
> ANGUS DALGLEISH

Training to fight only spike protein makes the body less able to fight the whole virus:

> In a 2017 article published in the Journal of Autoimmunity, the authors wrote "In the case of vaccines, if we only immunize to a single strain or epitope, and if that strain/epitope changes over time, then the immune system is unable to mount an accurate secondary response. In addition, depending (on) the first viral exposure the secondary immune response can result in an antibody dependent enhancement (ADE) of the disease or at the opposite, it could induce anergy. Both of them triggering loss of pathogen control and inducing aberrant clinical consequences."

That tiny passage from the 2017 *Journal of Autoimmunity*[275] packs a wallop and needs to be broken into layperson's terms to be understood, applying its concepts specifically to the Covid-19 shots.

"In the case of vaccines, if we only immunize to a single strain (Wuhan strain, or later Wuhan/Omicron boosters) **or epitope** (an epitope is a foreign protein, or antigen, capable of stimulating an immune response – in this case, the spike protein – just 1 of 29 SARS-CoV-2 proteins), **and if that strain/epitope changes over time** (if it mutates, which all coronaviruses do – rapidly – there had already been 12,000 catalogued mutations[276] in SARS-CoV-2 by September 2020) **then the immune system is unable to mount an accurate secondary response** (less able to fight the virus on second exposure because it responds as if fighting the previous strain).

In addition, depending (on) the first viral exposure (in this case, vaccination specified to the Wuhan spike protein), **the** secondary immune response **can result in an antibody dependent enhancement (ADE) of the disease** (the body overreacts to the virus on second exposure) **or at the opposite, it could induce anergy** (anergy is a condition in which the body fails to react to an antigen because it's primed to the first exposure, also known as original antigenic sin[277], or immune imprinting[278]). **Both of them triggering loss of pathogen control** (the virus rages largely unchecked by the immune system) **and inducing aberrant clinical consequences** (such as severe, even lethal reactions)."

The above passage from the *Journal of Autoimmunity* in 2017 shows that concerns about vaccinating to a single strain, or single protein, could cause problems for vaccine recipients – knowledge that appears to have been largely ignored when the Covid shots were developed.

Immune Escape, also known as viral escape or antigen escape:

"Immune escape (aka viral escape and antigen escape) is the ability of a virus to elude an individual's immune response," states a January 15, 2021 Covid-19 Q&A site[279]. The article goes on to explain that immune escape through drug resistance is more common than in vaccine resistance because "vaccines tend to induce immune responses against multiple targets on the virus whereas drugs tend to target very few."

The blog post, which is dismissive of concerns about immune escape being caused by mass vaccination during a pandemic, neglects to see its own point: the Covid-19 "vaccines" in fact target only one aspect of SARS-CoV-2 – the spike protein, which is continuously mutating.

Dr. Robert Malone wrote in June 2022[280] about **studies that were**

showing immune imprinting **in people who had received multiple Covid shots**. Malone noted that **post-vaccination there was a rapid loss of detectable** T cell **immunity against the spike protein, which is the main** antigen **in the vaccines**. In addition, "if you were first infected with Wuhan Hu-1, then vaccinated, then infected with Omicron, your antibody levels against the...Spike...were lower than those who had not been infected." This means **the unvaccinated experienced a stronger immune response during subsequent exposure to SARS than did the vaccinated.**

Malone said, "In the case of Omicron, these levels dropped after either two or three doses of vaccine...independent of prior virus infection history. Which is more evidence of immunologic escape of Omicron from B cell mediated (antibody) control." In other words, **the more Covid shots you take, the less your body is able to fight Covid, even if you have been infected with Covid and should be able to mount a robust response.** The Covid-19 shots interfere with, and even sideline, your immune system.

12. mRNA HAS NEVER BEEN SUCCESSFULLY USED IN VACCINATION

There has never been a successful vaccine for SARS or MERS, two viruses related to SARS-CoV-2 (the virus that can lead to COVID-19 disease.) Animal models revealed that although experimental SARS and MERS vaccines could create high amounts of antibodies, when challenged with exposure to the wild viruses they were supposed to be protected against, the animals instead developed severe, even lethal disease reactions.

INFORMEDCHOICEWA.ORG
May 18, 2020[281]

As explained, **mRNA had never been successfully used in vaccination**, not even in the animal trials[282] conducted on the first SARS in 2003, and MERS in 2012. Also, it was known that exposing the immune system to only one portion of a rapidly mutating virus had high potential of failure.

Another problem with the mRNA shots is that the lipid nanoparticle travels throughout the body, including crossing the blood-brain barrier, creating inflammation. As Dr. John Campbell[283] of the U.K. explains, that's not much of a problem in the shoulder, but it's a big problem in the heart and the brain.

As shown in the following excerpt from a CDC article published September 16, 2022, we have continually been told that the Covid shots stay in the arm, and that the mRNA would be eliminated by the body within 2-3 days. Nothing was explained to us about the lipid nanoparticles.

12/24/22, 9:48 AM Understanding How COVID-19 Vaccines Work | CDC

Español | Other Languages

COVID-19

Understanding How COVID-19 Vaccines Work

Updated Sept. 16, 2022

How mRNA COVID-19 vaccines work

- First, mRNA COVID-19 vaccines are given in the upper arm muscle or upper thigh, depending on the age of who is getting vaccinated.
- After vaccination, the mRNA will enter the muscle cells. Once inside, they use the cells' machinery to produce a harmless piece of what is called the spike protein. The spike protein is found on the surface of the virus that causes COVID-19. After the protein piece is made, our cells break down the mRNA and remove it, leaving the body as waste.

We now know that the mRNA does not stay in the arm muscle[284], and can continue in the body for weeks, continuously instructing to make toxic spike protein. We also know that the lipid nanoparticles are harmful[285]. (See Dr. Byram Bridle – Canada:" in Section 15)

As mentioned before, a study[286] accepted for publication on August 15, 2023, found that vaccine-induced spike protein was found in 50% of the biological samples analyzed even 187 days after vaccination. This means people were still producing the toxic spike protein six months after injection.

> This is the first time in human medicine that we have an uncontrolled exposure, for an uncontrolled duration and quantity, in the human body...To make matters worse, the vehicle that carries these genetic products into the human body (the lipid nanoparticle envelope) goes into vital organs. And it's unprecedented that we've ever exposed a single human, let alone hundreds of millions of people, to this form of technology.
>
> **DR. PETER MCCULLOUGH** | Cardiologist

Fauci "Sometimes a vaccine looks good in trials, but actually makes people worse:"

Dr. Fauci was aware of prior problems with mRNA technology for vaccines, including the problem of vaccine-induced antibody dependent enhancement (ADE). In the interview of March 2020, Mark Zuckerberg asked Fauci about the vaccines. **"What's the public health rationale of proving it's effective, if you've shown it is safe?...Why not push hard on rolling out more aggressively even if you don't know how effective it is?"** Dr. Fauci replied:

> There is another element to safety. That is, if you vaccinate someone and they make an antibody response and then they get exposed and infected, does the response that you induce actually enhance the infection and make it worse? The only way you'll know that is if you do an extended study not in a normal volunteer who has no risk of infection, but in people who are out there in a risk situation. **This would not be the first time, if it happened, that a vaccine that looked good in initial safety actually made people worse.**
>
> It was the history of the *Respiratory Syncytial Virus*[287] vaccine in children which paradoxically made the children worse. One of the *HIV vaccines*[288] that we tested several years ago actually made individuals more likely to get infected, so you can't just go out there and give it... **That's why you gotta do the trial.** (emphasis added)

In an interview years before, during the AIDS crisis, **Dr. Fauci explained**[289] the importance of the normal 10-15 year process for a new vaccine, and **especially the importance for the lengthy clinical trials before full approval for public use:**

> If you take it (the vaccine) and then a year goes by and everybody's fine, then you say okay that's good, now let's give it to 500 people. And another year goes by and everything's fine, and then you say now let's give it to thousands of people. **And then you find out that it takes 12 years for all hell to break loose, and then what have you done? (emphasis added)**

There was ample knowledge and reason to be cautious with the Covid-19 vaccines, but those reasons were discarded, along with the doctors and scientists who tried to raise a warning voice.

The problem is not just the Spike Protein, it's the mRNA platform: Evolutionary biologist Bret Weinstein points out[290] the **danger of the current mRNA technology, regardless of the protein it instructs the body to make**:

> Now you've got a huge fraction of the world that has already accepted mRNA vaccine technology. How are you going to rescue it from a 1 in 35 chance of damaging your heart? **How are you going to overcome the dawning awareness that this stuff is** lethally **dangerous and to tissues you cannot afford to have damaged**? If the problem is inherent to the spike protein [then]... all you've got to do is swap in a different protein that doesn't have some flaw like this and the mRNA platform is right back on track. On the other hand...**it doesn't matter what protein. Any foreign protein transcribed by your cells is going to cause your own immune system to go after your own heart cells, if that's where this thing ends up transfecting.** (Transfection is the altering of the properties of the cell through the insertion of foreign nucleic acid - DNA or RNA - into the cell.)

Weinstein believes that mRNA apologists, who are stating that the problem is just the use of the spike protein, without acknowledging all the other failings of the novel use of mRNA technology for vaccination, are engaged in a campaign to rescue the mRNA platform for future use:

> [They're] doing a limited hangout designed to rescue the mRNA platform from the natural consequences of what we have now discovered at the cost of who knows how many lives globally? Who knows how many years people lost? Who knows how much the cost of this destruction has been? **Yes the spike protein was not a good choice...but it's far from the worst problem here and the worst problems do indict this platform until proven otherwise.**

The biggest danger, aside from the harms caused by the Covid mRNA shots, is the fact that Big Pharma and the FDA have plans[291] to convert all traditional vaccines to the new, profitable new mRNA technology. They appear determined to proceed full speed ahead, regardless of safety signals, and will only be stopped by the public's refusal to partake and comply.

> The **mRNA technology, lipid nanoparticle technology, and adenoviral technology** were all essentially untried for human use before the development of the Covid-19 vaccines. In addition, **there were known problems with each in animal trials**, problems which were largely overlooked in the rush to roll out the Covid shots, and which are further being ignored as plans are made to expand mRNA technology.

Vaccine misunderstanding...or intentional deception?

> *They had a pandemic. They had a technology they couldn't bring safely to market, and the emergency allowed them to do it... allowed them to go through emergency use authorization, to get people on board with taking it because people were so scared of Covid. It allowed them to basically push aside all of the safeguards that should have prevented a prototype technology like this from reaching the market without demonstrating safety.*
>
> **BRET WEINSTEIN** | Evolutionary biologist | August 5, 2023[292]

Dr. David Gortler[293] points out that, "New **vaccine development has historically involved a meticulous, slow, decade-long discovery**, testing, review, and approval process. In contrast to that, those established epidemiological and time-honored standards seemingly evaporated under the justification of a Public Health Emergency (PHE). That scuttling was not for a "classic" type vaccine – but instead for brand-new, mRNA "vaccines" and their LNP components. Next, the entire expedited approval/review process was condensed to less than one year." But we had been taught to trust vaccines.

We have been told from the time we were children and receiving our first shots, that vaccines are a miracle of modern medicine. A little bit of discomfort leads to lifelong protection against serious diseases such as polio and measles. Sometimes a booster shot is required, as in the case of tetanus, to maintain protection.

Most of us also knew that **once you received the vaccine, or had the disease, you didn't get that illness again.** For example, those who had chickenpox as children did not receive the chickenpox vaccine when it became available years later, because they already had immunity against the disease.

This is what we understood. This is what doctors are taught in medical school where they spend roughly one week on vaccines – basically learning about the childhood immunization schedule, and proper handling and administering of the vaccines. Dr. Joseph Ladapo[294], the Florida Surgeon General, says that other medications are looked at objectively in medical school, but "with vaccines, they are treated instead as something that is inherently good or inherently benevolent...And **when individuals... choose not to participate in vaccine programs [they are considered] essentially bad people.** There's a value judgment that is absent from every other medication in medicine."

With Covid-19 we were told, in no uncertain terms, that vaccination was our only way out of the pandemic, that natural immunity[295] from prior infection was insufficient [296]protection, and that **the Covid vaccines were 95% effective.** "Safe and effective" was the mantra. **Great news! Except it wasn't, because "95% effective" was based on obfuscation.** The fact that Pfizer and the FDA wanted 75 years to release the data on the vaccine development and clinical trials is evidence of wrongdoing. It took multiple Freedom of Information Act requests, and a court order[297] to make the data available. What it revealed is shocking.

13. CLINICAL TRIALS DID NOT SHOW COVID SHOTS WERE "SAFE AND EFFECTIVE"

I was a bit familiar, not very deeply, with this mRNA class in my professional work, when these products were in development for other things—severe conditions such as cancer. I knew that these products were inherently dangerous, which isn't unusual in pharmaceutical research and development. We frequently work on things that are risky and can be toxic, such as chemo agents.

Yet, all of a sudden, our regulators were all gung-ho saying, "These are prophylactic vaccines. They can be given to children, pregnant women, and everyone else." I became extremely suspicious about this whole situation, and that's how I started looking into it.

SASHA LATYPOVA

Former Pharmaceutical Executive | June 17, 2023[298]

"Safe and effective" is not a medical term. It's a marketing phrase. The clinical trials for Covid-19 vaccines were not testing for safety or preventing transmission. **They were looking at one thing only: did the Covid-19 shots prevent infection?**

Every clinical trial has a specific hypothesis being examined, which is the endpoint, or purpose, of the trial. The trial is set up with the hypothesis, a study design, procedures for the study, analysis of outcomes, and a summary of results. A good trial is clear on each of these points and also adjusts for confounding factors that could affect the trial results.

The endpoint for Pfizer and Moderna's clinical trials of the mRNA Covid shots was not broad. **The clinical trials did not test the Covid shots for**

whether they prevented transmission[299] of Covid-19. They did not take into account adverse reactions to the vaccines in terms of trial results.

> The endpoint for the Pfizer clinical trial was this: How many of those who displayed Covid symptoms (no matter how mild), **and** tested positive on a PCR test, were in the vaccinated group versus the unvaccinated group?

That's it. **The endpoint of the trial was to determine who had symptoms that were PCR-test-confirmed to be Covid-19. They did not test for safety.** They didn't test if the vaccine prevented transmission[300] of Covid-19. They **did not test for how the vaccine interacted with the body long-term.** They **did not test the vaccines on pregnant women, or the elderly, or those with existing health problems**. They **combined the second and third phases of the already shortened clinical trials.** They dropped people from the trials who had serious reactions to the vaccines, largely ignored them[301], and **minimized or deleted their data.** They kept track of **adverse events,** but they were **not made known to the public,** and were **not a consideration in the endpoint of the trial.**

They tweaked a few numbers to come up with the claim that the shots were "95% effective," ended the clinical trial by offering the vaccine to the control group **(thereby eliminating an effective control group to follow in the months and years ahead),** and started their vaccination campaign. Thus began the largest clinical trial in history, as the experimental Emergency Use Authorized Covid shots were administered to the population of the world en masse.

"95% effective" was based on mathematical deception:

> *The public would have been far more skeptical of these products had the clinical trial results been translated into normal English.*
>
> **ROBERT BLUMEN** | Brownstone Institute author
> "Vaccine Was '95% Effective How?"[302]

To understand the misleading statement of "95% effective," we need to understand some clinical trial legalese. **In clinical trials they look at Absolute Risk Reduction and Relative Risk Reduction.**

Absolute Risk is your odds of getting Covid at all. Relative Risk is your odds of contracting Covid compared to the other group in the trial. In order for people to have informed consent about taking the vaccine, they needed a good cost benefit analysis. The Relative Risk number does not give good cost benefit information, but that is the number used by the drug companies.

For example, in a Canadian Covid Care Alliance analysis[303] of a slightly earlier release of the Pfizer clinical trial data, Pfizer claimed the data showed the vaccine to be "91% effective," derived from the total number of Covid infections that were confirmed by PCR testing during the trials. There were a total of 927 PCR-confirmed Covid cases in 40,000 people during the trials. Seventy-seven of those cases were in the vaccinated, and 850 were in the placebo group, the unvaccinated.

How Relative Risk is calculated:

The "91% effective" number is calculated by subtracting the number of cases in the vaccinated group, 77, from the number of cases in the placebo group, 850, then dividing by 850 and multiplying by 100 to get the percentage.

870 – 77 = 793
793 ÷ by 870 = 0.91
0.91 x 100 = 91%

The Relative Risk number that was repeatedly shared by government, media, and public health officials was essentially an insignificant number. Put another way, 927 people out of 40,000 contracted Covid during the clinical trial, which equaled just 2.3% of the total Pfizer Covid trial participants. When you consider that a Covid "case" was defined as symptoms (runny nose, sore throat, fever, body aches, cough, etc.) confirmed by a PCR test, you realize that serious disease was not what they were checking for in the clinical trials.

How Absolute Risk is calculated:

Absolute Risk, again that is your risk of getting Covid at all, is calculated by dividing the total number of Covid cases by the total number of participants, times 100.

Here is the Absolute Risk number for the placebo group:

850 ÷ 20,000 = 0.0435 x 100 = 4.35%

Here is the Absolute Risk number for the vaccinated group:

77 ÷ by 20,000 = 0.00385 x 100 = 0.385%

Absolute Risk *Reduction* is the comparison of the two arms of the trial, subtracting the one percentage from the other:

0.385% subtracted from 4.35% = 3.965%, rounded up to 4%

The actual relevant number was "only 4% effective," not "95% effective:

Vaccination was shown, in the Pfizer clinical trials, to reduce your Absolute Risk of contracting Covid-19 by 4% compared to the unvaccinated group. That's what is known as "clinically insignificant,"[304] in terms of trial outcomes, especially when dealing with the unknowns of a vaccine that was only given Emergency Use Authorization.

The actual significant numbers people needed to have to make an informed decision on whether or not to receive the Covid shots were withheld from the public. People needed to know the adverse events associated with the vaccines in the clinical trial, and they needed to know which of those adverse events required hospitalization, or caused death. **As shown in the following chart from the Canadian analysis, the adverse events, need-to-treat, and deaths were higher in the vaccinated group than in the placebo group.**

The vaccinated fared worse than the placebo group in Pfizer clinical trials:

PFIZER TRIALS DID NOT PROVE SAFETY
THEY PROVED HARM

ILLNESS

	BNT162b2	Placebo	Risk Change
Efficacy	77	850	-91%
Related Adverse Event	5,241	1,311	+300%
Any Severe Adverse Event	262	150	+75%
Any Serious Adverse Event	127	116	+10%

DEATHS

BNT162b2	Placebo
20	14

These are the results of Pfizer's own randomized control trial.

LEVEL 1 EVIDENCE OF HARM.

As shown in the above chart the vaccinated group (red numbers) performed worse in the clinical trials than the placebo group (blue numbers) in every metric except for specific Covid symptoms (cough, fever, runny nose, body aches, etc.).

No one would have been impressed with a 4% Absolute Risk Reduction, so the pharmaceutical companies chose the higher Relative Risk number to market their products, while withholding information about adverse events and deaths. Pfizer, Moderna, and Johnson & Johnson all focused on Relative Risk in order to get people to take their vaccines. It was a form of false advertising.

Blumen asks, 'Would you take a [potentially dangerous] drug that could reduce the incidence of a rare disease by 50%?" He suggests that sounds like pretty good odds until you realize that a 50% Risk Reduction can easily be skewed. For example, moving from 10 per 1 million to 5 per 1 million is a 50% Risk Reduction, but is only a 0.0005% Absolute Risk Reduction.

Blumen states, "Here's what '95% effective' did not mean: if you take the shots, then you will have a 95% lower chance of getting Covid. But that is how most people understood it because that is what the words mean in normal English."

Dr. John Farley, of the FDA, in advising doctors[305] on how to get patients vaccinated, suggested that if a patient was seeking ivermectin as a treatment for Covid-19, they should "tell that patient to get vaccinated." So someone who is already ill and seeking safe FDA-approved ivermectin as a treatment for Covid-19 is supposed to get vaccinated instead. Farley stated, "If they're sitting in your exam room, even if they're pushing back and giving you a hard time, they trust your recommendations. I would encourage them to get vaccinated for prevention." It would appear someone needs to school the FDA on the difference between Absolute Risk and Relative Risk, as well as the benefits of early treatment for Covid-19 disease.

This was not an honest mistake – it was a narrative driven by profits and power:

They knew. Pfizer and Moderna, Johnson & Johnson, Dr. Fauci, all those CDC and FDA spokespeople, many of the scientists and doctors, and government and public health officials who remained silent. They knew[306] that the vaccines hadn't been tested for stopping transmission[307]. They knew that the Covid symptoms in the trials could be anything from a sniffle to flu-like, and that no distinction was made between them. They knew that in a trial of about 40,000 people, a total of 927 cases was almost nothing and certainly not definitive. **They knew that the unquestioning media, and the general public would pounce on the almost irrelevant "95% effective" number and think we were saved from Covid, if only we'd roll up our sleeves.** They knew it was likely that some people would be seriously injured[308] by the Covid shots, yet their mantra was "safe and effective."

Highly Effective (2:26 min)[309]

The selfishness, greed, and irresponsibility of those who knowingly pushed for the experimental vaccination of every human being on earth, for only a 4% reduction in risk of contracting Covid, is nauseating. They claimed "95% effective," and "safe and effective," through twisting and withholding of data. When you take into consideration the harms that have been inflicted by the experimental vaccines, there is no question that medical malfeasance[310] was perpetrated on the world.

> At this point, powerful people in public health and Pharma are in full cover-up mode. They almost have to be, because it's hard to imagine how they could pivot from what they've done to what they ought to do. At this point, the negligence is criminal.
>
> **ED DOWD** | Wall Street analyst | Author of *Cause Unknown*[311]

14. THE MORE COVID SHOTS YOU TAKE, THE MORE LIKELY YOU ARE TO GET COVID

Informed consent must be based on an accurate risk-benefit ratio: Every medical procedure has some risk associated with it. It is always necessary to consider if the risk of the procedure is greater than the risk of the disease. With regard to the Covid-19 vaccines, the absolute risk of the disease is less for every population cohort than the risk from the vaccine, **other than the oldest cohort which is about equal. In a cost-benefit analysis for the Covid shots, the shots are not worth it.**

Professor Dalgleish explains[312] that **the Covid shots keep the immune system occupied, and less able to respond to other viruses and illnesses.** Dalgleish states that the Covid shots have "perturbated the immune system" and "used its resources to fight viruses that don't exist. **It's making lots of useless antibodies which are actually contributing to a phenomenon called** antibody dependent enhancement (ADE)." Professor Dalgleish continues:

> You've only got a limited capacity in your immune system. If you boost with another (Covid) vaccine to harness half the immune system to make antibodies to a virus that no longer exists on the planet...you are going to weaken it. You're going to reduce the front line...it also suppresses the T cell response.

"No coronavirus vaccine has been shown to be of any use at all and that's why we don't have one for the (common) cold," states Dr. Dagleish, "[T]he main reason they don't work is the coronaviruses **have this phenomenon - once the immune system sees it, it locks in. It's called ecological imprinting or** original antigenic sin. So when you do any variation (of the Covid shots) they're...letting all other variants in through the back door, and other people have said that **we probably wouldn't have had the problem with the variants if it hadn't been for the vaccine program in the first place."**

Cleveland Clinic study - more Covid shots equal more Covid infections: A Cleveland Clinic[313] study of 51,017 employees beginning on the day the bivalent booster became available, September 12, 2022, charted cumulative incidence of Covid-19 cases by vaccination status. **The following chart from the study demonstrates that the more Covid shots individuals receive, the more likely they are to become infected with Covid-19.** Participants were tracked from 12 September 2022 to 14 March 2023:

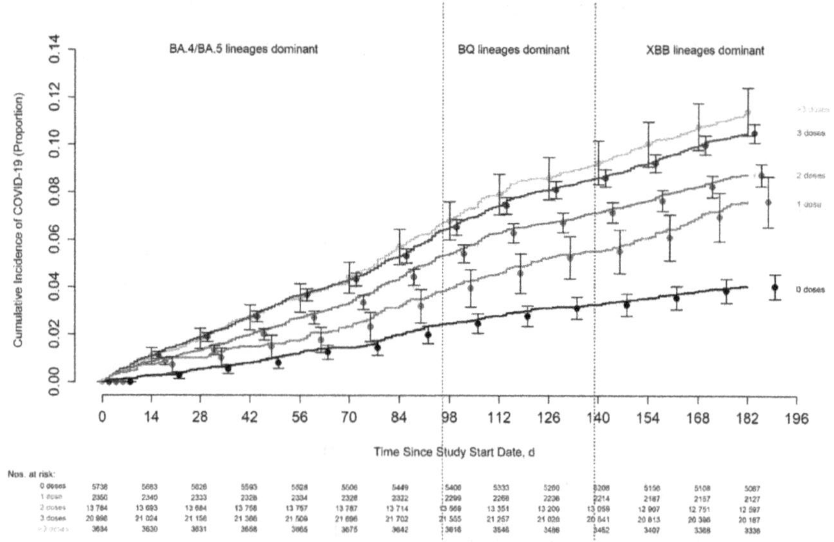

In the above chart, Covid cases are tracked from date of vaccination to date of infection. Black is no dose, red is 1 dose, green is 2 doses, blue is 3 doses, orange is more than 3 doses. The more doses received, the more susceptible the recipients were to contracting Covid-19. The unvaccinated, the black line, were the least likely to contract Covid-19.

The Cleveland clinic study also found, **"The more recent the last prior Covid-19 episode was the lower the risk of Covid-19,"** **which certainly speaks to the strength of natural immunity.**

> Ours is not the only study to find a possible association with more prior vaccine doses and higher risk of COVID-19....it is important to examine whether multiple vaccine doses given over time may not be having the beneficial effect that is generally assumed.
>
> **EXCERPT FROM THE CLEVELAND CLINIC STUDY**
> Open Forum Infectious Diseases | Volume 10, Issue 6, June 2023

The findings also beg the question: What is happening in the body that would cause someone to be more susceptible to disease after receiving a Covid shot? That answer lies within the immune system.

IgG4 antibodies – when the immune system is trained NOT to fight Covid: A study from Uversky and colleagues published in *Vaccines*[314] on May 17, 2023, states this:

> To date, 72.3% of the total population has been injected at least once with a COVID-19 vaccine. As **the immunity provided by these vaccines rapidly wanes**, their ability to prevent hospitalization and severe disease in individuals with comorbidities has recently been questioned, and increasing evidence has shown that,[unlike] many other vaccines, **they do not produce sterilizing immunity, allowing people to suffer frequent re-infections**. Additionally, recent investigations have found **abnormally high levels of IgG4 in people who were administered two or more injections of the mRNA vaccines**. (emphasis added)

Immunoglobulin IgG is part of the secondary immune response to an antigen. There are four IgG subclasses in humans. IgG1 through IgG3 antibodies are central in fighting various diseases, **IgG4 comprises only 4% of total IgG antibodies and is mostly associated with allergies, and immune tolerance.**

As stated in Uversky, **vaccinated individuals are experiencing a "class switch" in which the disease fighting antibodies are switching to IgG4 antibodies that do not fight antigens.** Uversky states:

It has been suggested that an increase in **IgG4 levels could have a protecting role by** preventing **immune over-activation**, similar to that occurring during successful allergen-specific immunotherapy by inhibiting IgE-induced effects. **However**, emerging evidence suggests that the reported **increase in IgG4 levels detected after repeated vaccination with the mRNA vaccines may not be a protective mechanism;** rather, it constitutes an **immune tolerance mechanism** to the spike protein that **could promote unopposed SARS-CoV2 infection** and replication **by suppressing natural antiviral responses."**

"Increased IgG4 synthesis due to repeated mRNA vaccination with high antigen concentrations may also cause autoimmune diseases, and promote cancer growth and autoimmune myocarditis in susceptible individuals." (emphasis added)

It would appear that we are doing more harm than good with every Covid shot administered. IgG is the only class of Ig that can cross the placenta in humans, and is largely responsible for protection of the newborn[315] during the first months of life. **Some infants born to mothers who have been Covid-vaccinated are presenting with low levels of IgG antibodies, leaving them less protected against disease.**

It's likely this IgG4 class switch is partly responsible for the appearance of turbo cancers, recurrent cancers, reactivated viruses, autoimmune disorders, and many other health issues that have increased since the Covid vaccine rollout began. **It appears the Covid shots interfere with the body's natural immune system.**

> It is grossly naïve, and frankly it's hubris, to believe that we have such a sophisticated understanding of virology and immunology and viral evolution that we can predict and administer something like this into the whole population and mitigate the natural process and come up with something that's better than what has naturally evolved over millennia...
>
> **ROBERT MALONE** | June 13, 2022[316]

2023: Emergency's over, but $1.4 billion is allocated to creating Covid vaccines:

Evidence continues to mount that Covid shots do not prevent Covid infection, but an additional $1.4 billion in taxpayer money[317] is being directed toward creating new Covid vaccines. Because, you know, all those variants.

In fall 2023, the "updated vaccines" were formulated to XBB.1.5, a strain of Omicron that has all but disappeared. XBB.1.5 was replaced by E.G.5 (Eris), which had at least 36 mutations from XBB.1.5 in the spike protein alone. Also emerging was variant BA.2.86 (Pirola). **Both Eris and Pirola are variants of Omicron, and they cause sore throat, runny nose, headache, blocked nose, sneezing, and cough– so basically, the common cold.**

Unsurprisingly, another variant entered the media attention in fall 2023 even as the **booster campaign was floundering[318], with around 14%[319] of U.S. citizens taking the shot.** HV.1 (possibly named to sound like HIV and add an additional fear factor?) is yet another Omicron variant causing cold-like symptoms. In the following CDC chart from November 20, 2023, HV.1 had replaced EG.5 as the dominant Omicron variant:

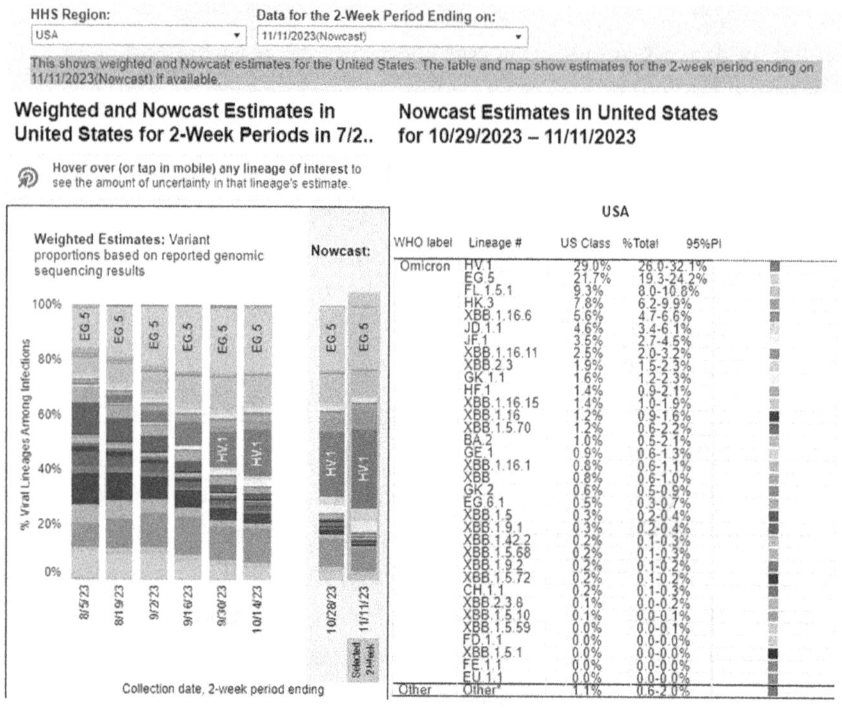

WHO label	Lineage #	US Class	%Total	95%PI
Omicron	HV.1	29.0%	26.0-32.1%	
	EG.5	21.7%	19.3-24.2%	
	FL.1.5.1	9.3%	8.0-10.8%	
	HK.3	7.8%	6.2-9.9%	
	XBB.1.16.6	5.6%	4.7-6.6%	
	JD.1.1	4.6%	3.4-6.1%	
	JF.1	3.5%	2.7-4.5%	
	XBB.1.16.11	2.5%	2.0-3.2%	
	XBB.2.3	1.9%	1.5-2.3%	
	GK.1.1	1.6%	1.2-2.3%	
	HF.1	1.4%	0.9-2.1%	
	XBB.1.16.15	1.4%	1.0-1.9%	
	XBB.1.16	1.2%	0.9-1.6%	
	XBB.1.5.70	1.2%	0.6-2.2%	
	BA.2	1.0%	0.5-2.1%	
	GE.1	0.9%	0.6-1.3%	
	XBB.1.16.1	0.8%	0.6-1.1%	
	XBB	0.8%	0.6-1.0%	
	GK.2	0.6%	0.5-0.9%	
	EG.6.1	0.5%	0.3-0.7%	
	XBB.1.5	0.3%	0.2-0.4%	
	XBB.1.9.1	0.3%	0.2-0.4%	
	XBB.1.42.2	0.2%	0.1-0.3%	
	XBB.1.5.68	0.2%	0.1-0.3%	
	XBB.1.9.2	0.2%	0.1-0.2%	
	XBB.1.5.72	0.2%	0.1-0.2%	
	CH.1.1	0.2%	0.1-0.3%	
	XBB.2.3.8	0.1%	0.0-0.2%	
	XBB.1.5.10	0.1%	0.0-0.1%	
	XBB.1.5.59	0.0%	0.0-0.1%	
	FD.1.1	0.0%	0.0-0.0%	
	XBB.1.5.1	0.0%	0.0-0.0%	
	FE.1.1	0.0%	0.0-0.0%	
	EU.1.1	0.0%	0.0-0.0%	
Other	Other*	1.1%	0.6-2.0%	

CDC Chart of Covid-19 Variants 20 Nov 2023 [320]

Assurances from Moderna and the government[321] that the shots formulated for an almost extinct version of the virus, will work against the new variants, seem rather thin - especially since the Covid shots have *never* been effective against preventing disease or transmission.

CDC: Vaccinated are susceptible to the new variant, so get vaccinated:

The CDC[322] stated in a Risk Assessment report on August 24, 2023, "BA.2.86 (Pirola) may be more capable of causing infection in people who have previously had Covid-19 or who have received Covid-19 vaccines."

> **The Covid shots do not prevent disease or transmission, and the CDC states the new variants may be more capable of causing infection in people who have received Covid-19 vaccines. It would appear there is no viable health reason for taking a Covid shot.**

The CDC *did* say people with previous Covid infection may also be at risk of infection with the new variants. However, the CDC failed to mention that people who have been infected with Covid-19 have developed a broad immune response[323] to all parts of the SARS-CoV-2 virus, and their immune system is prepared to fight new variants. **The Covid vaccines train only to fight an extinct version of the spike protein.** The vaccinated who have also had Covid may have some innate protection, but **concerning evidence shows that the Covid vaccines interfere with the body's natural immune system, making it less able to fight all types of disease.** (see Section 14 The More Covid Shots You Take, the More Likely You Are to Get Covid)

I refer you to Section 10 These Are Not your Father's Vaccines, and Section 15 Early Warning Voices about the Dangers of the Covid-19 Shots.

A December 9, 2023 review of CDC's Covid Data Tracker finds that the "scariants," as some people call the variants, continue to mutate rapidly, as coronaviruses are known to do. As shown in the following table, E.G.5 (Eris), XBB.1.5, BA.2.86 (Pirola) together account for 11.1% of infections. HV.1 accounts for 29.6%, and is being replaced by JN.1 at 21.4%. But the fall 2023 Covid mRNA shots continue the same, instructing your body to mount an attack against a spike protein that no longer exists.

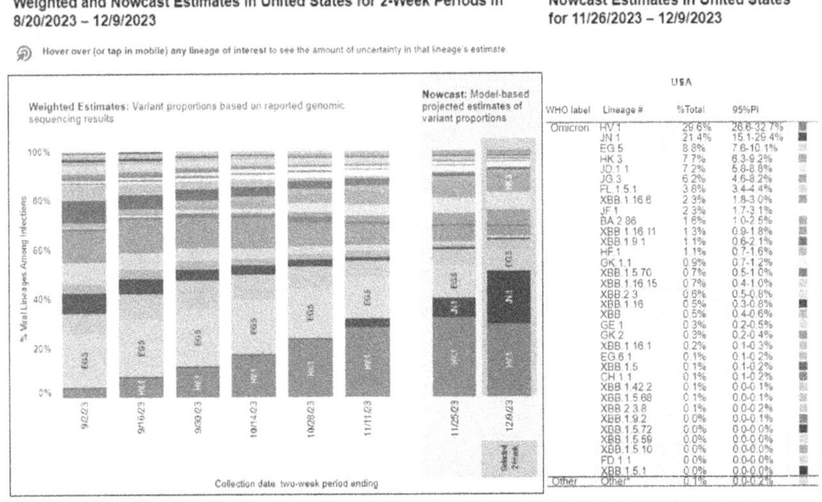

Nowcast Estimates in United States for 11/26/2023 – 12/9/2023

USA

WHO label	Lineage #	%Total	95%PI
Omicron	HV.1	29.6%	26.6-32.7%
	JN.1	21.4%	15.1-29.4%
	EG.5	8.8%	7.6-10.1%
	HK.3	7.7%	6.3-9.2%
	JD.1.1	7.2%	5.8-8.8%
	JG.3	6.2%	4.6-8.2%
	FL.1.5.1	3.8%	3.4-4.4%
	XBB.1.16.6	2.3%	1.8-3.0%
	JF.1	2.3%	1.7-3.1%
	BA.2.86	1.6%	1.0-2.5%
	XBB.1.16.11	1.3%	0.9-1.8%
	XBB.1.9.1	1.1%	0.6-2.1%
	HF.1	1.1%	0.7-1.6%
	GK.1.1	0.9%	0.7-1.2%
	XBB.1.5.70	0.7%	0.5-1.0%
	XBB.1.16.15	0.7%	0.4-1.0%
	XBB.2.3	0.6%	0.5-0.8%
	XBB.1.16	0.5%	0.3-0.8%
	XBB	0.5%	0.4-0.6%
	GE.1	0.3%	0.2-0.5%
	GK.2	0.3%	0.2-0.4%
	XBB.1.16.1	0.2%	0.1-0.3%
	EG.6.1	0.1%	0.1-0.2%
	XBB.1.5	0.1%	0.1-0.2%
	CH.1.1	0.1%	0.1-0.2%
	XBB.1.42.2	0.1%	0.0-0.1%
	XBB.1.5.68	0.1%	0.0-0.1%
	XBB.2.3.8	0.1%	0.0-0.2%
	XBB.1.9.2	0.0%	0.0-0.1%
	XBB.1.5.72	0.0%	0.0-0.0%
	XBB.1.5.59	0.0%	0.0-0.0%
	XBB.1.5.10	0.0%	0.0-0.0%
	FD.1.1	0.0%	0.0-0.0%
	XBB.1.5.1	0.0%	0.0-0.0%
Other	Other*	0.1%	0.0-0.2%

Also, a little reminder of what the original Covid vaccines were supposed to do – remember them? Six or seven shots ago? They were supposed to be 95% effective against you getting Covid. But, as discussed in Section 13 Clinical Trials Did NOT Show Covid Shots Were "Safe and Effective" they were only 4% effective against contracting disease in clinical trials, and never should have been unleashed on the world.

15. EARLY WARNING VOICES ABOUT THE DANGERS OF THE COVID-19 SHOTS

Crucial safety steps are being skipped on a product that surely will be marketed to every person on the planet. The rush to market appears to be driven by potential profits not a desire to save lives. Several effective therapeutics exist and more are emerging that can prevent cases and effectively treat patients... A rushed vaccine product that has not undergone extensive safety testing could do more harm than good.

ICWA ARTICLE[324]

March 19, 2020

As the vaccine rollout started throughout the world, eminent doctors in different countries noticed alarming trends and called for a halt to the Covid vaccine program. In March of 2021, I stumbled across an interview with Dr. Sucharit Bhakdi, retired professor and former Chair of the Institute of Medical Microbiology and Hygiene, Johannes Gutenberg University of Mainz in Germany.

> I felt horrified. If Dr. Bhakdi was properly describing the harms that the mRNA (Pfizer, Moderna) and adenoviral vector shots (J&J, AstraZeneca) would cause people, **we were going to see an increase in heart attacks, strokes, blood clots, neurological diseases, autoimmune disorders, and deaths across all ages** – some right at the time of injection, but many more in the months and years ahead.

Dr. Sucharit Bhakdi - Germany:

Excerpts from the March 29, 2021 Journeyman Pictures interview[325] with Dr. Sucharit Bhakdi:

> **Humans are now being the test animals.** Humans. Millions! And now we're seeing the outcome, and the outcome is horrible and frightening...such a nightmare.
>
> This is a disastrous situation because the spike protein itself is now sitting on the surface of the cells facing the bloodstream. And it is known that the spike proteins, the moment they touch platelets, they activate them, and that sets the whole clotting system on... **Clot formation in brain veins can give any symptoms. The symptoms of these people who are getting their second shot of the vaccine are so diverse, but they would all fit into that.** Even these people who have jerking, limb movements that they cannot control anymore.
>
> Whether it's mRNA or vector, the fact is that a gene for the spike is entering the bloodstream and reaching cells of the blood vessel wall, at locations that are actually forbidden because, if they are then used, these genes are then put into the machinery of the cell to produce these spikes, these spikes are going to be produced at locations where they never are produced normally.
>
> **[People] won't have died [from Covid] but you are killing them with the vaccine,** and **secondly maiming others for life** because even in the vaccination trials it turned out that several hundred people who were vaccinated got such severe side effects they had to be treated in the hospital. On the one hand you might have...9 severe cases, you prevented 140 mild cases, but in return, you got 100-200 side effects so severe they had to be taken care of in the hospital.
>
> Interviewer: If people have gotten the shot and they haven't had any bad symptoms, hopefully they'll be all right going forward, but you do believe there is a danger of long-term impact going if they are re-exposed to something – a similar virus?

Dr. Bhakdi: Yes of course, because the immune system is now primed to fight, super aggressive. So if the real virus comes in tomorrow or day after or next year, or a related virus, because there are so many coronaviruses flying around us, any one will do. And if they come and infect the lung, the moment those cells start making the slightest bit of the virus, they're going to mount an attack and kill that lung. This is called overreaction, and this would be **the immune dependent enhancement of disease,** which is very, very bad potentially.

This has been shown to take place in animal experiments where people were vaccinating against SARS-CoV-1 or MERS and so this sort of thing is able to take place. Second, and this is to me perhaps even more frightening. If people get revaccinated, those side effects that will take place in the brain are going to be enforced and magnified so those guys who escaped the first time may not escape another time, and a third time, or fourth time, I don't think so.

* * *

Dr. Bhakdi and 11 others directed an open letter[326] to the European Medical Association (EMA) on February 28, 2021, calling for an **immediate halt to the vaccine campaign based on "a wide range of side effects… being reported following vaccination of previously healthy younger individuals with the gene-based Covid-19 vaccines."** The EMA responded[327] on March 23 with a 'There's nothing to worry about' attitude that culminated in, "We hope that the above reassures you that …stringent safety monitoring is in place for all Covid-19 vaccines, to ensure that the benefits always outweigh the risks." For raising the alarm, Dr. Bhakdi was eventually threatened with loss of his credentials and pension, and also with criminal charges[328] for continuing to speak out[329].

Dr. Peter McCullough - U.S.A:

At about the same time, Dr. Bhakdi became concerned about the Covid vaccines, cardiologist Dr. Peter McCullough, cited earlier in this paper, also became alarmed[330].

> **By February 2021, about 90 days into the mass vaccination campaign against Covid-19, Pfizer's clinical trial documents showed that there had been 1,129 vaccine-related deaths, and over 40,000 vaccine injuries.**

Dr. McCullough, who had served on multiple vaccine safety review boards for the FDA in the past, knew from experience that **even a possible causal relationship between the vaccines and injury reports was concern enough for the shots to be pulled**. For example, in 1976 during a swine flu[331] scare, a mass vaccine rollout was abruptly canceled after more than 500 vaccinated people contracted Guillain-Barré syndrome, and more than 30 died. (See this article[332] by a physician who writes under the pseudonym of A Midwestern Doctor for a history of government vaccine-injury cover ups from 1976 to the present.)

Dr. McCullough called for an end to the Covid-19 vaccine rollout, but was discounted and vilified in the press and by government and public health officials, eventually losing his hospital privileges and coming under attack from his former employer[333], various boards[334], and colleagues. Despite the persecution, he has continued to advocate for patient rights and evidence-based medicine.

Dr. Geert Vanden Bossche - Belgium:

On March 6, 2021 in Belgium, Dr. Geert Vanden Bossche, a veterinarian and an independent virologist and vaccine expert, formerly with GAVI and The Bill & Melinda Gates Foundation, issued an open letter to the WHO[335], **warning against vaccinating into an active pandemic.** Vanden Bossche explained that the current vaccines would actually make people more susceptible to Covid-19. **Because the SARS-CoV-2 was actively circulating and constantly mutating, the vaccines being administered would cause the virus to work around them (immune escape), and create stronger variants.** The body would have a less adequate response to new variants than if no vaccination had occurred, due to immune imprinting.

Vanden Bossche explained that the immune system would be outcompeted by the antigen-specific antibodies to the Wuhan spike created by the vaccine, and **the body would respond to the wrong virus** (pathogenic priming) **and be unable to adequately fight the new mutation.** "That is a long-lived suppression," said Vanden Bossche. "The [vaccine induced] antibodies don't work anymore. And your innate immunity has been completely bypassed, and this while highly infectious strains are circulating." He called for an end to the Covid vaccine drive. Dr. Vanden Bossche was mocked and attacked in the press and by peers, but has continued to speak out[336].

Dr. Luc Montagnier - France:

In France in May 2021, Luc Montagnier[337], who was awarded the Nobel Prize in Medicine in 2008, warned that mass vaccination during the pandemic was creating immune escape variants that would cause deaths. "**The new variants are a production and result from the vaccination**," said Montagnier. "You see it in each country, it's the same; **in every country death follows vaccination**." (See Ed Dowd's *Cause Unknown*,[338] p. 259 for charts of 17 countries that had low Covid death rates before vaccination, and high Covid death rates after mass vaccination.)

Montagnier said the mass vaccine rollout into an active pandemic was "unthinkable" and a historical blunder. **"The history books will show that... it is the vaccination that is creating the variants," said Montagnier.** Not being able to ignore Montagnier's decades of competence, his attackers implied that he was losing his professional acumen in his old age.

In one of his last public appearances[339], before his death in February 2022, the 89-year-old virologist stated, **"The spike protein, which is synthesized by the messenger RNA contained in the vaccine, is toxic to the cells. It's poison." Montagnier warned that the coronavirus vaccines were "likely to cause neurological disease, very serious brain diseases."** Montangier's last paper[340], in peer review when he died, connected the Covid vaccines to a new form of Creutzfeldt-Jakob disease, an incurable brain disorder which leads to rapid onset dementia and death. Montangier's paper, written with colleagues, **studied 26 Creutzfeldt-Jakob cases all diagnosed in 2021, "with the first symptoms appearing within an average of 11.38 days after a Pfizer, Moderna, or AstraZeneca Covid-19 injection."**

I am asking all my colleagues to stop vaccinating people with this vaccine. The physicians who keep vaccinating people are absolutely aware of this. If they weren't before, they are now. Beware, beware, they are responsible for the future of mankind."

LUC MONTAGNIER | January 2022[341] | Winner of Nobel Prize in Medicine, 2008

Dr. Byram Bridle – Canada:

In May 2021, Dr. Bryam Bridle of the University of Guelph in Canada gained access to a biodistribution study[342] that the Japanese Regulatory Agency had conducted on the Pfizer mRNA Covid shots. Dr. Bridle, a virologist and immunologist, was an Associate Professor of Viral Immunology at Guelph, in the Department of Pathobiology.

Pfizer had not released the results of the Japanese bio-distribution study, but what Bridle found as he analyzed the data was alarming. Bridle said the **study refuted the premise that the Covid-19 vaccine stays in the shoulder muscle. "The spike protein gets into the blood, circulates through the blood in individuals over several days post-vaccination,"** said Dr. Bridle. "Once it gets in the blood, it accumulates in a number of tissues such as the spleen, the bone marrow, the liver, the adrenal glands. And **of particular concern for me is it accumulates in the ovaries, in quite high concentrations."**

Bridle explained the **danger of the spike protein crossing the blood-brain barrier, where it causes neurological damage and possibly fatal blood clots in the brain, as well as causing damage to the circulatory system, leading to strokes and heart issues.** Dr. Bridle also cited an alarming study showing that spike proteins transfer through breast milk, **passing Covid-19 vaccine vectors into breastfeeding infants.**

A few months later, in October 2021, Dr Bridle found his job in jeopardy for refusing the Covid vaccine. Bridle, normally a vaccine proponent, would not take the experimental vaccine both because of what he knew about its many problems, and because he had been infected with Covid-19 and had naturally acquired immunity as confirmed by blood tests. He pointed out to the University administration that not only were the Covid shots not preventing infection or spread, but **people who had been infected**

with Covid prior to vaccination, were experiencing more severe side effects from the Covid-19 vaccines.

Bridle was harassed[343], ostracized, prevented from entering campus[344], and eventually lost his job.[345]

Naomi Wolf, PhD – U.S.A:

Dr. Bridle's concerns about the accumulation of spike protein in high concentration in the ovaries became **Naomi Wolf's concern, as she began to hear accounts of disrupted menstrual cycles in women post-vaccination. Among other things, these irregularities involved missed periods, or conversely, intense vaginal bleeding, including bleeding in young girls and in post-menopausal women.**

Wolf writes, **"You didn't have to be a rocket scientist to understand that a dysregulated menstrual cycle was going to negatively affect fertility."** Wolf talked about this issue in a Fox News interview, and on the political podcast War Room, as well as posting on social media. As she puts it: "The reaction? My Twitter account was suspended; the reason given was that I had committed 'medical misinformation.'" (*The Bodies of Others, p. 165*)

As time went on, Wolf continued[346] to raise alarm about the Covid vaccines[347]. She also raised concerns about the safety and efficacy of face masks, and the totalitarian nature of the increasingly stringent Covid policies and mandates. For speaking out she was ejected from her synagogue and circle of professional and personal contacts, relationships that had been built over a lifetime of reporting, university teaching, and being highly involved in the Democratic party.

In her book *The Bodies of Others*[348], Wolf writes:

> The vaccines were not the endgame. The endgame was to transform our countries. We had been people who respected privacy; now we had become people who monitored the bodies of others. We had been kind, generous people with public morals that were truly decent; now we had become people who tolerated and supported cruelty. We had been people who accepted individual needs and requirements; now we had become people who could not accept or tolerate individual need, individual variation, indeed people who rolled over it without compassion." (*The Bodies of Others* p. 190)

When the Pfizer data were finally released, due to court order[349], Naomi Wolf's independent news group Daily Clout[350], along with Steve Bannon's The War Room, organized a team of 3,250 medical expert volunteers to pore over the hundreds of thousands of documents. Much of what we know about the clinical trials and the perfidy in the vaccine development and approval process is due to their work[351].

16. THERE HAS NEVER BEEN AN EFFECTIVE VACCINE FOR A RESPIRATORY VIRUS

"The key to ending the pandemic has always been the immune system."

STEVE TEMPLETON

Associate Professor | Microbiology & Immunology[352] | Indiana University

The Covid vaccine campaign was based on the faulty premise that a respiratory virus could be controlled through human behavior, including mass vaccination. Airborne respiratory viruses are not good candidates for vaccine treatments. Before Covid-19 there were four recirculating coronaviruses[353] that regularly infect humans, including the one that causes the common cold. Now there are five, thanks to SARS-CoV-2.

Respiratory viruses and other microbes are constantly in our environment, with people always developing immunity, and the virus always trying to evade immunity. Vaccine development for respiratory viruses will always be behind the current strain. For example, the December 2020 Covid vaccines were designed to address the original Covid-19 Wuhan strain. By the time the vaccine rollout began, the Delta variant and B117 variants had replaced the Wuhan strain. By the time the Omicron strain became dominant in the fall of 2021, when the first booster shots were being pushed and the vaccine mandates arrived, the vaccines were almost completely outdated, formulated for strains that were no longer circulating. **CDC Director Rochelle Walensky acknowledged in August 2021 that the Covid shots could not prevent the spread of Covid-19[354].**

The bivalent boosters that were released in the fall of 2022 were geared for the Wuhan spike and the original Omicron spike. Wuhan

was long gone, and the original Omicron had been outcompeted by other mutations. The Covid booster shots had no significant impact on preventing infection or spread of Covid-19, which is why almost everyone from Dr. Fauci, to the arrogant talking heads on TV, to your friends and family members, got Covid despite vaccination.

Diseases have not been eradicated through vaccination:

The only diseases that have been eradicated[355] through vaccination are smallpox and a cattle disease, rinderpest, which has never impacted humans. Neither is caused by an airborne respiratory virus. **We don't have effective vaccines for** endemic **respiratory diseases; that is, diseases that continually circulate in the population, usually seasonally.** Each year when the flu shots roll out, there is around a 30 percent chance that the shot will impact infection with influenza, in part because the shot is always based on strains from the previous year.

Some might say, well, we have to try things if we're going to have any advances in medicine and science. There is some truth to that. After all, it's called "the practice of medicine." However, when it comes to new medications and vaccines, and novel medical technologies, there are strict rules. These **rules and procedures, put in place to protect people from dangerous, untested products, were violated completely with the Covid-19 vaccines.**

The push[356] for future capacity to design and inject mRNA vaccines within 100 days[357] of the appearance of a new pathogen, ignores two essential points: 1) mRNA technology has not been proven to be safe or effective, and 2) humans have immune systems that have successfully learned to fight diseases since the beginning of time, without being given constant injections.

The 100-days goal for mRNA vaccines ignores the key point that the first mRNA vaccines tested on people – the Covid-19 vaccines - were a complete failure. They neither prevented illness nor prevented spread, but they have caused widespread and long-lasting harm.

"Wrong shot, wrong protein, wrong virus"

> *We are literally doing the largest experiment on humanity ever done, not knowing the long-term outcomes.*
>
> **DR. RYAN COLE** | Pathologist | June 13, 2022[358]

When the vaccine rollout began, Dr. Ryan Cole was the owner of a large medical lab that provided doctors the results of patients' blood samples. **His lab began to notice** unusual pathologies, including an **increase in sudden and rapidly progressing cancers** (given the name "turbo cancer" by some, but it's not an official medical term), a large rise in **cancers coming out of remission,** and **reactivation of viruses**, such as the chicken pox herpes zoster being reactivated as shingles. He began to see a **skin condition** that is usually only in young children, before they develop immunity to that virus, **reactivating in adults**. Over time **Cole concluded that patients' immune systems were being suppressed by their reaction to the Covid-19 shots.**

Dr. Cole explains, "**We're giving a sequence to make individuals' bodies a spike protein toxin factory**. Now any cell that makes that spike becomes the target of your own immune system. Now your natural killer cells come in to destroy your own cells. **The spike is a toxin and we made a shot that makes your body the toxin factory.**" Dr. Cole states:

> This is a dangerous product...being used on humanity for a virus that no longer exists (the Wuhan strain), that does nothing but cause increased disease in those who get a series of these shots.

Observing that the Covid shots were leading to more medical product-related deaths than ever seen before, Dr. Cole summarized the Covid shots this way: Wrong shot. Wrong protein. Wrong virus." **As James Lyons-Weiler, and Dr. Angus Dalgleish and others had warned in April 2020, the spike protein was the wrong part of the SARS-CoV-2 virus on which to base the vaccine.**

Wrong shot: neither mRNA in a lipid nanoparticle envelope, known to cause inflammation and other problems, nor adenovirus vector shots have a targeting mechanism to reach certain cells, meaning they have no control over where and how long the spike protein will be produced.

Wrong protein: messaging cells throughout the body to create the toxic spike protein, which is homologous to DNA and likely to produce autoimmune diseases or antibody dependent enhancement (ADE), is not a good idea.

Wrong virus: the human body is always fighting coronaviruses which are always trying to evade the immune response. This leads to constant mutations which is one of the reasons why coronaviruses are not good candidates for treatment with a vaccine.

Dr. Ryan Cole becomes another target for the censorship complex:

As a consequence of speaking out about what he has observed, Dr. Cole lost insurance approval and was removed from one of Idaho's biggest health networks, leading to lost business and reduced income from his pathology lab. Cole is also facing disciplinary action[359] against his medical license.

Some who still want to give benefit of the doubt to the authorities, might say, "Well, we had to try something. A pandemic was raging, and the vaccines were supposed to help." But let's not forget that even if that were true, and vaccines were somehow necessary, we did not need to use untested mRNA, lipid nanoparticles, and adenovirus vector technologies with known safety issues.

Dr. Robert Malone states[360], "I do not know how to write this more strongly. This technology is immature." He notes that the World Health Organization approved six Covid-19 vaccines that were more traditional, all of which the U.S. government could have licensed. "These genetic vaccines [were] not the only option."

2016 study outlined some of the problems with lipid nanoparticle technology:

A May 2016 study[361] of mRNA vaccine delivery using lipid nanoparticles found that they induced inflammation. The study stated, "IV injections of LNP (lipid nanoparticle)–mRNA vaccines are less common because of the potential of systemic side effects. Indeed, injecting immunogenic material in the bloodstream may lead to massive cytokine production, also known as cytokine storm, that can lead to shock and death. Additionally, vital organs, including the liver and lungs, are transfected by mRNA vaccine delivery using LNPs. Expression of the antigen by these organs could recruit T cells that induce tissue damage and inflammation."

> Inflammation and organ and tissue damage were previously known problems associated with lipid nanoparticle technology that were not resolved during the development of the Covid-19 vaccines. Massive amounts of vaccine injuries and deaths have been the result.

17. ISRAEL: PFIZER'S GUINEA PIG, AND THE WORLD'S CANARY IN THE COAL MINE

In January 2021 leaders in the Israeli government made an exclusive deal with Pfizer:[362] vaccines for data. As reported by NPR in January of 2021, "Israel's small size and technologically advanced public health system offer an attractive model for Pfizer to demonstrate the impact of the vaccine on an entire population." Despite some concerns expressed about privacy of data, and caution in using a new product on an entire country all at once, the mass vaccination rollout began.

Myocarditis in young men post-vaccination:

Over the next few months **there was an increase in myocarditis[363], which is inflammation of the heart, observed in younger men especially in those age 16-19,** with 148 cases occurring around the time of vaccination, more often after the second dose. The Ministry of Health reported "Most cases have been in the hospital for up to 4 days, and 95% are considered to be mild cases." It is doubtful the parents of those previously healthy young men felt that a hospital stay, followed by limited physical activity of unknown duration, was a "mild" situation for their children.

As heart muscle does not repair itself, it is difficult to predict how a "mild" case of myocarditis may impact someone down the road. Venkatesh Murthy, Professor of radiology and cardiovascular medicine at the University of Michigan, states[364], "People with myocarditis are usually counseled to limit activity, placed on one or more meds and are at lifetime increased risk of cardiac complications. This can have profound consequences."

The vaccinated become the bulk of Covid hospitalizations:

As over 90 percent of the Israeli population became fully vaccinated, the expectation was that Covid cases would decrease. Instead there continued to be outbreaks, with the majority of Covid hospitalizations occurring in the vaccinated[365]. There was also a **sharp rise in ER visits[366] for the 16-39-year-old population, due to cardiovascular adverse conditions**, when compared to the years 2019-2020.

Pfizer cited the breakthrough Covid cases in Israel and acknowledged that "efficacy in preventing both infection and symptomatic disease has declined six months post –vaccination," especially in light of the Delta variant. The solution? **The logical solution would have been to halt the vaccination program and conduct more research.** But no. Pfizer stated, "[I]t is likely a third dose may be needed within 6 to 12 months after full vaccination." By October 2021, booster shots were required [367] in Israel in order to be considered "fully vaccinated" and obtain the Green Pass[368] that allowed participation in public life. As the following charts show, there is no correlation between masking, vaccination, and the rise and fall of cases and Covid deaths. The first chart shows daily new cases from January 2020 through January 2022:

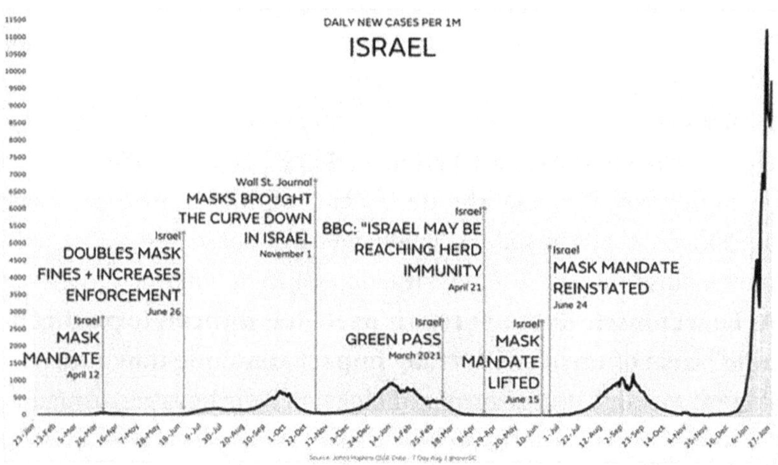

The above chart shows daily new cases per 1 million from January 2020 through January 2022. Note the phenomenal rise in cases in January 2022, long after 90% of Israelis were vaccinated against Covid-19.

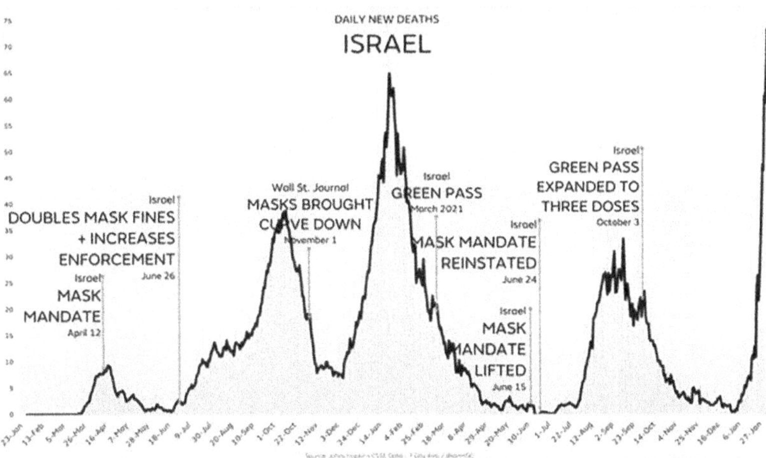

The above chart shows daily new deaths in Israel from January 2020 to January 2022. By September of 2021, Israelis were required to show their Green Pass, verifying they had received the required three doses of a Covid shot, in order to participate in daily life. The mandated masking and vaccines did not prevent the largest surge in Covid deaths, which occurred in early 2022.

Israeli Ministry of Health suppresses adverse vaccine events:

Anecdotal word from Israel[369] was that people were experiencing health problems after receiving the Pfizer shots, but the Israel Ministry of Health (IMOH) suppressed the reports. **It was not until the end of December 2021 that the data were analyzed finding the most common post-vaccine side effects, beyond heart issues, included neurological injuries, menstrual irregularities, musculoskeletal system disorders, and kidney/digestive system and urinary system problems.**

As explained by Yaffa Shir-Raz, PhD, of the University of Haifa and Reichman University, and Professor Retsef Levi of MIT[370], upon learning of these long-term side effects, the IMOH withheld the findings, even from their own expert committee, which in the meantime approved the vaccine for infants as young as six months.

Shir-Raz states, "**While Israel is a relatively small country, it was dubbed 'the world's laboratory.'** The eyes of much of the world were on it, and the FDA and other regulators have repeatedly cited its experience with the vaccine as a basis for policy-making, including for boosters and mandates and much else...**So if Israel did not in fact have a functioning adverse event monitoring system in place and its data was fiction[371]... what was the FDA really relying on?"**

Rather than learning from Israel's experience, the U.S. and other developed nations that were focused on the experimental mRNA shots suppressed the data about vaccine harms and continued their vaccine campaigns.

18. MYOCARDITIS HARMS SHOW UP EARLY IN THE U.S. VACCINATION CAMPAIGN

The CDC [Centers for Disease Control and Prevention] maintains the VAERS [Vaccine Adverse Event Reporting System] database, and there are others, such as Yellow Card in the UK and EudraVigilance in Europe, and all kinds of other databases. They all started showing huge numbers of adverse events and deaths right away. The government was denying that those were associated with the vaccine, and they continue denying it to this day.

SASHA LATYPOVA

Former pharmaceutical executive | July 20, 2023

In addition to early warning signs from Israel, there were also developing data in the U.S. that linked myocarditis and pericarditis (inflammation of the fibrous sac surrounding the heart) to the Covid shots.

Myocarditis cases rise in 16 to 17 years olds in U.S.:

In June 2021 Stat News reported[372], "**Among 16 to 17 year olds**, who had received 2.3 million doses**, there had been 79 cases of myocarditis or pericarditis** reported through VAERS. Based on baseline frequency of the myocarditis and pericarditis, there **would have been an expected 2 to 19 cases in that group**."

VAERS[373] is the Vaccine Adverse Event Reporting System, established by the CDC[374] in 1990, as "a national early warning system to detect possible safety problems in U.S.-licensed vaccines."

Adverse Event[375] (AE) is defined as, "a harmful and negative outcome that happens when a patient has been provided with medical care." A Serious

Adverse Event[376] (SAE) is an adverse event that leads to medical or surgical intervention, hospitalization, disability or permanent damage, or death.

This chart compiled from VAERS data shows the alarming increase in reports of myocarditis and pericarditis in 2021, the starting year of the Covid-19 vaccine campaign:

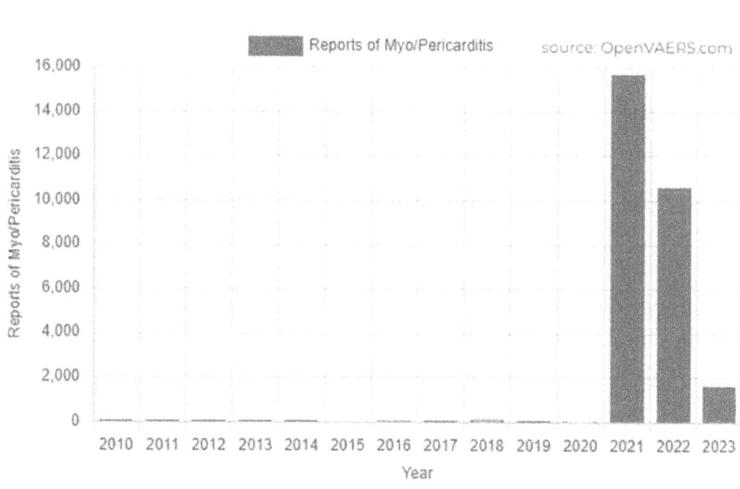

As the above chart shows, to go from almost no reports during 2010 through 2020, to close to 16,000 myocarditis/pericarditis reports in 2021, is a safety signal that cannot be missed or dismissed. Yet the CDC chose to ignore its own safety signal system.

VAERS shows a safety signal, so the government questions VAERS data: The **CDC**, throughout the pandemic, **disparaged its VAERS system as being "self-reported," and therefore not an accurate method** for tracking Covid vaccine injuries. Yet there have been **more Covid-vaccine related deaths reported to VAERS since the rollout began, than there had been for all other vaccines combined in the previous 30 years.** Other vaccine databases, such as the WHO's VigiAccess and the U.K.'s Yellow Card system also received unprecedented adverse event reports regarding the Covid shots.

Following is a chart of VAERS data, from 1990 to the present, showing the dramatic increase in vaccine-related deaths corresponding with the rollout of the Covid shots:

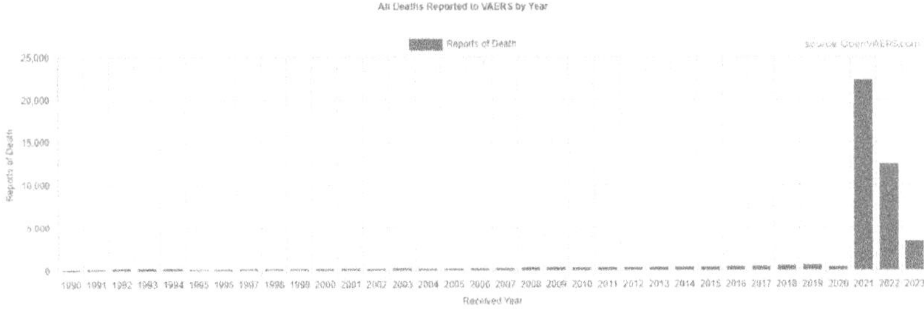

The above chart shows the number of vaccine deaths reported to VAERS with a dramatic increase in death reports shown in the red bars on the right for the years 2021, 2022, and 2023.

Majority of VAERS-reported deaths occurred within 2 days of Covid vaccination:

As of December 18, 2023 **there have been over 1.6 million adverse event reports to VAERS**[377] including 27,832 reports of myocarditis/pericarditis. Other adverse events include heart attacks, miscarriages, Guillain Barré syndrome, Bell's palsy, anaphylaxis, blood clots, shingles, menstrual disturbances, neurological problems, and permanent disabilities. In addition, there have been 36,726 vaccine deaths reported to VAERS, the majority of which occurred within two days of Covid vaccination.

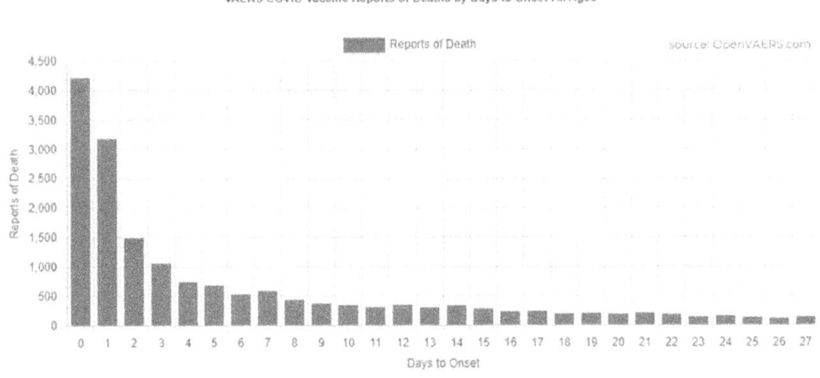

The above chart shows deaths reported to VAERS based on the number of days since vaccination. The majority of deaths reported to VAERS after Covid vaccine were within 2 days, as shown in the red bars on the left.

VAERS is thought to represent only 2% of total adverse events: As shown in the above charts, there has been an alarming rise in vaccine-related adverse events, including deaths, reported to VAERS since the advent of the Covid vaccines. Also concerning is the longstanding estimate that **only about 2% of adverse events are ever actually reported to VAERS. It's a clunky system, with difficult paperwork.** Although anyone can report to VAERS, it's **mostly used by medical professionals** who are required by law[378] to report "moderate to severe" suspected vaccine injuries.

Attorney Warner Mendenhall, who specializes in bringing lawsuits against the government for overreach and abuse of power, is currently representing **nurses** who were **fired for reporting suspected Covid vaccine injuries to VAERS.**

> We believe that every hospital in this country has ignored its legal obligation to submit injury reports to the VAERS system.
>
> **ATTORNEY WARNER MENDENHALL**
>
> American Thought Leaders interview | June 4, 2023[379]

Analyst Ed Dowd writes, "Media and government leaders tried to dismiss VAERS out of hand, implying that anti-vaxxers were using it to submit false claims...**The idea that significant numbers of these reports are fake is ludicrous. Aside from it being a Federal offense to file a false VAERS report, it's impenetrably difficult** to enter a VAERS report." - Dowd, Ed. "Cause Unknown": The Epidemic of Sudden Deaths in 2021 & 2022 (Children's Health Defense) (p. 123). Skyhorse. Kindle Edition.

Dr. Jessica Rose[380], immunologist and independent researcher, has tracked VAERS data extensively since December 2020. She states,

> It's very important that everybody understands...that by the end of January 2021 there was way more of a signal than would be required to at least do an investigation. Because that's what VAERS is for, it's a pharmacovigilance system which detects safety signals by way of adverse events that are reported, that weren't detected during pre-market testing, like clinical trials. The system was actually

> functioning beautifully. There were safety signals left, right and center going all the way to death...We knew right away that these things were not safe by any definition of the word...

Rose concludes, "**Something should have been done in the way of trying to assess whether these shots were the cause of all of the deaths**." Instead, the CDC, FDA and pharmaceutical companies proceeded full speed ahead.

Increased Adverse Event reports to VAERS are NOT because More Shots were given:

Some people have claimed that the increased safety signal to VAERS was nothing alarming, and was a natural function of so many shots being administered at the same time. The idea being, "Of course there are more reports to VAERS; there are more shots being given – it's proportional." Dr. Rose dismisses **that idea by providing data comparing[381] the VAERS data on flu shots and Covid shots. Adverse events reported per 1 million doses of the flu vaccine versus 1 million doses of the Covid vaccines show that the reports to VAERS are not "proportional." The adverse events reported for the Covid shots are astronomically higher:**

2/11/23
JESSICA ROSE, PHD

NUMBER OF VAERS REPORTS FOR FLU (2020)/1M DOSES VS. COVID (2021)/1M DOSES

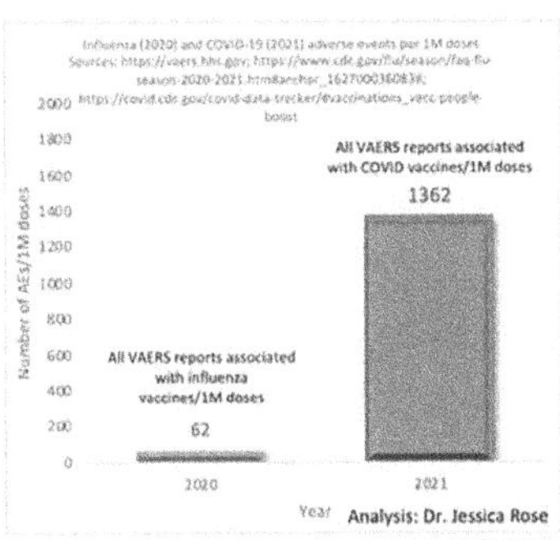

https://jessicar.substack.com/p/lets-put-the-its-cuz-there-are-so
https://www.cdc.gov/flu/season/faq-flu-season-2020-2021.htm#anchor_1627000360838
https://covid.cdc.gov/covid-data-tracker/#vaccinations_vacc-people-boost

The above chart compares adverse events reported to VAERS for the flu vaccine with adverse events reported to VAERS for the Covid vaccines. The blue bar represents adverse reports to VAERS per 1 million doses of the flu vaccine in

2020. The red bar shows adverse reports to VAERS per 1 million doses of the Covid vaccines. Comparing 1 million doses of the flu vaccine with 1 million doses of the Covid vaccines shows that proportionally, there are significantly more adverse events reported to VAERS associated with the Covid shots.

Whether or not someone values the VAERS reporting system, there is **no denying the experiences of the thousands of people who have been Covid vaccine-injured**. Following are a few quotes from the documentary "Anecdotals:"

- "This is not political. This is a human issue."
- "And so it's usually someone who finds out I'm vaccine-injured says, 'Oh, really? Which one did you have?' and I'll say, 'Oh I got Pfizer.' They'll reply, 'Oh really? My husband got that one and he's fine.' I'll reply, 'Yeah, and so did a lot of my family members and they're all totally fine, but some people are *not* fine.'"
- "Paresthesia is a word I'd never heard of a year ago, but now it wakes me up every morning."
- "One word describes how I felt the first few months after my diagnosis – abandoned."
- "I did what was asked, got harmed, and there's no program in place for me."
- "I was told that I was ethically and morally irresponsible because sharing my story could sway people away from getting vaccinated."
- "They say we (the vaccine-injured) have a small percentage, but how do they truly know what the percentage is when they won't even acknowledge that we exist? "

These people are the Anecdotals – those whose stories are not considered necessarily true or reliable because they're based on personal accounts:

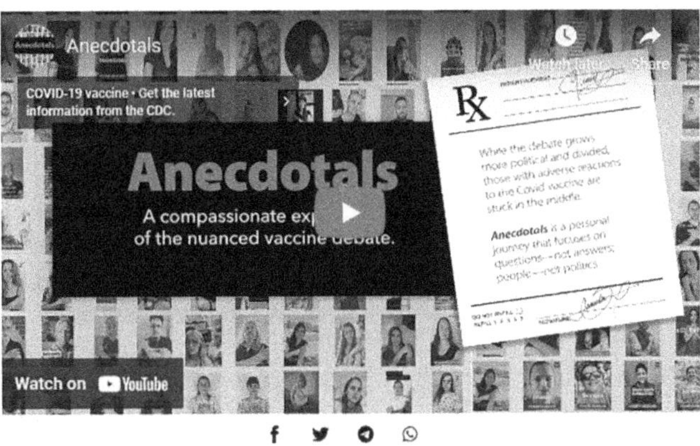

"Anecdotals" - Accounts of the Vaccine Injured[382]

A logical person is led to ask: If 30 deaths and 500 injuries were enough to pull the vaccines during the 1976 flu pandemic[383], why has the Covid-19 vaccine campaign continued?

D-Med, the Military medical record system, shows a rise in myocarditis:

> *Our soldiers are being experimented on, injured, and sometimes...killed....This is corruption at the highest level. We need investigations. The Secretary of Defense needs investigated. The CDC needs investigated.*
>
> **ATTORNEY TOM RENZ** | U.S. Senate Roundtable | January 24, 2022[384]

The **U.S. Military** was also seeing an unusual number of myocarditis cases in healthy servicemen. A June 2021 report in *JAMA Cardiology* stated[385], "In this case **series of 23 male patients, including 22 previously healthy military members, myocarditis was identified with four days of receipt of a Covid-19 vaccine,"** most after the second dose. D-Med, the Department of Defense medical record system, began showing a **"catastrophic increase in illnesses and injuries,"** after the Covid vaccines rollout, according to whistleblower Lt. Col. Theresa Long, MD, MPH, FS.

Lt. Col. Long states that reportable medical events, defined as **"severe life-threatening clinical manifestations that disrupt military training**

and deployment," almost doubled from 110,985 in 2020 to 205,651 in 2022. Long is convinced that the risks of the Covid vaccines far outweigh any benefit. (See U.S.Senate[386] December 2022, at 29:30)

Testimony during a January 2022 Senate roundtable regarding the Covid-19 response stated that D-Med reports of **miscarriages and cancers increased by almost 300% above the 5-year average after the Covid vaccines were required in 2021. Neurological issues, which especially affect pilot readiness, increased over 1,000%.**

Human Rights Attorney Leigh Denison, working with Tom Renz on the whistleblower lawsuit against the Department of Defense, said **acute disease reports** in all areas in the five years previous to the Covid vaccine rollout totaled 1.7 million. **"They introduced and mandated a Covid-19 vaccine for our U.S. Military when they'd only lost 12 service members total to the disease,"** said Denison, "and in the 10 months of 2021 after that it (acute disease reports) **jumped from 1.7 million to darn near 22 million.** That was a **20 million increase**." (See U.S. Senate Roundtable, Jan 24, 2022, at 5:09:50[387])

When Lt. Colonel Long approached upper command with her concerns, the alarming safety signals in D-Med were dismissed as a "computer glitch." D-Med was then taken offline for "maintenance." When it came back online, the database had been altered[388] to reflect fewer vaccine safety signals.

Before FDA Emergency Use Authorization, they already knew about myocarditis:

Epoch Times reporter Zachary Stieber has compiled a detailed timeline[389] of Covid-19 vaccines and myocarditis. In reviewing the compilation, it becomes **undeniably clear that Pfizer, Moderna, the FDA, the CDC, and other entities involved in the vaccines development knew the shots were associated with multiple adverse health conditions.**

In **September 2020, the CDC gave an internal presentation** on Vaccine Adverse Events titled, "Enhanced safety monitoring for Covid-19 vaccines in early phase vaccination." **The preliminary list of VAERS AESI's (Adverse Events of Special Interest) in connection with the Covid-19 vaccines included** myocarditis **and** pericarditis, along with a host of other concerns, as shown in Slide 8 from the CDC presentation:

Preliminary list of VAERS AESIs

- COVID-19 disease
- Death
- Vaccination during pregnancy
- Guillain-Barré syndrome (GBS)
- Other clinically serious neurologic AEs (group AE)
  - Acute disseminated encephalomyelitis (ADEM)
  - Transverse myelitis (TM)
  - Multiple sclerosis (MS)
  - Optic neuritis (ON)
  - Chronic inflammatory demyelinating polyneuropathy (CIDP)
  - Encephalitis
  - Myelitis
  - Encephalomyelitis
  - Meningoencephalitis
  - Meningitis
  - Encepholapathy
  - Ataxia
- Seizures / convulsions
- Stroke
- Narcolepsy / cataplexy
- Autoimmune disease
- Anaphylaxis
- Non-anaphylactic allergic reactions
- Acute myocardial infarction
- Myocarditis / pericarditis
- Thrombocytopenia
- Disseminated intravascular coagulation (DIC)
- Venous thromboembolism (VTE)
- Arthritis and arthralgia (not osteoarthritis or traumatic arthritis)
- Kawasaki disease
- Multisystem Inflammatory Syndrome in Children (MIS-C)

In an **FDA meeting on October 30, 2020**, the FDA presented a **draft of "possible adverse event outcomes"** to the Covid-19 vaccines. Again, the list was long and included **myocarditis and pericarditis:**

FDA Safety Surveillance of COVID-19 Vaccines :
DRAFT Working list of possible adverse event outcomes
*****Subject to change*****

- Guillain-Barré syndrome
- Acute disseminated encephalomyelitis
- Transverse myelitis
- Encephalitis/myelitis/encephalomyelitis/meningoencephalitis/meningitis/encepholapathy
- Convulsions/seizures
- Stroke
- Narcolepsy and cataplexy
- Anaphylaxis
- Acute myocardial infarction
- Myocarditis/pericarditis
- Autoimmune disease
- Deaths
- Pregnancy and birth outcomes
- Other acute demyelinating diseases
- Non-anaphylactic allergic reactions
- Thrombocytopenia
- Disseminated intravascular coagulation
- Venous thromboembolism
- Arthritis and arthralgia/joint pain
- Kawasaki disease
- Multisystem Inflammatory Syndrome in Children
- Vaccine enhanced disease

Yet the Covid-19 vaccines were still given Emergency Use Authorization, and the information about adverse events was kept from the public. In fact, it was kept from the clinical trial participants, who were only told to expect common vaccine side effects such as redness and swelling at the vaccination site, a fever, and nausea.

As the vaccine rollout to the public began in December 2020, the incoming number of Covid-19 adverse event reports exceeded the

expected maximum of 1,000 reports per day, leading to a delay in processing and follow-up by the outside contractor hired[390] to manage VAERS.

Despite the safety signals coming in from VAERS, the Department of Defense, and Israel, **concerns were minimized or dismissed, the public was not informed, and the vaccine rollout continued.** In fact, there were multiple communications[391] and meetings between different government players about how to keep the information from the American people. It wasn't until after several months, and reports flowing in from around the world, that the FDA finally put a warning label on the Covid vaccines.

FDA adds myocarditis warning to Covid shots – recommends all receive them:

On June 25, 2021 the FDA added to its Covid-19 provider information a warning that there is a possible link between the shots and myocarditis and pericarditis, especially in young males between the ages of 15-30. Nonetheless, their issued statement was, "The CDC continues to recommend vaccines for everyone 12 years of age and older, given the risk of COVID-19 illness and related, possibly severe complications." That update is no longer available with the link from June 2021.

The link now reads[392] **"CDC continues to recommend that everyone ages 6 months and older get vaccinated for Covid-19. The known risks of Covid-19 illness...far outweigh the potential risks of having a rare adverse reaction to vaccination**, including the possible risk of myocarditis or pericarditis." No mention is made of the age stratification in Covid risk, nor of the fact that children have a much higher risk of death from drowning[393] and auto accidents[394], and even being struck by lightning[395], than death due to Covid[396].

> There were almost no incidents of a healthy young person being seriously injured by Covid-19 infection in 2020. The same cannot be said for what has happened to young people since the introduction of the Covid-19 shots.

19. WHY CHILDREN, ESPECIALLY, SHOULD NOT BE GIVEN THE COVID-19 SHOTS

Today, the U.S. Food and Drug Administration authorized the emergency use of the Pfizer-BioNTech COVID-19 Vaccine for the prevention of COVID-19 to include children 5 through 11 years of age.

FDA NEWS RELEASE

October 29, 2021[397]

Today, the U.S. Food and Drug Administration authorized emergency use of the Moderna Covid-19 and the Pfizer-BioNTech Covid-19 Vaccine for the prevention of Covid-19 to include use in children down to 6 months of age.

FDA NEWS RELEASE

June 17, 2022[398]

The CDC and FDA have never emphasized that children are not susceptible to serious illness from Covid, but instead they continually **try to frighten parents into thinking their children are at risk[399].** The fact is, the younger a person is, the fewer ACE2 receptors[400] they have in their nose. Because the spike protein binds with the ACE2 receptor in the replication process, most children have natural protection against serious Covid-19.

> The case fatality [for Covid] is not one that should have caused us to vaccinate literally billions of people...At the point that we're seeing an undeniable pattern of [Covid vaccine] damage, and we know that damage is disproportionately experienced by younger people, and

> that the risk of Covid is disproportionately to the old, the idea that it doesn't stop...means, I think, that somehow whatever is driving this policy is absolutely comfortable with the death of other people.
>
> **BRET WEINTSTEIN** | Evolutionary biologist | January 2023 (1:18:30)[401]

Another tactic to try to get children vaccinated is the guilt trip that they need to take the shot to protect the elderly. However, children with Covid infection are often asymptomatic, or have mild symptoms, and are largely not infectious to adults**. It was known early on that children are not the drivers of Covid infection, even in close interactions**, (see here[402], here[403], and here[404]) which has remained the case. But beside that, adults are supposed to protect children - not use them as shields to protect themselves. To inject children with an Emergency Authorized Use drug, that has possible serious side effects and no benefit for the child, is unethical.

European countries discontinue the Covid mRNA shots for young people:

Throughout the entire pandemic, hardly any children have been seriously ill with, or have died of, Covid-19. Yet the CDC to this day recommends Covid vaccination in everyone 6 months and older. Most in the U.S. are not aware that starting as early as 2021, multiple European countries[405] stopped giving the Moderna Covid vaccine to those under age 30, all citing studies pointing to possible risks of heart inflammation in young recipients of mRNA vaccines. In October 2022, Sweden[406] stopped recommending Covid shots for healthy children between the ages of 12 and 17, and Denmark ended Covid shots for anyone under the age of 50, noting that the shots do not prevent infection.

The mRNA Covid-19 shots may permanently harm infants and toddlers:

Dr. Geert Vanden Bossche and Dr. Robert Malone have both called the injection of infants and children with Covid-19 mRNA vaccines "a crime," and "unconscionable." Why? Dr. Vanden Bossche is a specialist in vaccinology, and Dr. Malone was a pioneer in mRNA development. Why would both these doctors speak so strongly against mRNA vaccines for children? It has to do with how the immune system develops.

Young babies are protected by their mother's antibodies, which they get from the colostrum in breast milk. After a few months, that protection wanes and the baby begins to develop its own immune system through continued exposure to pathogens. **Many viruses are coated in a sugary coating of molecules called glycans, which are similar to components in the human host so as to mislead and subvert the immune system.**

Dr. Vanden Bossche explains **that the developing immune system is designed for broad response to the glycan-coated viruses** such as coronavirus, measles, mumps, rubella, RSV, and influenza. The developing immune system learns to recognize these foreign compounds as intruders, distinguishing the pathogen from self, and developing innate antibodies against each pathogen. Antibodies wain, but the immune system stores the memory in T cells and B cells, and is able to pull forward an antibody response when the pathogen is encountered again.

Now enter the Covid-19 mRNA shots. Vaccinal anti-spike antibodies bind more strongly to the virus than the innate antibodies can. The innate antibodies, which are designed for a broad response, are outcompeted and sidelined by the dominant spike-specific protein induced through Covid vaccination.

Frequent exposure to circulating Omicron variants combined with vaccine-induced antibodies may continually suppress innate antibodies. This constant emphasis on SARS spike protein may inhibit a child's innate immune system from learning how to recognize RSV, influenza, and other childhood diseases. Vanden Bossche said there is **a concern that illnesses such as RSV, and influenza could become serious, even life-threatening diseases for children whose innate immune systems were interfered with by the Covid shots.**

As mentioned, **children have natural protection against Covid-19,** in part because of their low ACE2 receptor count. **Giving them an mRNA injection bypasses this natural defense and directly exposes children to unlimited amounts of spike protein,** leading to autoimmune complications, and all the other harms identified with the shots including damage to the cardiovascular, neurological, and reproductive systems.

> The price you pay for a few weeks of protection is severe disease. You have messed up the education of your innate immune system because the body hasn't learned to recognize other viruses. Young children (who receive the Covid-19 vaccines) are at high risk for suffering severe immune pathology.
>
> **DR. GEERT VANDEN BOSSCHE** | June 13, 2022[407]

Loss of Bifidobacteria – another possible risk to infants due to Covid-19 shots:

Gastroenterologist and CEO of Progenabiome, Dr. Sabine Hazan is an expert on gut microbiome. The microbiome[408] is the collection of all microbes, such as bacteria, fungi, viruses, and their genes, that naturally live on our bodies and inside us.

Dr. Hazan found[409] during the Covid pandemic that people who had severe Covid lacked a bacteria called Bifidobacteria. Bifidobacteria in the human gastrointestinal tract are **one of the major bacteria beneficial to human health, comprising an important aspect of the immune system.** Mothers transfer[410] Bifidobacteria to newborns during vaginal delivery, and also through breastfeeding[411].

Dr. Hazan undertook a study of the bifidobacteria count in patients pre- and post-Covid vaccination, and found a dramatic drop post-vaccination. This was concerning as the Covid shots were intended to improve immunity, not reduce it. Hazan followed four patients, "who were in amazing shape," for 90 days and found that **post-vaccination their Bifidobacteria dropped "from a million to zero."** Not only that, but the damage persisted in the months to follow.

As Dr. Hazan analyzed the gut microbiome of breastfeeding newborns born to vaccinated mothers, she found that there was no Bifidobacteria present, where normally Bifidobacteria would comprise up to ninety percent. Hazan speculates that the Covid spike protein has a negative impact on the infant's developing gut microbiome.

The interplay between the gut microbiome and the immune system is a relatively new field of research; however a June 2022 study[412] found a **significant loss of biodiversity in the gut microbiota one month after**

the second dose of Covid-19 vaccine. An analysis[413] of the study notes, "On a theoretical basis, vaccine-induced loss of diversity would increase the opportunity of pathogens to thrive in the intestine." The study noted that while "certain gut microbes could enhance vaccine efficacy," Covid-19 vaccines "could impact the gut microbiota, enhancing the growth of some microbes but suppressing the growth of many others."

The analysis concludes that further study is needed to "understand and possibly harness such interactions and convert their potential therapeutic value into reality." But the real point is that we don't know.

> What are we doing giving mRNA shots to the entire world's population when there are so many body systems that are impacted, and we know so little about the complexities of those interactions?

Medical ethics require extensive testing and evaluation of new medications and vaccines before administering them to **pregnant women, obviously in part because the developing fetus is also impacted. This ethical norm was largely discarded in the case of the Covid-19 vaccines.** Although no pregnant women were included in the original Covid-19 vaccine clinical trials, they were encouraged[414] and even coerced[415] to take the EUA shots anyway.

Dr. Robert Malone issues a warning against Covid shots for children:

> *The most alarming point about this is that once these damages have occurred, they are irreparable...* ***"[T]his genetic vaccine, based on the mRNA vaccine technology I created...has not been adequately tested".*** *(emphasis added)*
>
> **DR. ROBERT MALONE** | October 29, 2021

After the FDA approved the Covid shots for children ages 5 through 11 on October 29, 2021, Dr. Malone issued a video statement[416] to parents warning them not to vaccinate their children against Covid. He explained that the toxic spike protein, created in response to the messenger RNA in the shots, **can cause permanent damage in children's critical organs**, including:

- brain and nervous system
- heart and blood vessels, blood clots
- reproductive system, and
- possibly causing negative fundamental changes to their immune system

Multiple doctors warn against the Covid shots for children. A few quotes:

> *"It is a crime to administer this vaccine to children."*
> **Dr. Luc Montagnier** | January 15, 2022[417]

> *"The risk for children with Covid is exceedingly low, but we now know there is a real risk for myocarditis from the vaccine."*
> **Dr. Kirk Milhoan** | Dec 7, 2022, 2:15:53[418]

> *"Putting this needle with this gene in their arm is nothing short of child abuse."*
> **Dr. Ryan Cole** | June 13, 2022[419]

> *"Other reputable countries have banned this shot for kids. We need to pause. What are we doing?"*
> **Dr. Renata Moon** | Dec. 7, 2022, 2:51:49[420]

> *"That's not how this is supposed to work; our kids should not be the guinea pigs. Stop using our children as shields in this battle that is an adult battle."*
> **Dr. Byram Bridle** | May 28, 2021[421]

> *"Why are they pushing this lethal vaccine and risking the future of all humanity?"*
> **Dr. James Thorp, OB/GYN** | Dec 7, 2022, 2:29:34[422]

> *"To me it is unconscionable that a society uses its children as shields for adults. Children do not have significant risk from this illness. Are we a society...where we are using our children, even if they did spread it, as shields? We're going to inject our children with an experimental drug, that they don't have a significant benefit*

from, to shield ourselves? ...That's really just a heinous violation of all moral principles, in my view."
Dr. Scott Atlas | Sept. 23, 2021[423]

Some doctors who were at first proponents of the Covid shots are now adamant that they should not be given to anyone, and especially not to children. Over time, many medical professionals began speaking out, despite the risk to their professional credentials and personal lives.

U.K. Doctors Express mRNA Vaccine Concerns (19 min)[424]

Evidence of suppressed immune systems in children emerges in fall 2023:

As this compendium was nearing completion, there began to be reports from China of children becoming seriously ill with common childhood illnesses. **The warnings of the doctors cited in Section 19** "Why Children, Especially, Should Not Be Given the Covid-19 Shots," **have been continually dismissed.** The CDC and FDA have been pushing hard for everyone 6 months and older to take the new-for-fall-2023 Emergency Use Authorized Covid-19 shots.

In late fall 2023 reports[425] are coming in that **children are becoming very ill and hospitalized with "familiar viruses such as influenza, rhinoviruses, respiratory syncytial virus (RSV), adenovirus, and bacterial mycoplasma pneumonia."** Forbes[426] reported on November 25, 2023 that in South Korea, 80% of the hospitalized patients with acute bacterial respiratory infections are children under the age of 5.

Mainstream media headlines say things like "Mystery child pneumonia outbreak reported in China hospitals," and "Full list of countries affected by Chinese respiratory illness as cases spreading in EU," and "WHO questions China about a mystery illness as hospitals fill with sick children ." (see here[427], here[428], and here[429])

It's not a mystery to doctors who kept their Hippocratic Oath to "do no harm," and refused to promote or give experimental mRNA shots to children who didn't need them. China attributes the rise in children hospitalized with common childhood illnesses to "the lifting of Covid-19 restrictions," and the factor of their population now being "vulnerable to regular bugs after limited exposure in recent years." Certainly that could have an impact, but limited exposure doesn't seem like the only explanation.

A November 30, 2023 article[430] in the *Messenger* has the headline "Mysterious Pneumonia Outbreak Emerges in US – Days After Similar Illness Reported in China." The article reports, **"a surge of mysterious pneumonia cases" in the U.S., "with 142 children falling ill with the disease since August."** Specifically, the "outbreak" emerged in Warren County, Ohio, with the average age of affected children" being 8 years old. The article states, "At least three pathogens have been detected in the outbreak, including Mycoplasma pneumoniae, Streptococcus pneumoniae and the adenovirus. However**, these bugs regularly circulate in the U.S., making it unclear why a sudden surge is appearing now."**

Unsurprisingly, but infuriatingly, the increase in serious respiratory illnesses in children, rather than leading to introspection about the past four years of the Covid response, is leading to calls[431] for a return to ineffective and damaging social distancing, masking, closures, and travel bans.

> If kids' immune systems are suppressed because they didn't get normal community exposure because of lockdowns and school closures, [and face-masking], then the CDC's nightmarish incompetence has heaved into sight once again. But there's another possibility, the one keeping PHE (public health establishment) up at night and making them invoke the "baffled" defense: *what if it's the jabs?*
>
> **JEFF CHILDERS** | Attorney and Author | December 1, 2023[432]

20. PHARMACEUTICAL COMPANIES HAVE NO LIABILITY FOR VACCINE HARMS

Why are they vaccinating our children [for Covid-19]? Because once it's on the childhood vaccine schedule, they are no longer liable for injury. So they're going to get off that EUA (Emergency Use Authorization), put it right on the vaccine schedule and then have no liability going forward.

DR. CHRISTINA PARKS

January 24, 2022 - 3:53:04[433]

Establishment of the National Childhood Vaccine Injury Act - 1986

In 1986, the National Childhood Vaccine Injury Act[434] (NCVIA) was signed into law. At the time there had been expensive lawsuits against vaccine manufacturers and healthcare providers due to vaccine injuries in children. Those involved in developing vaccines were threatening to cut back or discontinue research, causing concern this would lead to vaccine shortages and reduced vaccination rates.

The **Act created the National Vaccine Injury Compensation Program (NVICP), with the stated intention that victims of vaccine injury would be compensated through the government**. People injured by vaccines would not be able to sue the pharmaceutical companies, which theoretically would free up the companies to ensure an adequate supply of vaccines and stabilize vaccine costs.

The number of vaccines on the childhood schedule exploded after 1986:

Now comes the most insidious part: For a vaccine to be covered by the National Childhood Vaccine Injury Act, the vaccine must be part of the routine vaccine schedule[435] for children or pregnant women.

Dr. Christina Parks sums it up, **"When they did the 1986 Vaccine Injury Act and said that manufacturers no longer have liability for any vaccine that's on the childhood schedule, the childhood [vaccination] schedule exploded**...That's when they said, 'We have the perfect business model. Every kid has to take these vaccines if we put it on the schedule, and we have no liability'...Safety corners were cut...[and] no one pushed back because of this sort of idea that vaccines were always, always, a positive health intervention." (U.S. Senate 01-24-22, 03:53:04[436])

Dr. Peter McCullough points out that the **child and adolescent immunization schedule just continues to grow, without any of the vaccines being removed.** (See "Shot Dead" 3:45 to 4:47[437]) Before Covid-19, there were 72 injections for 16 different diseases, which children are advised to get before age 18. Now with Covid-19 there are over 100 shots on the childhood immunization schedule, for 17 diseases, including the RSV shot[438] that was just approved.

McCullough is appalled that the gene therapy Covid-19 vaccines, that are still only under Emergency Use Authorization and have not been stringently tested for safety in children, have been added to the schedule (see "Shot Dead" 0:18 to 2:18[439]). Prior to Covid-19, Dr. McCullough, a cardiologist and one of the most published doctors in his field, never questioned vaccines. At the first of the Covid vaccines rollout, McCullough even recommended the shots for some of his elderly patients. Due to what he has seen and experienced during the Covid-19 pandemic, his entire view about the childhood vaccination schedule, and vaccines in general, has changed.

Dr. Anthony Fauci has been the architect of diminishing childhood health: Robert F. Kennedy, Jr. points out that Dr. Anthony Fauci is a key player in both the burgeoning childhood vaccination schedule, and an overall diminishing of health for all Americans. Kennedy states,

> Under Dr. Fauci's leadership, the allergic, autoimmune, and chronic illnesses which Congress specifically charged NIAID to investigate and prevent, have mushroomed to afflict 54 percent of children, up from 12.8 percent when he took over NIAID in 1984. Dr. Fauci has offered no explanation as to why allergic diseases like asthma,

> eczema, food allergies, allergic rhinitis, and anaphylaxis suddenly exploded beginning in 1989, five years after he came to power.
>
> On its website, NIAID boasts that autoimmune disease is one of the agency's top priorities. **Some 80 autoimmune diseases**, including juvenile diabetes, rheumatoid arthritis, Graves' disease, and Crohn's disease, **which were practically unknown prior to 1984, suddenly became epidemic under his watch.**
>
> **Autism**, which many scientists now consider an autoimmune disease, **exploded from between 2 to 4 per 10,000 Americans** when Tony Fauci joined NIAID, **to one in thirty-four** today. Neurological diseases like ADD/ADHD, speech and sleep disorders, narcolepsy, facial tics, and Tourette's syndrome have become commonplace in American children.
>
> (Intro to: *The Real Anthony Fauci* - emphasis added)

Kennedy acknowledges that **other factors including those related to foods, pesticides, various types of radiation, and other environmental factors are also likely contributors to the overall diminishing of health** in the United States. However, **Dr. Fauci is not interested in finding the sources of these diseases through epidemiological research**. Fauci is interested in pharmaceuticals as the answer to every health problem.

> Tony Fauci has more power than any other individual to direct public energies toward solutions. He has done the opposite. Instead of striving to identify the etiologies of [chronic diseases]...Dr. Fauci has deliberately and systematically used his staggering power over Federal scientific research, medical schools, medical journals, and the careers of individual scientists, to derail inquiry and obstruct research that might provide the answers.
>
> **ROBERT F. KENNEDY, JR.** | Intro, *The Real Anthony Fauci*

Putting the Covid-19 shot on the childhood schedule shelters Big Pharma: Because the Covid-19 shots were Emergency Use Authorization only, the pharmaceutical companies had no liability for vaccine injury. But once the emergency was officially over, Big Pharma would no longer be protected from vaccine-injury lawsuits, unless the Covid shots became part of the childhood vaccination schedule[440].

On February 9, 2023 the CDC's Advisory Committee on Immunization Practices (ACIP) voted unanimously to add Covid-19 to the recommended immunization schedule for children 6 months and older, although there had not been any Covid-19 vaccine clinical trials for infants and young children.

A seroprevalence study[441] by the **CDC** conducted from September–December 2021 **found that approximately 75% of children and adolescents in 50 states, plus D.C. and Puerto Rico, already had evidence of previous infection with SARS-CoV-2.** Pathologist Ryan Cole[442] notes the incongruity of giving a vaccine to someone who has already been infected. "You never give a vaccine to someone who has recovered from a disease," says Dr. Cole, "It's scientifically illogical." Certainly by 2023, almost all children had either been exposed to, or infected with, Covid-19.

> Given all that we have learned about the dangers and ineffectiveness of COVID shots over the last two years, it is horrifying to see the CDC now recommend this as a routine shot to children. Although it is unsurprising given the agency capture, it is nonetheless tragic.
>
> **MARY HOLLAND** | Feb. 10, 2023 | President and General Counsel Children's Health Defense Fund[443]

Dr. Richard Urso points out that **giving children the Covid shots**, when most of them have likely already had Covid, **ignores natural immunity** as well as the possible harms of the vaccines. In addition, he states that infants and children were excluded from the vaccine trials, so there was no clinical trial data when ACIP recommended the shots for children. (3:50:34[444])

It's important to remember that many members of the CDC's Advisory Committee on Immunization Practices (ACIP) have financial ties[445] to one or more of the pharmaceutical companies that produce vaccines.

Corruption – the collusion of Big Pharma, the Government, and the Media: Dr. Pierre Kory makes no excuses for the government's mishandling of the Covid-19 response. He states:

> These departures from our policies, from what we know are to be scientific truths, the denial of natural immunity...Why are they doing this? There could be multiple reasons. The simplest and most easily understandable and provable is that every vaccine - these novel, patented, high-cost drugs - is profit. **They are putting profits ahead of patients.**
>
> You know we can call attention to all of these policies. They are non-scientific...yet they're being carried out and distributed across the country, and doctors, and states, and health departments are willingly accepting these without question, without critical thinking... **This is corruption, plain and simple. It's corruption!** (see *here* at 3:52:20[446]) (emphasis added)

PREP Act 2005: additional liability protections for Big Pharma and others
As if the 1986 Childhood Vaccine Injury Compensation Program were not enough protection for Big Pharma, the Public Readiness and Emergency Preparedness Act (PREP Act[447]) was signed into law in 2005. The PREP Act "authorizes the Secretary of the Department of Health and Human Services to issue a Declaration that **provides immunity from liability for any loss caused, arising out of, relating to, or resulting from administration or use of countermeasures to disease, threat and conditions** determined in the Declaration to constitute a present or credible risk of a future public health emergency."

The Declaration states that **liability immunity "applies to entities and individuals involved in the development, manufacture, testing, distribution, administration, and use of medical countermeasures.**" The only statutory exception to this immunity is for actions or failures to act that constitute willful misconduct.

The PREP Act provides for a **Countermeasures Injury Compensation Program (CICP), which compensates people through government funds, if they can prove they were injured** by "products delivered during certain public health emergencies."

Vaccine injury compensations through the PREP Act are almost nonexistent:

On March 15, 2022 Utah Senator Mike Lee and Mississippi Senator Cindy Hyde-Smith introduced legislation[448] to help individuals who have suffered adverse reactions due to Covid-19 vaccination. "It is extremely difficult to obtain awards under the CICP, particularly related to Covid-19 countermeasures," states the explanation of the bill. CICP records reflect that truth. The filing process is complicated and time consuming, and likely discourages many vaccine-injured from even attempting the boondoggle.

The October 1, 2023 CICP report for fiscal years 2010-2023[449] shows only 36 claims have been compensated during that time. Of the 12,775 claims filed from 2010-2023, only 542 are for products *other* than Covid-19 products. Only eight people have been approved by CICP for compensation of Covid-19 vaccine injury, receiving payouts ranging from $1,033 to $8,961 for a total of $30,855. Lee and Hyde-Smith's bill, S.3810, did not pass[450] the 117th Congress and was wiped from the books.

21. VACCINE MANDATES, AND HATRED TOWARD THE UNVACCINATED

"As soon as you resort to mandates, you've failed."

JAY BHATTACHARYA

Professor of Medicine | Stanford University

I guess to do certain stuff, to work and all that, I guess you don't own your body...they make the rules of what goes into your body and what you do.

ANDREW WIGGINS

NBA star forward | October 5, 2021[451]

Unfortunately, once collective welfare is allowed to override individual rights, the government acquires unlimited unchecked power. Whole societies ended up with every aspect of their lives micromanaged by bureaucrats and technocrats with no philosophical grounding in public health ethics.

RAMESH THAKUR[452]

Former U.N. Assistant Secretary-General

As evidence mounted[453] that many people were being injured[454] by the vaccines, and that the vaccines were not preventing Covid-19 disease or transmission, about one quarter of Americans resisted taking the Covid shots. This was not acceptable to the powers that be.

On August 23, 2021 the FDA gave full approval[455] to Pfizer's vaccine, calling it Comirnaty. Pfizer's bulletin explained that the only difference between the emergency authorized mRNA shot, and the full approved Comirnaty, was a legal difference, and that the two products are interchangeable. Pfizer also said it had no plans[456] to actually distribute Comirnaty. In other words, legally Pfizer could be sued for vaccine injuries from a fully FDA-approved vaccine, so it was sticking with the Emergency Use Authorized vaccine. Meanwhile, the Biden Administration now had its excuse for vaccine mandates.

Pres. Biden's infamous and inflammatory September 2021 address to the Nation:

On September 9, 2021, Pres. Joe Biden addressed[457] the American people about the pandemic. As shown in the following excerpts, Biden scolded and demeaned[458] the country as he announced vaccine mandates and blamed the spread of Covid-19 on the unvaccinated:

> Many of us are frustrated with the nearly 80 million Americans who are still not vaccinated, even though the vaccine is safe, effective, and free. The vast majority of Americans are doing the right things. Nearly three quarters of the eligible have gotten at least one shot. But one quarter has not gotten anything...That 25% can cause a lot of damage and they are. **The unvaccinated overcrowd our hospitals, they're overrunning ER units, hospitals and intensive care units, leaving no room for someone with a heart attack, or pancritis (sic), or cancer.**
>
> As your president I'm announcing a new plan to require more Americans to be vaccinated to combat those blocking public health. Many said they were waiting for approval from the FDA. Last month the FDA granted that approval. So the time for waiting is over.
>
> This is not about freedom or personal choice, it's about protecting yourself and those around you. (emphasis added)

Biden then made the unprecedented announcement that every executive branch employee, all federal contractors, and all healthcare workers would need to be fully Covid-vaccinated in order to continue working. Then he incited others to invoke mandates:

> I issue this appeal to those of you running large entertainment venues, to sports arenas, to concerts venues, movie theatres, **please require folks to get vaccinated or show a negative test as a condition of entry.**
>
> My message to unvaccinated Americans is this: What more is there to wait for? What more do you need to see? We've made vaccinations free, safe, and convenient. The vaccine is FDA approved. ..**We've been patient, but our patience is wearing thin, and your refusal has cost all of us.** So please do the right thing.
>
> **I** understand **your anger at those who haven't gotten vaccinated,** understand your anxiety about getting a breakthrough case. But as the science makes clear, if you're fully vaccinated, you're highly protected from severe illness, even if you get COVID-19.
>
> **Tonight I'm** calling **on all governors to require vaccination for all teachers and staff.** Some already have done so. We need more to step up. Vaccine requirements in schools are nothing new. They work. They're overwhelmingly supported by educators and their unions. (emphasis added)

The divisiveness and ugliness that Biden unleashed toward anyone who didn't do just what the government said, set the tone for months and years to come, in America and around the world. When the leader of the free world stomps on the rights enshrined in the U.S. Constitution, others follow his lead. (See Austria[459], New Zealand[460], and Australia here[461] and here[462], and also my own local newspaper here[463], for a few examples)

The Unvaccinated are the Problem (3:43 min)[464]

In a Twilight Zone kind of world, HIPAA[465] was knocked off its privacy pedestal, and the Nuremberg Code[466] was forgotten. Evidence of vaccination became necessary for many people to enter public spaces, attend school, and keep their jobs. Uncomfortably, far too many people seemed to enjoy policing and snitching on their neighbors, and throwing their weight around in telling other people what to do.

No One is Safe Until We're All Safe (11:24 min)[467]

The private sector pushes back. Daily Wire refuses to enforce vaccine on employees

Americans by and large are law-abiding citizens, accustomed to thinking of themselves as free, and seeing their government as being "of the people, by the people, for the people." During Covid, the government became a tyrannical force. If it weren't for brave individuals who pushed back, we would be in much worse shape than we are today.

The Daily Wire says "No!" (3 min 29 sec)[468]

The *Daily Wire* hit back hard on September 9, 2021, filing a lawsuit against the federal government. Cofounder Jeremy Boreing issued a video reply, shown in the above link, to the government vaccine mandate:

> "No! The Daily Wire *does* have more than 100 employees but we won't be enforcing Joe Biden's unconstitutional and tyrannical vaccine mandate. That's it. We'll use every tool at our disposal, including legal action, to resist." Which they did[469], and ***Daily Wire*** won[470] the court case.

However, on December 11, 2023, the U.S. Supreme Court[471] basically wiped the slate clean on Covid litigation, preventing the decisions of lower courts from holding the weight of precedent in the future. The Court reasoned that the state of medical emergency no longer exists, therefore, the cases are moot. This is a blow for the dissident community, but the foundation has still been laid for pushback in the future.

22. CARDIAC EVENTS, DEATHS RISE IN ATHLETES AND YOUNG PEOPLE

If those who are the healthiest among us are suddenly dying, what would that mean for the rest of the population? In other words, what if healthy young athletes are the canary in the coalmine? To my trained eye, what others might consider sad anecdotes became something more: A trend change was underway.

ED DOWD, WALL STREET ANALYST

Author of *Cause Unknown*, p.11[472]

Data worldwide in 2020 showed it was mostly the elderly and those with health problems who were adversely affected by Covid disease. Still, because of reports of cardial events in the very ill, the NFL, college teams, and some high school sports teams set up screening programs for their athletes, worried about Covid-caused myocarditis. The college athletes, and young people in general, were not particularly impacted by Covid disease, and the screening programs ended. (see here[473]and here[474])

CDC data shows that **through February 13, 2021, 81% of all deaths due to Covid-19 in the U.S. were in the 65+ age groups.** However, in 2021, as **the vaccine campaigns began,** and especially as vaccine mandates were enforced, something alarming began to occur with unusual frequency: **healthy young people and fit professional athletes began dropping[475] on the field, the court, and the track with heart problems – some of them fatal.** Other young people were passing away in their sleep. The incidents of "died suddenly" and "cause of death unknown" soared.

Written and video compilations of news reporters, politicians, athletes, celebrities, and entertainers collapsing suddenly started to circulate. These on-film collapses, often due to heart attack or stroke, are labeled as a "sudden illness," or "medical event," and we are often never told what caused the event. For others heat and dehydration, or stress, or anxiety are blamed for the sudden heart attack or collapse. Sometimes the cardial problem is attributed to a previously unknown genetic defect. Dr. Peter McCullough points out that it is rare for a genetic heart defect to go undiagnosed in athletes in our times because the physical exam required for participation in sports picks up most of them.

Dr. David Gortler, a former senior advisor to the FDA Commissioner, is concerned that the experimental lipid nanoparticles (LNPs) that encased the mRNA in the Pfizer and Moderna shots could be causing the heart issues. (see Novel lipid nanoparticle technology made the mRNA shots possible: in Section 10) Gortler states:[476]

> Could positively charged LNPs be attracted to negatively charged tissues/cells in the body? Could that lead to LNPs forming in one's vasculature leading to occlusive stroke (also prominently reported in VAERS and V-safe)?
>
> Unfortunately, there is abundance of data showing positively charged lipids to be inherently toxic. One area in the human body with a lot of electrical activity is the heart, and many adverse events reported to VAERS and V-safe (myocarditis, pericarditis, heart attack, arrhythmia, stroke) are heart-originated. Could positively charged LNPs electrically attaching themselves to cardiac tissue be a source of reported adverse events?

Unfortunately, Covid vaccination status is not being examined in connection with these dramatic and tragic events.

The "fact-checkers" and legacy media try to normalize abnormal medical events:

"Fact-checkers," governments and health officials, the media, and sometimes even the individual who experienced the event essentially have tried to normalize what is obviously not normal.

For a few examples, look up Damar Hamlin[477], Jamie Foxx[478], Grant Wahl[479], Heather McDonald[480], Nick Nemeroff[481], Jessica Sutta[482], Sergio Aguero[483], Oscar Cabrera Adames[484], David Renne[485], Bronny James[486], and Brooke Shields[487]. Note that these are prominent people in the public eye, and consider for a moment if you have observed similar medical events among your own family, friends, and acquaintances.

For a larger scale example, observe this report from India[488], where over 1,000 emergency calls were made related to heart problems and breathlessness during the energetic Garba dances in India's Western state of Gujarat during the Hindu festival of Navaratri in October 2023. Within 24 hours 10 people died of heart attacks, including a 13-year-old and 17-year old. Indian authorities called on event organizers to have ambulances and medical personnel at the ready, and attributed these alarming events to people being out of shape. A media headline asks, *"Why are so many heart attack cases happening during Garba? Experts Explain."* One shrewd observer of Covid-shot analogous incidents replies:[489]

> No, the experts did not explain. It was all just a bunch of hand-waving about the exertions of dancing, literally that was their very best idea, and without any context offered about the previous six thousand years of human dancing history during which teenagers did NOT get heart attacks and they did NOT need to keep defibrillators on hand and they did NOT need to require CPR training for dancing events.
>
> Totally unrelated, over 2 billion vaccine doses were delivered into Indian arms as of September 1, 2022 (mostly AstraZeneca).

The incidents of vaccinated people being sidelined, or dying, from "sudden illness" are glaring, despite the exhortations to "pay no attention to that man behind the curtain."

> **Certainly not every death or other adverse event that happens these days is a direct result of the Covid-19 shots, but how would we know? Because the Covid shots were Emergency Use Authorized only, part of any investigation of adverse medical events should be to establish whether or not the individual has received one of the**

Emergency Use Authorized Covid-19 shots, but that information is not a required data point.

Fatalities after Covid shots "many folds higher" than after flu shots: In April 2023, Dr. Eyal Shahar, Professor Emeritus of Public Health at the University of Arizona explained that **short-term fatality after Covid shots is "many folds higher" than fatalities after a flu shot.** (As discussed previously in this paper, CDC records of serious adverse events after Covid vaccination, including deaths, are higher than for all other vaccines combined in the past 30 years. (see Section 18 Myocarditis Harms Show Up Early in the U.S. Vaccination Campaign)

Shahar asks:[490]

> What about long-term consequences, fatal and non-fatal? Will we be able to estimate the rates in a decade or two? How would we know, for example, if a future Covid death was caused by changes in natural immunity? How would we know if a sudden death was caused by subclinical myocarditis? How would we know if an auto-immune disease was triggered by Covid vaccine?
>
> This is not going to be an easy task. There had been good reasons, before the Covid era, for not rushing a new drug, or a new vaccine, into the market. We have learned the lesson the hard way, or have we?

Professor Mark Crispin Miller posts[491] an ongoing record of sudden deaths and disabilities from around the world, many of which are listed as "cause unknown." **It used to be that when the cause of death was not listed, it was likely suicide. These days they just don't say anything, or refer vaguely to a sudden illness, or the previously unheard of SADS** (Sudden Adult Death Syndrome). Attorney Jeff Childers shares cases of SADS almost daily on his Substack[492].

Medical professionals, who are paying attention and speaking out at great personal risk, state that the increased heart issues in young people are the result of the Covid-19 shots, "until proven otherwise."

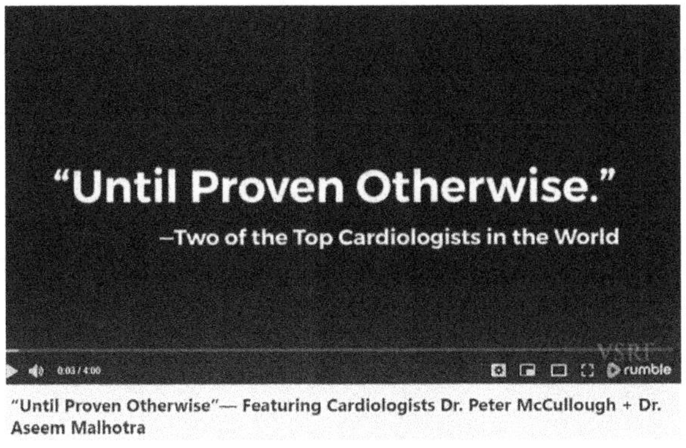

"Until Proven Otherwise"— Featuring Cardiologists Dr. Peter McCullough + Dr. Aseem Malhotra

Until Proven Otherwise (4 min) [493]

Life insurance reports highest excess mortality ever recorded:

In January 2022, major life insurance companies in the U.S. dropped a bombshell.

Life insurance CEOs reported that starting in the third quarter of 2021 excess mortality in working-aged people 18 to 64, suddenly spiked overall by 40 percent.

"Excess mortality" is a term referring to the difference between the number of deaths that were expected to occur during a given time period, and the number of deaths that actually occurred.

The 40% increase was the highest ever seen[494] in the history of life insurance, and the majority of excess deaths were **mostly not attributed to Covid**. There was also a sharp rise in disability payouts[495] in these same age groups.

Ed Dowd, a former Wall Street growth fund manager and analyst, has spent his career analyzing statistics and identifying trends. In 2021, Dowd noticed not a Wall Street trend, but a trend in unexplained sudden deaths among healthy young people.

> Dowd describes it this way, "From February 2021 to March 2022, millennials experienced the equivalent of a Vietnam war, with more than 60,000 excess deaths. The Vietnam war took 12 years to kill the same number of healthy young people we've just seen die in 12 months."

Analyst finds excess deaths correspond with vaccine mandates in Q3 of 2021:

Dowd worked with insurance industry executives, and expert analysts from around the world to compile his findings about the "alarming and unusual stories of young, exceptionally fit athletes dropping dead on the field of play." Dowd published his findings in his book, *Cause Unknown*[496], which **in addition to charts and analysis, links to articles on hundreds of young people who have "died suddenly" since the Covid shots rollout began.** (p. 11 Cause Unknown)

What Dowd determined through the CDC's All-Cause Mortality Data in early analysis was confirmed by the Society of Actuaries report released August 17, 2022.

Table 5.7
EXCESS MORTALITY BY DETAILED AGE BAND

Age	Q2 2020	Q3 2020	Q4 2020	Q1 2021	Q2 2021	Q3 2021	Q4 2021	Q1 2022	4/20-3/22	% COVID	% Non-COVID	% Count
0-24	116%	124%	104%	101%	119%	127%	110%	91%	111%	3.3%	8.1%	2%
25-34	127%	132%	121%	118%	131%	178%	131%	125%	133%	13.3%	19.6%	2%
35-44	123%	134%	128%	129%	133%	200%	156%	136%	142%	23.1%	19.2%	4%
45-54	123%	127%	129%	133%	119%	180%	151%	143%	138%	27.4%	10.8%	9%
55-64	117%	123%	130%	130%	114%	153%	141%	137%	131%	24.0%	6.7%	18%
65-74	117%	115%	133%	130%	108%	131%	125%	122%	122%	18.6%	3.9%	17%
75-84	114%	114%	133%	123%	106%	119%	121%	121%	119%	14.0%	4.6%	20%
85+	112%	103%	124%	111%	92%	104%	105%	103%	107%	10.3%	-3.5%	27%
All[11]	116%	115%	129%	123%	107%	134%	126%	122%	121%	17.1%	4.3%	100%

As shown in the above chart, excess mortality skyrocketed in the third quarter of 2021, especially in the younger age groups.

At the time of the announcement of the **catastrophic rise in excess deaths,** OneAmerica's CEO[497] Scott Davison pointed out that the **majority of deaths were not Covid deaths**. Davison encouraged the shareholders to get their vaccines, saying at least they wouldn't be at risk for Covid.

> Dowd writes, "Federal public health officials and media had developed ways to ignore and explain away the increase in All-Cause Mortality: It must be more suicides or overdoses, or missed cancer screening during lockdowns. **But the rate of change in Fall 2021 was particularly striking as it coincided with the corporate mandates – and <u>it simply wasn't statistically possible that suicides, overdoses and</u>**

> **deaths from delayed treatment of rapid-onset fatal cancers all spiked in that very same 3-month period.** The only thing that had changed was mass vaccination forced upon the millennial generation via government and corporate mandates... We know that hundreds of millions of doses were administered, and there is no other factor that affected nearly all working-age people simultaneously. (p. 151-152, 222 - Cause Unknown)

The fact-checkers are also compromised by ties to Big Pharma:

> *As an editor once said to me: "Just because someone says something doesn't mean you have to put it in the paper." If the media followed this one simple rule, there would be no need for "fact-checking" at all. But the media does not and will not follow this rule because printing lies – as long as they are said by a government official the media likes or about an official they don't like – is now an integral part of the industry.*
>
> **THOMAS BUCKLEY**[498] | Former newspaper reporter

In the world of news, one would think the highest increase in all-cause excess mortality in the history of life insurance might be an important story, but no. **Fact checkers are quick to dismiss anything that could lead to questions about the Covid vaccines**. A favorite trope is that the high number of reports to VAERS doesn't prove causality. True, but it should be looked into. It's anti-science to say, "It's not because of the vaccine," while failing to conduct studies that could confirm or refute that statement.

However, **fact-checking is another compromised industry[499].** For example, Reuters has become quite prominent in the fact-checking world, consistently supporting the official narrative about the Covid-19 response. However, Thomson Reuters' Chairman of the Board[500], Jim Smith, is also James C. Smith, Director at Pfizer Inc[501]., having served on Pfizer's Board since 2014.

Dr. Peter McCullough notes that factcheck.org is part of the Annenberg Public Policy Center, University of Pennsylvania, which is heavily funded[502] by the Bill & Melinda Gates Foundation.

> The fact checker programs now are intentionally deceiving. They're making direct false claims against what's observed in the scientific literature.
>
> **DR. PETER MCCULLOUGH** | American Thought Leaders Now December 20, 2023[503]

With regard to policies in compromised fact-checking organizations, Thomas Buckley states:

> Lies from government officials and lies from nonprofit and advocacy groups and non-governmental organizations (who directly pay news outlets for the "coverage" of an issue they are involved in) are all waved through as gospel… **It's dangerous because a "true" rating is just that: something has been determined to be true and can therefore never be questioned** again or something is mostly true so any error can be tied back to an accidental misspeak. And then this "truth" can be spread as 100 percent Grade-A verified fact, no matter whether it actually is or not. It has received an imprimatur from on high and that's that.

Ed Dowd has continued to compile data, and in a recent analysis[504] found that **in 2023 all-cause mortality and disabilities continue to be elevated in the U.S. and U.K.** He believes we are seeing the long-term effects of the damaging Covid shots, and the attempts by governments and Big Pharma to cover up the truth.

Insidious lie: "Long Covid caused by natural infection, not by the vaccines"

The excess deaths are explained away or ignored, and the long-term disabilities are blamed on Covid-19. The CDC's article on Long Covid, updated July 20, 2023, states:[505] "Long Covid is broadly defined as signs, symptoms, and conditions that continue or develop after acute COVID-19 infection. This definition of Long Covid was developed by the Department of Health and Human Services (HHS) in collaboration with CDC and other partners…

People who are not vaccinated against Covid-19 and become infected may have a higher risk of developing Long Covid compared to people who have been vaccinated."

There are no quality long-term studies that show vaccination prevents Long Covid, or that Covid infection causes it. Dr. Peter McCullough called out the fallacy [506]of Covid infection causing long Covid in testimony before a subcommittee of the European Parliament on September 13, 2023:

> We are seeing now a third false narrative. The 1st false narrative was, "The virus is unassailable we have to stay in lockdown and be fearful." The 2nd false narrative is "Take a vaccine – it's safe and effective." The **3rd false narrative now is "It's not the vaccine causing these problems. It's Covid.** It's Covid that we saw back in 2020 causing all these problems in 2023." Don't fall for the false narrative.

Dr. McCullough said that serious Covid-19 infection promotes cardiovascular disease, but studies in V.A. patients showed the risks to be time limited to about six weeks post-infection (6:50[507]). However, **the Covid vaccines have had an FDA warning on them since June 2021 stating that they are associated with myocarditis.**

Dr. McCullough says there have been a number of papers on pediatric vaccine-induced myocarditis, showing that "in some unlucky children, the heart develops a permanent scar." **The weakened heart is then susceptible to heart attack when it's hit with a surge of adrenalin, such as during exercise, or in the hours just before waking.** (12:54[508]).

The Society of Actuaries Research Institute released a May 2023 report acknowledging the unprecedented, and continuing, increase in deaths among younger people, but denying a connection[509] to the Covid-19 shots. **As Dowd anticipated, life insurance executives blame delayed doctor appointments, and the higher rates of depression, suicide, and increased substance abuse.**

The Society of Actuaries Group Life Covid-19 Mortality Survey Report[510] November 2023 states, "One of the observations from both the Group Life Survey and the U.S. Population results is that the deaths were higher than expected through the first quarter of 2022 for all age groups and have remained higher than expected for the 15-44 age group, even after excluding explicit Covid-19 deaths. (p. 43)…Aside from Covid-19 we see that All Other/Unknown is the largest contributor to excess deaths in all three years (2020 – 2022) and for both populations (Group Life and overall U.S. population)…The next largest contributor to excess deaths has generally been major cardiovascular disease…(p. 49)."

In light of the Society of Actuaries report, it seems appropriate to repeat the warning[511] of Dr. Sucharit Bhakdi from March 2021:

> This is a disastrous situation because the spike protein itself is now sitting on the surface of the cells facing the blood stream. And it is known that the spike proteins, the moment they touch platelets, they activate them, and that sets the whole clotting system on… **Clot formation in brain veins can give any symptoms. The symptoms of these people who are getting their second shot of the vaccine are so diverse, but they would all fit into that.** Even these people who have jerking, limb movements that they cannot control anymore…
>
> Whether it's mRNA or vector, the fact is that a gene for the spike is entering the bloodstream and reaching cells of the blood vessel wall, at locations that are actually forbidden because, if they are then used, these genes are then put into the machinery of the cell to produce these spikes, these spikes are going to be produced at locations where they never are produced normally.
>
> **[People] won't have died [from Covid] but you are killing them with the vaccine,** and **secondly maiming others for life…**

23. COVID VACCINES KILLED 114,134 AMERICANS WITHIN 3 MONTHS IN 2021

There probably wouldn't have been any wave at all if we hadn't kept on first-dosing people when the Delta variant was predominant.

FABIAN SPIEKER

Life insurance executives have not attributed the unprecedented and ongoing excess deaths to the Covid shots, and the government claims the Covid shots saved lives. However, the detailed work of Fabian Spieker shows[512] that in fact, **the second half of 2021 was a Mass Vaccine Casualty Event.**

Spieker has spent hundreds of hours analyzing dozens of U.S. datasets from both county- and state-level data during July-September of the Delta outbreak. He charted time-series data, meaning data points collected over an interval of time. Spieker also ran millions of regressions[513] on the data, which is a statistical method designed to determine the strength and character of the relationship between one dependent variable and a series of other variables.

One must assume that Spieker's meticulous work has not received more attention because 1) Legacy media will not touch it, 2) government and Big Pharma do not want to be exposed, and 3) most people are not very adept at reading and interpreting charts.

Because reading charts is not my forte, I asked Spieker for clarification, so as to be able to "interpret" his charts for others who also aren't very "chart literate."

Scatterplot charts that show correlation between Covid vaccination and Covid deaths:

Spieker created scatterplots for all the age groups. Along the x-axis (horizontal line) he charts the number of first doses administered. Along the y-axis (vertical line) he shows the total number of Covid deaths per 100k. Spieker used data from each of the 50 states in the U.S. to show the correlation between number of first doses and number of Covid deaths per 100k population in August 2021. Here is a small rendering of the chart, which will be addressed further in the next pages:

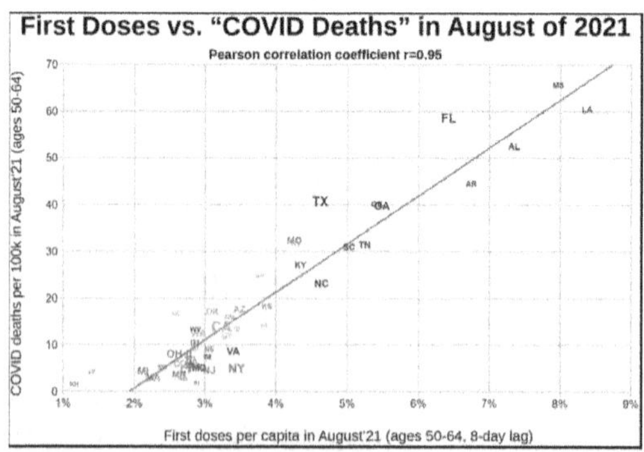

Statistical tools of correlation and regression are used to show trends on a scatterplot. A scatterplot[514] usually looks like a line or curve moving up or down from left to right along the graph with points (dots) "scattered" along the line. A scatterplot helps to uncover information about any data set, including:

- The overall trend among variables (to quickly see if the trend is upward or downward.)
- Any outliers from the overall trend.
- The shape of any trend.
- The strength of any trend.

The closer the dots are to being located on a diagonal line, the stronger the correlation and the higher the probability of a direct causal link between the two variables. A correlation of 1 or -1 means that all dots are located on a diagonal line.

Here are some examples[515] of what scatterplots look like when there is a positive, a negative, or no correlation between variables:

When the y variable tends to increase as the x variable increases, we say there is a **positive correlation** between the variables.

When the y variable tends to decrease as the x variable increases, we say there is a **negative correlation** between the variables.

When there is no clear relationship between the two variables, we say there is **no correlation** between the two variables.

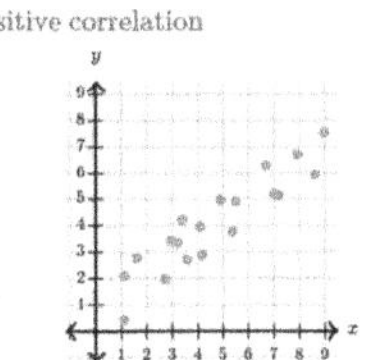

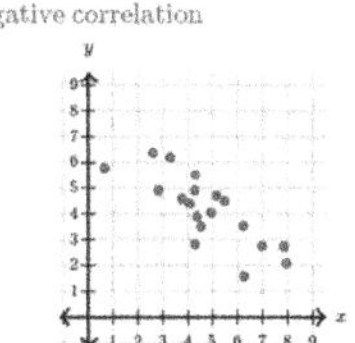

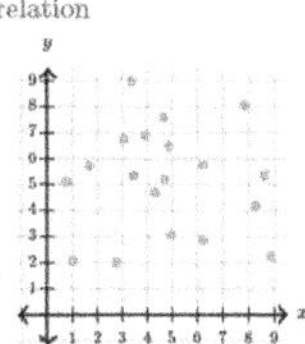

In the scatterplot below, note that the "dots" (the abbreviations for the States) are all quite close to the diagonal line, showing a positive correlation. This scatterplot is designed to determine if there is a correlation between the number of people who received a first dose of a Covid vaccine, and the number of deaths per 100k. The x-axis shows "First doses" per capita (meaning per person) in August 2021. The y-axis shows "Covid deaths" per 100k in August 2021.

The millions of regressions Spieker ran on the data from the third quarter of 2021, consistently show a strong correlation between administration of the first dose, and deaths attributed to Covid-19. The data comes from death certificates where Covid-19 is designated as the Underlying Cause of Death (UCOD). (The diagnostic code for Covid-19 on insurance records is U07.1)

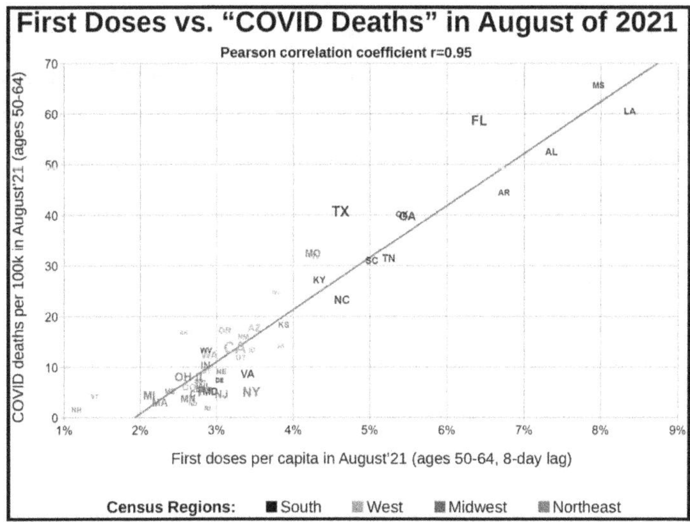

The x-axis shows the percentage of the 50-64 age group, in each state, who received the first dose of a Covid vaccine in August 2021. The y-axis shows Covid deaths per 100k during that same time period. The clustering along the diagonal line shows a positive correlation between first doses and Covid deaths.

The above chart shows the data for those aged 50-64. **Spieker said the correlation between first doses and deaths is strongest in the 50-64 age group, but the correlation is almost as strong in all the other age groups.** (All of Spieker's charts are available on his Substack[516].)

Note that the higher percentage of the population that received a first dose, the higher the number of "Covid deaths." For example, Louisiana and Mississippi at the top far right (showing the most deaths) were states where around 8 percent of the population aged 50-64 received the first dose of a Covid vaccine in August 2021. New Hampshire and Vermont, in the lower left corner (the fewest deaths) were states where only 1% of the population aged 50-64 received a first dose of a Covid vaccine.

Spieker explains that the correlation between the x-axis (deaths coded as Covid-19 deaths) and y-axis (first doses administered) reaches 0.95 on the ninth day, meaning almost a correlation of 1. A perfect correlation of 1 is almost unheard of in population-level data.

CDC policy was to NOT count people as partially vaccinated until two weeks had passed after they received a first dose of the Covid shots. This means people who tested positive within 14 days of their first Covid shot were **counted as unvaccinated**, when, in fact, **they should have been in a separate category of "one Covid vaccination received."**

As Spieker points out, **"The average delay between first vaccine dose and positive test result of those individuals later declared as a "COVID casualty" seems to be short enough (<14 days) for most decedents to be considered unvaccinated along with those who never received a dose."** The requirement by the CDC, to not list anyone as having received a first dose of vaccine until two weeks after the fact means that the hospitalized who had received one dose of vaccine were counted as unvaccinated in CDC reports.

> It would be extremely unusual to witness anything close to a correlation of 1 in population-level data, where countless factors are usually influencing each other and no single cause is ever solely responsible for whatever is being measured.
>
> FABIAN SPIEKER

Quarter 3/2021 Vaccine Deaths Attributed to Covid-19:

In the following chart, Spieker sums up the total number of vaccine deaths attributed to Covid-19 in Q3 of 2021 by age group. Spieker notes, "It looks like Covid mortality would have been negligible without first doses being administered at the time:"

Q3/2021 Vaccine Deaths Attributed to COVID

	0-17	18-49	50-64	65-74	75+	Total
Jul	15	1,360	2,658	2,246	3,419	9,698
Aug	87	6,719	12,560	10,710	15,276	45,352
Sep	92	8,268	16,961	14,403	19,359	59,084
Total	**194**	**16,347**	**32,179**	**27,359**	**38,054**	**114,134**

Following is another way of looking at Spieker's findings. He has charted the total number of deaths per 100k population, for all ages, in each state from July 2021-September 2021. The deeper the shade of pink, the higher the incidence of Covid Deaths per 100k:

Covid Deaths per 100k - All Age Groups July- September 2021

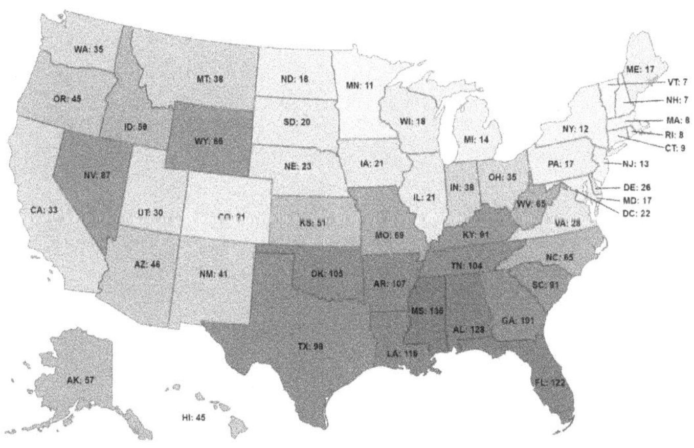

The following chart shows the percentage, for each state, of residents who received a first Covid shot in the same time period of July to September 2021. Compare the pink chart above with the blue chart below. Note the strong correlation between the percentage of the population that received the first dose and higher Covid deaths in that state, such as in Florida, Alabama, Mississippi, Louisiana, and Arkansas:

Percentage (All Age Groups) Received 1st Covid Shot July –September 2021

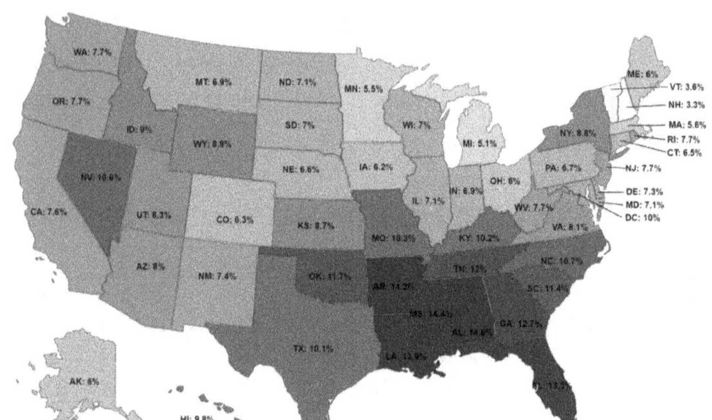

Delta became the predominant variant throughout the world at almost the same time, but did not immediately cause outbreaks in all regions. The large "outbreaks" happened when first doses of Covid vaccines were being administered. In Europe, this same correlation emerged in October 2021.

First Doses P.C. VS. COVID Deaths Per 100k Across Europe Week 2021/43

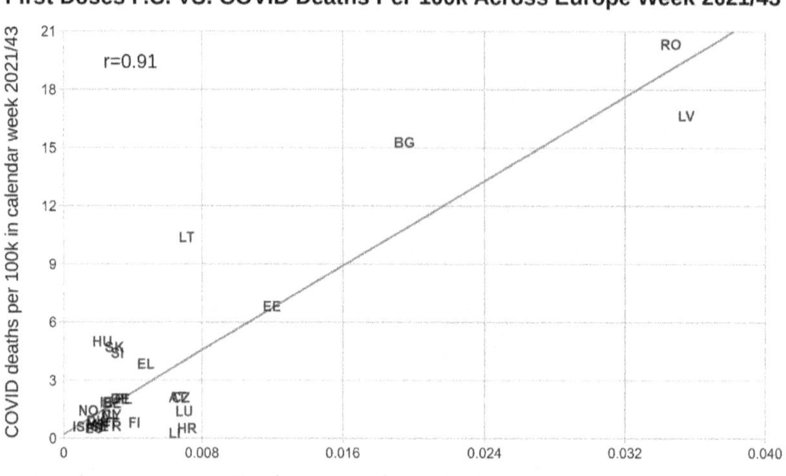

The almost perfect correlation between the number of first doses given, and the number of deaths attributed to Covid-19, is highly suggestive of first doses killing by mechanism of "vaccine-mediated enhanced disease."

Vaccination makes people more susceptible to Covid infection:

The Covid vaccines made people more susceptible to circulating Covid infection immediately after vaccination, due to its impact on the immune system. This increased susceptibility caused more disease to circulate in the population overall, which **means the vaccine also led to the deaths of unvaccinated people because the vaccine exacerbated Covid outbreaks.** More people were infected with the Delta variant, and more people died. If a region was vaccinating heavily during Delta infection, Spieker notes, **the "innate immune system was overwhelmed by spike protein from two sources (injection and infection), or was suppressed by the vaccine."** Meaning, the immune system was occupied by the effects of the Covid shot, which made it less effective at fighting other illnesses, including the circulating Covid virus.

Spieker continues:

> So we really have two groups of infected individuals: Those who were recently first-dosed and those who were not. **The recently first-dosed individuals seem to be more infectious.** For case rates to increase and to form a "wave," infected individuals have to infect more than 1 person. If the infected individuals are not infectious enough, there won't be a wave. There probably wouldn't have been any significant Delta wave at all if we hadn't kept on first-dosing people during this time.

Spieker observes, **"There are many factors influencing other types of mortality, but in the third quarter of 2021 in the U.S., vaccines were almost solely to blame for Covid mortality."**

Spieker believes that there were people involved in the vaccine rollout who knew that infections, and consequently deaths, would increase with Covid vaccination, but they proceeded anyway. He notes:

> In an interview with the *NY Times* in May of 2021, CDC director Dr. Walensky stated that she was preparing for a "Pandemic of the Unvaccinated." When this anticipated situation - which should be understood as "Epidemics of the Recently Vaccinated" - did finally manifest in mid-July of 2021, the NYT proceeded to release her

statement in an attempt to "prebunk" anyone stumbling over what would happen in the months that followed."

Spieker's data analysis provides a **strong indication that the massive deaths in the latter half of 2021 were the result of the Covid vaccines, and the subsequent waves of Covid-19 infection that they caused.** His conclusion that there likely would not have been a continuation of the pandemic, if we had not been mass vaccinating into an active pandemic, supports and confirms what Dr. Geert Vanden Bossche and Dr. Luc Montagnier warned of at the beginning of the vaccination campaign.

Spieker's analyses also corroborate the findings of Ed Dowd with regard to insurance actuary tables and the excess deaths in Quarter 3 of 2021. (see Life insurance reports highest excess mortality ever recorded:)

> A vaccine that prevents infections prevents epidemics. A vaccine that makes the vaccinated person more susceptible to infection exacerbates outbreaks.
>
> **FABIAN SPIEKER**

24. WORLDWIDE DATA POST-COVID VACCINATION SHOWS REDUCED LIVE BIRTHS

There is strong support of a causal link between LNP/mRNA injections and acceleration of declining births nine months later.

ROBERT W. CHANDLER, MD, MBA[517]

Harm to women and the unborn due to the Covid-19 shots is evident in data from around the world. Miscarriages, stillbirths, and preterm births have increased. For example, on August 25, 2022, the Swiss Hagemann group published a statement[518] regarding the decline of live births in Europe:

> [E]very single examined European country shows a monthly decline in birth rates of up to more than 10% compared to the last three years. It can be shown that this very alarming signal cannot be explained by infections with Covid-19. However, one can establish a clear temporal correlation to Covid vaccinations incidence in the age group of men and women between 18 and 49 years. Therefore, in-depth statistical and medical analyses have to be demanded.

One might ask, where is the U.S. data regarding birth rates in the U.S.? Dr. Robert Chandler, who completed extensive research[519] on this topic, wrote on January 16, 2023:

> US, UK, and Australian data may not meet high standards. American data has yet to appear and may be suspect when it does, given the

control exerted by the Department of Defense and the US intelligence agencies. The UK data has ceased to flow.

Dr. Chandler states, "**The evidence is growing that both male and female reproductive functions and organs are adversely affected by LNP/mRNA products with lowered sperm motility and counts, menstrual irregularities, and reproduction organ dysfunction...**The pattern that began years to decades ago of declining birth rates in the developed world appears to be accelerating after the introduction of LNP/mRNA gene products suggesting at least a temporary reduction in fertility as a result of interference with reproductive function in both men and women."

Reproductive system adverse events in women post-Covid vaccination: Dr. Chandler provides the following compilation using VAERS records from **December 14, 2020 to July 29, 2022, noting that there were almost 20,000 reported cases of menstrual problems after receipt of the Covid shots**. Dr. Chandler states, "The Vaccine Adverse Events Reporting System (VAERS) **reporting is arduous** and unfamiliar to some or many health care providers. **So, 20,000 could easily represent exponentially more.**"

Chart 2: Menstrual Irregularities after LNP/mRNA

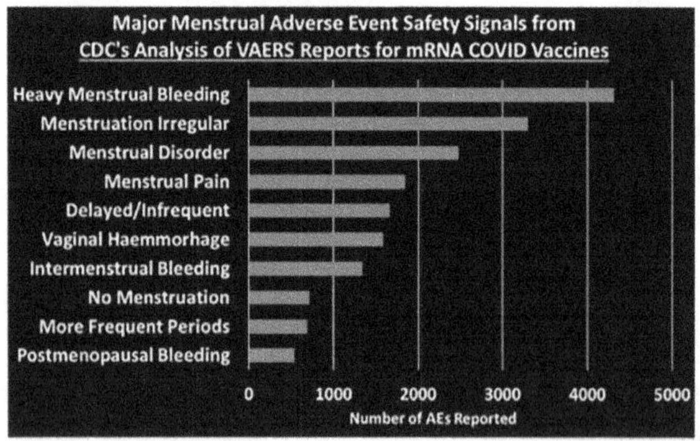

The above chart shows the cumulative numbers of major menstrual adverse events reported to VAERS post-Covid vaccination from 14 December 2020 to 29 July 2022. Based on past VAERS reporting, it is assumed that this data represents only about 1-2% of actual events.

Maternal and fetal complications post-vaccination:

As a practicing OB/GYN for 40 years, Dr. James Thorp has seen a striking increase in maternal and fetal complications in his practice since the rollout of the Covid shot. Using U.S. government data, Thorp notes that nationally, the stillbirth rate per 1,000 births was between 5.75 and 5.89 from 2017 through 2020. But starting in July 2021 it increased to 29.3/1,000 births (36:08[520]). Thorp says the 2022 data has not yet been released, and questions why.

Michelle Gershman, a registered nurse in Fresno, California, said before March of 2021 her hospital would see one or two fetal demises (stillbirths) every couple of months. After March of 2021, they started having one or two per week, many of them full-term (38:34[521]). After the Covid shots rollout began, Gershman saw multiple incidents where the expectant mom would go to her OB, get the Covid shot, and within one week they were delivering a dead baby. Gershman states,

> I kept seeing these fetal demises. I kept seeing these mothers with problems...high blood pressure, bleeding from their eyes, blood clots coming out of them. All of these horrific things that you would only see in a horror movie and this was every time I would come to work. And then I'd see these babies that were having severe cases of like, jaundice, and they're having respiratory issues – all these things that didn't used to happen.

In September 2022, an email was issued to the medical staff at the hospital where Gershman was working, stating that there were 22 fetal demises in August. The email described confusion with handling the remains of a miscarried baby and requested, "please follow the procedure in the fetal demise binder." **Gershman was appalled that the emphasis was on how to handle the extra work, and not on figuring out why the number of stillbirths remained elevated.** Gershman said the number of stillbirths continued to increase, one time getting up to 8 in one day.

Thorp notes that the Pfizer documents show pregnant women were not included in the Pfizer trials. However, in the early rollout of the vaccines, contrary to medical ethics, 270 pregnant mothers received the Pfizer shots. Thorp points out that there were 25 spontaneous abortions reported, and no outcome was provided for 238 pregnancies. Pfizer simply didn't follow up with 238 pregnant mothers to determine if they had any side effects from the experimental Covid shots. (44:35[522]) This information was not made available to the public as pregnant women were being pressured to take the shots and were being told that the Covid vaccines were "safe and effective" in pregnancy.

> Why study the disrupted menses of thousands of women, as we see in the Pfizer documents, and not then disclose these risks to women, or why [not] report on the deaths of two babies in utero due to "maternal exposure" to the vaccine, unless you wish to disrupt the menses of women and harm babies in utero? There is no way these injuries are accidental if they are studied and documented in detail and then rolled out on the public with no informed consent.
>
> **NAOMI WOLF, PHD** | December 14, 2023[523]

Trillions of dollars kept the vaccine rollout in motion:

"This was a trillion dollar cash cow," says Thorp of the Covid shots (45:45[524]). He believes the money involved is one of the reasons the vaccine rollouts continued, despite the fact that during the first 90 days there were 1,223 vaccine deaths, as shown in the following chart from the Pfizer report:[525]

BNT162b2
5.3.6 Cumulative Analysis of Post-authorization Adverse Event Reports

Table 1 below presents the main characteristics of the overall cases.

Table 1. General Overview: Selected Characteristics of All Cases Received During the Reporting Interval

Characteristics		Relevant cases (N=42086)
Gender:	Female	29914
	Male	9182
	No Data	2990
Age range (years): 0.01 -107 years Mean = 50.9 years n = 34952	≤ 17	175[a]
	18-30	4953
	31-50	13886
	51-64	7884
	65-74	3098
	≥ 75	5214
	Unknown	6876
Case outcome:	Recovered/Recovering	19582
	Recovered with sequelae	520
	Not recovered at the time of report	11361
	Fatal	1223
	Unknown	9400

a. in 46 cases reported age was <16-year-old and in 34 cases <12-year-old.

As shown in Figure 1, the System Organ Classes (SOCs) that contained the greatest number (≥2%) of events, in the overall dataset, were General disorders and administration site conditions (51,335 AEs), Nervous system disorders (25,957), Musculoskeletal and connective tissue disorders (17,283), Gastrointestinal disorders (14,096), Skin and subcutaneous tissue disorders (8,476), Respiratory, thoracic and mediastinal disorders (8,848), Infections and infestations (4,610), Injury, poisoning and procedural complications (5,590), and Investigations (3,693).

Thorp says it's unprecedented that a vaccine causing that kind of damage in the first 90 days would still be on the market. Yet the federal government spent $13 billion funding the Covid-19 Community Corps, which was used to hire various organizations and individuals to promote the vaccines:

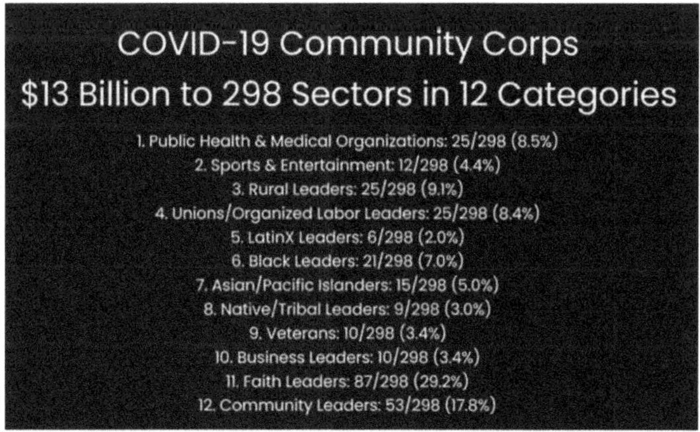

A Freedom of Information Act request discovered that the American College of Obstetrics and Gynecology took at least $11 million in exchange for agreeing they would not deviate from the narrative of the HHS and CDC, and if they did, they would have to pay back the money. (46:40[526]) Dr. Thorp reports that fertility rates are down throughout the world and all-cause mortality is elevated.

> Our federal agency that's tasked with protecting Americans from harms of drugs and biological products appears to be willfully blind to disabilities, injuries, and deaths. As a clinician I'm gravely concerned about the federal agencies and the lack of consumer protection for Americans.
>
> **DR. PETER MCCULLOUGH** | American Thought Leaders
> December 20, 2023[527]

25. DANGEROUS VARIATIONS IN COVID-19 VACCINE BATCHES

[T]oward the fall of 2021, I knew for sure they were not good manufacturing practice compliant. So, we have a not good manufacturing practice compliant product being produced, shipped in millions of doses, and injected in millions of people, including pregnant women and children. We have the CDC and FDA lying and saying, "There is no signal at all," and this continues for a long time.

SASHA LATYPOVA

Former pharmaceutical executive | June 17, 2023[528]

Sasha Latypova, who worked for years in pharmaceutical research and development, as well as clinical trials, observed that with Covid-19 **the vaccine manufacturers and government regulators were not following established practices for clinical research and public health protocols.**

Latypova explains that the **quality control for the mRNA shots was abysmal.** Batches of vaccine were not consistent with regard to how much mRNA was in each shot. The website How Bad Is My Batch[529] allows anyone to enter in their Covid batch number, found on the vaccination card, and see how many adverse events have been reported. Some batches of mRNA have few, others have thousands.

Dr. Marik[530] agrees about the variability of content in batches stating, **"We don't know what's in the vaccines.** Many of them may not have RNA or may have low quantities of messenger RNA, so that may explain the enormous difference between the lots...What's in the vaccine may make a big difference...There seem to be very strong genetic factors increasing your predisposition to complication."

Dr. Renata Moon, a practicing pediatrician for more than 25 years, demonstrated in a December 2022 Senate roundtable (2:25:45 [531]) that **the product information insert for the Covid shots was a plastic-sealed multi-folded large sheet of paper. When removed from the wrap and unfolded, the paper was blank except for one phrase in the middle "intentionally left blank."** Dr. Moon asked, "How am I to give informed consent to parents when this is what I have?...[Yet] I have a government that's telling me that I have to say 'safe and effective,' and if I don't, my license is at threat." (Dr. Moon was fired from her position[532] as professor of medicine at Washington State University due to her Senate testimony.)

Dr. David Wiseman, a PhD Research Bioscientist who formerly worked at Johnson & Johnson, has headed his own R&D consulting businesses since 1996. Dr. Wiseman spoke to the lack of information about the Covid vaccines in a January 2022 Senate roundtable (5:03:53[533]). **Wiseman explained that every patient insert contains the chemical structure of the drug in picture form, but the chemical structure of the mRNA vaccines has never been released by the FDA.** As of December 2023, an internet search still does not bring up a chemical structure image of Pfizer's Comirnaty or Moderna's Spikevax, the two shots that received full FDA approval.

The lack of information about the shots, and the known variability in batches is unacceptable in terms of long-held FDA safety standards. As Latypova explains, "[I]f you're buying Advil today and you're buying it a month from now, your experience...shouldn't be 1,000 percent different, because that's going to be really dangerous. When you see a variability like this between batches of what is supposed to be consistently produced... it means that the product is not good manufacturing practice compliant."

Bait and Switch:

> *Having a regulator that lies to you and papers over problems is worse than having no regulator at all. It's the equivalent of thinking you have a seatbelt on when you don't. It's no basis for sound decisions.*
>
> **EL GATO MALO** | October 5, 2023[534]

Joshua Guetzkow, PhD, Senior Lecturer at Hebrew University of Jerusalem, explains that the participants in the Pfizer clinical trials were given Process 1 vaccines, which is a relatively clean manufacturing process, but it's expensive and produces small quantities. They needed another manufacturing process to roll out billions of shots to the world. Process 2 batches used E.coli bacteria essentially as a medium for growing the mRNA and that was the product that was sold and injected into billions of arms around the world. Guetzkow explains[535] why this is a problem:

> **When we're talking about biological medical products, biologics, the process is the product.** You can't change the process, definitely not in as dramatic a fashion as they changed their production process, and then not run a clinical trial on that new product. It's a totally different product. **They advertised one product...and then they totally switched it with a different product that had a very different safety profile.**

Kevin McKernan is a scientist with 25 years' experience in the genomic field, and is a leading expert in sequencing methods for DNA and RNA. As explained in a May 2023 Brownstone article[536]**, McKernan sequenced vials of the Covid vaccines, and found that they contain large amounts of contaminants** that have no place in a final product. Specifically, McKernan was shocked to find that **"They actually had a lot of DNA in the background."** The article explains that there are strict guidelines about DNA contamination levels in mRNA products, and these vials tested by McKernan exceeded that amount by 1,000 times.

DNA is the hereditary material in humans and almost all other organisms. The Pfizer and Moderna vials of bivalent vaccine that McKernan tested were contaminated with plasmid DNA fragments. "Plasmid DNA is the 'complete recipe'[537] to program bacterial cells to mass produce mRNA," needed for the mRNA shots, but **we were assured by the CDC and FDA that there wasn't any DNA in the *actual* Covid shots.**

Dr. Robert Malone explains[538] that a central dogma of biology is this: "DNA makes RNA, RNA makes protein." **In a proper manufacturing method, after the RNA is manufactured from the DNA, a type of filter is used to remove the degraded DNA fragments** and the small unused chemical subunits from the large RNA molecules. The remaining product is "basically pure RNA dissolved in water." Dr. Malone states:

> [O]nce you have the negatively charged purified RNA in water, you can make it more or less concentrated, mix it in fancy ways with other stuff like self-assembling positively charged fats to produce lipid nanoparticles, store it in a glass vial, and inject it into people. And that is the manufacturing process for pseudo-mRNA vaccines in a nutshell.

The mRNA vaccines erroneously contain DNA fragments:

But something is terribly wrong because there are DNA contaminants in the vaccines, as McKernan and colleagues discovered, including fragments of the DNA virus SV-40[539] which is a cancer causing virus. **Other plasmid DNA sequences were also identified, including antibiotic resistant gene fragments.** Dr. Malone speculates that, "at a minimum such fragments are likely to impact on gene expression in the human cells that take up the DNA. One ***possible*** (sic) impact could involve development of cancers..." Malone states,

> [T]he DNA fragments (in the mRNA vaccines) meet the "formal criteria for pharmaceutical 'adulteration,'" which is strictly prohibited by US Federal law. **The prevention of drug, device and food "adulteration" is one of the central missions of the FDA - basically, a central reason that the FDA was created in the first place.**
>
> The FDA's job is to insure that drugs, vaccines, medical devices and foods are not adulterated. **The remedy for adulteration is immediate recall and seizure** if necessary. (emphasis added)

Immunologist Jessica Rose, PhD, states "Now with this DNA issue... It's integration. That's the biggest concern for me, besides immunological activity against this...This is why we have safety measures...This is why we test for DNA because we really don't want this kind of contamination. In a nutshell, it could be disastrous in terms of cancer." (Integration refers to the foreign DNA becoming part of the human genome.)

A September 2003 article[540] in *Gene Therapy* highlighted the **concern of foreign DNA being incorporated into chromosomal DNA.** "Two main classes of integration mechanisms exist: those that draw on sequence similarity between the foreign and genomic sequences to carry out homology-directed modifications, and the nonhomologous or 'illegitimate' insertion of foreign DNA into the genome. Gene therapy procedures can result in illegitimate integration of introduced sequences and thus pose a risk of unforeseeable genomic alterations."

Dr. Rose explains that the **Covid shots are causing immune suppression**. There are inherent problems in the shots themselves, including the effects of the spike protein, says Rose, and when you **"combine that with this issue of potential integration, it makes you come to the conclusion that this is why we're seeing all these reports of turbo cancers.** It's the combination of all of these factors...People also have preexisting conditions." For example, "If you have some kind of impairment in your BRCA gene, you're more predisposed for breast cancer. It makes you wonder if it's not akin to **throwing gasoline on a fire** [to give these biologic Covid mRNA shots]. It just seems like it's making any current situation going on in the body just explode. Not just cancer, but also immunological stuff."

"Further investigation by McKernan[541] showed the plasmid DNA contained in the vaccines was indeed viable and capable of transformation in bacterial cells...**The question is, does this DNA have the potential to become part of the genome of a human organism and if so what might be the consequences?"** (emphasis added)

We don't fully know the answer to that question, but highly qualified and credentialed doctors and scientists are concerned about the possibilities. What we know for certain is that the vaccine development and approval processes were rushed, good manufacturing practices for the Covid shots were not followed, and known problems with mRNA technology were not addressed before injecting the world's population.

"Why weren't we told? Where is the informed consent?"

In July 2023, John Campbell, PhD, a retired nurse educator, interviewed Dr. Vibeke Manniche[542], one of the authors of a study[543] analyzing suspected adverse effects (SAEs) connected with the Pfizer Covid vaccines in Denmark from 27 December 2020 to 11 January 2022.

Campbell explained that clinical data on individual vaccine batch safety levels had not been reported by Pharma or government health, but that Manniche, Max Schmeling, and Peter Riis Hansen had conducted a self-funded study on the topic.

Using publicly available data the authors examined the results from 52 different BNT162b2 (Pfizer) vaccine batches that were given to 4,026,575 persons. They linked individual SAEs to batch numbers by using the batch label the vaccinated individuals had received.

SAEs were categorized as "non-serious, serious (hospitalization or prolongation of existing hospitalization, life-threatening illness, permanent disability or congenital malformation), or SAE-related death." The study findings were startling.

As discussed previously, vaccine quality is normally monitored for consistency and safety, but Manniche and colleagues' work found that, "unexpectedly, rates of SAEs per 1000 doses varied considerably between vaccine batches."

The study notes, "Three predominant trendlines were discerned, with noticeable lower SAE rates in larger vaccine batches and additional batch-dependent heterogeneity in the distribution of SAE seriousness between the batches." In other words, the Pfizer shots given in Denmark were not good manufacturing practice compliant.

Manniche said the rate of adverse events was as high as one in twenty people for some batches. As shown in the following chart, 4.2% of the doses were associated with the majority of adverse events:

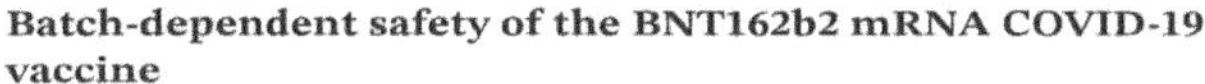

Batch-dependent safety of the BNT162b2 mRNA COVID-19 vaccine

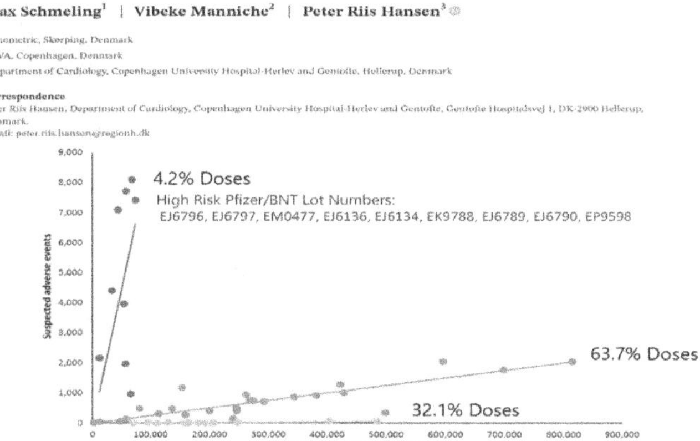

Max Schmeling[1] | Vibeke Manniche[2] | Peter Riis Hansen[3]

[1]Innometric, Skørping, Denmark
[2]LIVA, Copenhagen, Denmark
[3]Department of Cardiology, Copenhagen University Hospital-Herlev and Gentofte, Hellerup, Denmark

Correspondence
Peter Riis Hansen, Department of Cardiology, Copenhagen University Hospital-Herlev and Gentofte, Gentofte Hospitalsvej 1, DK-2900 Hellerup, Denmark.
Email: peter.riis.hansen@regionh.dk

This chart was compiled from the Pfizer vaccine rollout in Denmark between December 2021 and January 2022. Each dot on the chart represents a batch. Suspected adverse events were much higher in earlier batches (blue line).

Campbell notes that these batches were given all over the world outside of the U.S., which had its own manufacturing plants. Campbell, said he **would not have taken the Covid shots had he known – the risk of a bad adverse reaction was too high.** Campbell asks, "Why were we not told? Where is the informed consent? It's outrageous!"

Questions for Pfizer, regulators, and politicians:

Dr. Manniche says the wide variability in adverse reactions related to batch numbers raises these questions for Pfizer, the regulators, and government decision-makers: **Has the product changed? When did it change? Why did it change? Why were the safety signals ignored and obscured?** Manniche says the population did not have proper information to give informed consent about receiving the Covid shots.

Manniche explains that she doesn't think it was bad that Pfizer changed the product, if they could see there was a problem early in the rollout, but it was a problem not to make the changes public. In light of the failure to inform, Manniche says the next question would be, "If they changed the product, did it actually work? Was there any reason to get the vaccines?"

Manniche commented that the vaccines did nothing to prevent transmission, and in addition, in Denmark, Covid cases were higher in the vaccinated group than the unvaccinated. Manniche asks, "So was it worth the trouble, so to speak, was it worth the side effects?"

> "This is a product problem - something with the product" states Mannice, "whether it's the product by itself, or the transportation, or the distribution, whatever, but there is a product problem.... People were pushed into these vaccines. They lost their jobs and were bullied...Lots of people now may sit back and wonder, what happened? Was this dangerous? And what will happen in the future?" Manniche worries that the further we get out from vaccination, the less patients and their doctors will recognize the possible connection between future health problems and the vaccine.

Dr. Eyal Shahar, also expressed concern[544] about the **difficulty of determining cause of illness and death due to the Covid vaccine**. Shahar's conservative **analysis estimated serious adverse events at a rate of 10-15 per 10,000, which is much higher than the flu shot.**

> Fatality after a flu shot is extremely rare: one death per five million. [The] short-term fatality after Covid shots is many folds higher: dozens of deaths per one million. That's unacceptable, or at least was unacceptable for any vaccine before 2021.
>
> **DR. EYAL SHAHAR**
>
> Professor emeritus of public health in epidemiology and biostatistics

Censorship of Science:

> *For potential scientific data to be rejected out of hand before it's been analyzed, before it's been critiqued, because it doesn't fit with a particular narrative, is a form of intellectual fascism...*
>
> **JOHN CAMPBELL, PHD**

Manniche expresses frustration about the censorship and inability to have a scientific debate about the vaccines. **"We have not been allowed to have this discussion. We have not been allowed to raise concerns about safety issues with the vaccines. If we did so, we were called names, bullied, and so on.** I think that's part of the sickness - that you are not supposed to speak against the narrative."

Manniche said **when they completed the study and submitted it to the large medical journals, it was rejected** by the *Lancet, Journal of American Medicine, New England Journal of Medicine*, and *Annals of Internal* Medicine, all large respected journals, **without peer review or explanation. Eventually the less prominent *European Journal of Clinical Investigation* printed[545] the research.**

Manniche is grateful to the *European Journal of Clinical Investigation* for having the courage to print the study, but was also disappointed it **took a year to get the information out to the public.** Campbell expressed dismay at the unprofessional conduct of the larger journals – rejecting the study outright, within hours of receipt, without any kind of peer review. **Campbell states, "If they're deciding what medical information we get and what medical information we don't get, that is a fundamental problem with all medical epistemology."**

Manniche asks, "Why has this not been in the mainstream media?" It was new information, yet it has not been reported. Manniche said **any other product would have been removed from the market immediately upon seeing the wide variance in adverse events between batches.**

Following is an excerpt from the FDA's website on December 1, 2023 regarding Current Good Manufacturing Practices. Based on Manniche's and colleagues' research, among other things, it is accurate to claim that **the FDA failed to employ oversight that "assures the identity, strength, quality, and purity of drug products** by requiring that manufacturers of medications adequately control manufacturing operations."

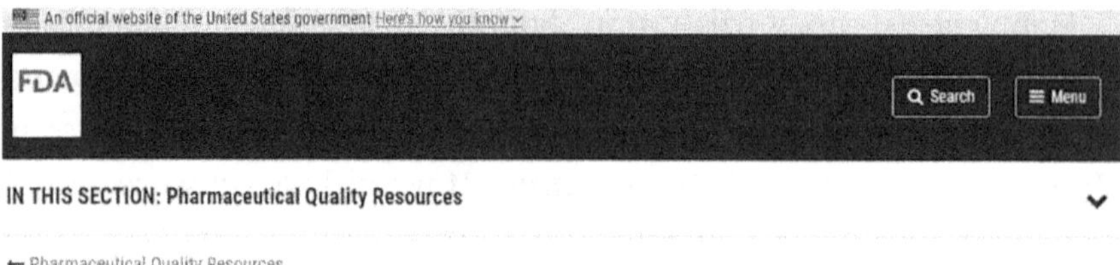

Facts About the Current Good Manufacturing Practices (CGMP)

What is CGMP?

CGMP refers to the Current Good Manufacturing Practice regulations enforced by the FDA. CGMP provides for systems that assure proper design, monitoring, and control of manufacturing processes and facilities. Adherence to the CGMP regulations assures the identity, strength, quality, and purity of drug products by requiring that manufacturers of medications adequately control manufacturing operations. This includes establishing strong quality management systems, obtaining appropriate quality raw materials, establishing robust operating procedures, detecting and investigating product quality deviations, and maintaining reliable testing laboratories. This formal system of controls at a pharmaceutical company, if adequately put into practice, helps to prevent instances of contamination, mix-ups, deviations, failures, and errors. This assures that drug products meet their quality standards.

> And so it's usually someone who finds out I'm vaccine-injured says, "Oh, really? Which one did you have?" and I'll say, 'Oh I got Pfizer.' They'll reply, "Oh really? My husband got that one and he's fine." I'll reply, "Yeah, and so did a lot of my family members and they're all totally fine, but some people are *not* fine."
>
> **FROM THE FILM "ANECDOTALS"**[546] | Testimonies from the Vaccine Injured

26. V-SAFE: CDC'S (SKEWED) COVID-19 VACCINE INJURY REPORT SYSTEM

Although intended to prevent viral infection, replication, spread and COVID disease or death, and to enable "herd immunity," these products were developed at "Warp Speed" and labeled "vaccines" but differed remarkably from all other currently available vaccine products. That haste and the associated regulatory compromises enabled by lax EUA requirements yielded products with high rates of avoidable existing and emerging treatment-associated serious adverse events including hospitalizations, permanent disabilities and deaths, as reported in both VAERS, and the CDC's now surreptitiously shuttered V-safe reporting system.

DR. DAVID GORTLER | former FDA senior advisor
ROBERT MALONE, MD | September 13, 2023[547]

The CDC often proclaimed[548] that the Covid-19 shots were monitored by the "most intense safety monitoring in U.S. History." Most likely one monitoring element they are referring to is V-safe, which was a smartphone app developed specifically to track the Covid-19 vaccines by asking recipients in real time about their post-vaccine experiences. V-safe was highlighted as being the first vaccine monitoring app ever deployed. Meanwhile, the 30-year-old VAERS system was downplayed as flawed and ineffective. However, V-safe was structured to gather the results the CDC wanted to see, not what the system *could* show.

CDC knew early studies reported that the mRNA shots caused serious harm:

As explained in Section 18, the CDC and others knew about expected medical harms from the Covid shots. Published studies[549] in the *New England Journal of Medicine* (July 2020[550]), and JAMA (October 2020[551]), both highlighted concerns. And the CDC presentation on October 30, 2020 reported that adverse events associated with the early injections of Covid-19 mRNA shots included:

- Eye disorders
- Gastrointestinal disorders
- Convulsions/seizures
- Encephalitis
- Transverse myelitis
- Nervous system disorders
- Anaphylaxis (allergic reaction)
- Guillain-Barré syndrome
- Immune thrombocytopenia
- MIS-C
- Myocarditis/pericarditis
- Multisystem inflammatory syndrome in children
- Musculoskeletal and connective tissue disorders

Yet V-safe specifically only asked users to check if they had experienced one or more of the following symptoms in the first seven days after taking the shot: chills, headache, joint pain, muscle or body aches, fatigue or tiredness, nausea, vomiting, diarrhea, abdominal pain, and rash. These conditions are not considered serious adverse events, but "reactogenicity,"[552] meaning a normal adverse reaction to a vaccine.

V-safe users were also asked to pick, if applicable, one or more of the following three "health impacts:"

- unable to perform normal daily activities
- missed work/school
- needed medical care

If a user answered yes to one of the above three impacts, they were asked to select one or more of the following options:

- hospitalization
- emergency room
- urgent care
- telehealth

V-safe users were asked to tick these same boxes once a week for six weeks, then at six months and one year. There were also limited free text fields where users could report other symptoms, which means all the actual adverse events were unnecessarily difficult to track. **Aaron Siri, an attorney who has filed multiple Freedom of Information requests to get information from the CDC, FDA, and other government agencies, explains:**[553]

> Reflecting that the CDC knew these serious adverse events were critical to track, and that the CDC sought to obfuscate reports of these harms, the CDC created an incredibly complex system to deal with text field reports of these conditions. **If a v-safe user reported one of these conditions, someone at the CDC would have to agree that what was written in a free text field actually reflected one of these conditions**, then someone from the CDC was supposed to reach out to the v-safe user by telephone (which, as discussed in a future part, often did not occur or occurred months or years later), and if the CDC ever actually reached out and thought the condition described was on the list, then **the CDC employee could assist the user in completing a VAERS report.**
>
> And then, once in VAERS, the CDC, as it does, would say that VAERS reports (i) cannot ever be used to show a vaccine causes a harm and (ii) cannot be used to determine a rate at which it may cause a harm because VAERS receives reports from an unknown population size. Meaning, the CDC says it doesn't know the denominator needed to calculate a rate using VAERS data.

> [H]ad the CDC simply had a check-the-box field in v-safe for each of these conditions, it would have had a denominator. It could simply divide the number of v-safe users reporting the condition by the total number of v-safe users. And boom, there it would be! The rate. Instead, the CDC knowingly, consciously, chose to not create check-the-box options for these serious adverse events, even though it had itself identified them as safety issues to track prior to the launch of v-safe. (emphasis added)

Legal action against the CDC was required before it would release V-safe data:

Although 10 million of the vaccinated public voluntarily enrolled in v-safe, the CDC would not release the V-safe data. Aaron Siri explains, **it took "multiple legal demands, appeals, and two federal lawsuits" to get the court order that required the CDC to release the v-safe data to the public** in November 2022, 464 days[554] after V-safe commenced. This was deceitful[555] on the part of the CDC, which knew vaccine-harm conditions such as myocarditis and blood clots (thrombosis) usually take longer than one week to manifest.

> Once the CDC finally released the data, **it still withheld the adverse events listed in the free text fields, for several months.** During the year and a half before being forced to make V-safe public, the CDC only reported data collected in the first 7 days after vaccination. Siri states[556], "CDC actively chose to cherry-pick data in order to claim the vaccine is safe when it knew that the only data point it really had from the check-the-box data, to determine whether the vaccines are safe, had a blaring red alarm screaming 'unsafe.' It just chose to hide that from the public."

The CDC ended V-safe enrollment on May 19, 2023:

Despite the fact that "updated" Covid vaccines are still only Emergency Use Authorized, the **CDC closed enrollment in the V-safe program on May 19, 2023.** The system disappeared from public view on June 30, 2023, accessible only through an archived zip file[557] that is not user-friendly.

On the discontinued V-safe page the CDC explained it was "developing a new version of v-safe which would allow users to share their post vaccination experiences with new vaccines," and encouraged users to contact VAERS to "report any possible health problems or adverse events following vaccination." The CDC stated, as shown in the screenshot below from this same page, "Covid-19 vaccines are safe and effective. CDC continues to recommend that everyone ages 6 months and older should stay up to date with Covid-19 vaccines."

Vaccine Safety

Vaccine Safety Home

V-safe After Vaccination Health Checker

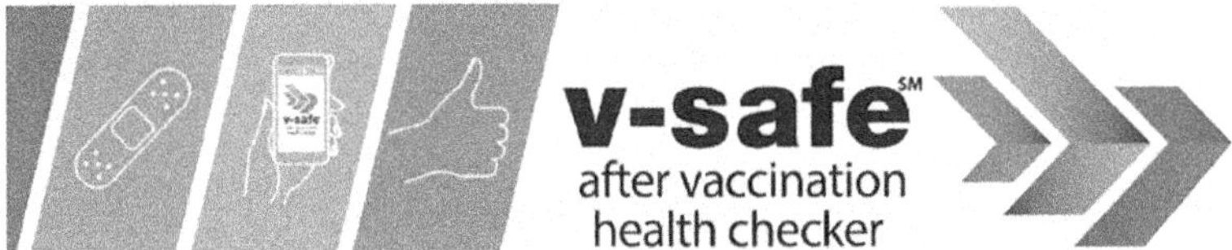

On May 19, 2023, CDC closed enrollment in v-safe for COVID-19 vaccines. V-safe was developed specifically for COVID-19 vaccines and has been an essential component of the pandemic vaccine safety monitoring systems that have successfully characterized the safety of the COVID-19 vaccines used in the United States. CDC is developing a new version of v-safe which will allow users to share their post-vaccination experiences with new vaccines.

CDC will continue to monitor the safety of COVID-19 vaccines through its other vaccine safety monitoring systems. V-safe users or others who get vaccinated can report any possible health problems or adverse events following vaccination to the Vaccine Adverse Event Reporting System (VAERS).

Since its launch in December 2020, 10.1 million v-safe participants completed more than 151 million health surveys about their experiences following COVID-19 vaccination, and v-safe data have been included in more than 20 scientific publications.

COVID-19 vaccines are safe and effective. CDC continues to recommend that everyone ages 6 months and older should stay up to date with COVID-19 vaccines.

After discontinuing V-safe as a Covid-19 vaccine tracker in June 2023 the CDC then switched, at the beginning of October 2023, to only reporting VAERS data once a month, instead of weekly.

Here is the last screenshot[558] of data compiled by an independent group, ICAN, from the V-safe dashboard before it was taken down:

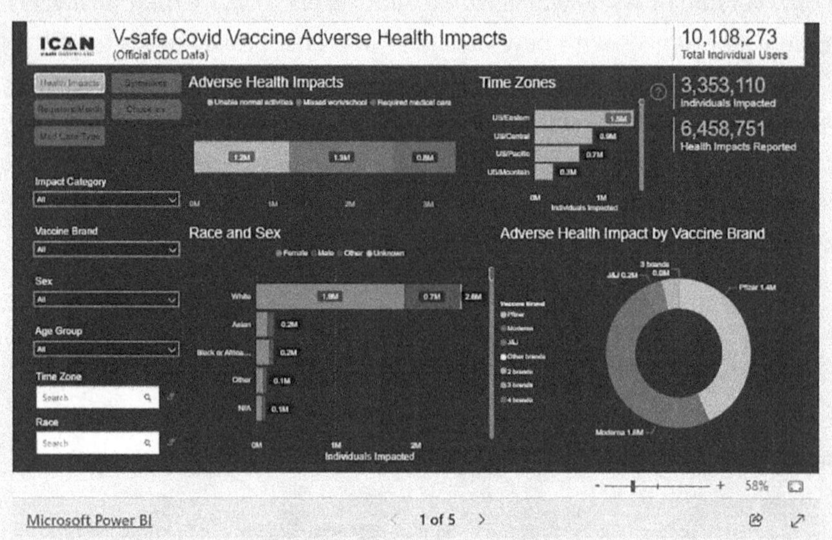

David Gortler[559], a former senior advisor to the FDA Commissioner, explains that the V-safe data showed huge amounts of adverse events, and wonders at the choice to end the reporting system:

> Existing data from the V-safe site showed around 6.5 million adverse events/health impacts out of 10.1 million users, with **around 2 million of those people unable to conduct normal activities of daily living or needing medical care**, according to a third-party rendering of its findings. In other words, despite mRNA shots still being widely available and the CDC promoting its continued use, it's "case closed" with regards to collecting new safety reports, under today's federal public health administration.

A check of V-safe[560] on November 1, 2023, showed it was now tracking "Adults ages 60 years and older who receive a respiratory syncytial virus (RSV) vaccine." It appears it is not being used to track the fall 2023 Covid vaccine, or any of the previous Covid shots.

> **It's mystifying as to why the CDC would close the adverse event reporting app on a vaccine product that still only has Emergency Use Authorization.**

27. LAWSUITS AGAINST THE FEDERAL GOVERNMENT

It seems that in any way possible, the federal bureaucracy was weaponized against American citizens during the pandemic. An obscure[561] phrase from Section 264(a)[562] of the Public Health Service Act of 1944 was stretched to justify mask mandates on public transportation, in airports, and on airplanes. The Centers for Disease Control and Prevention, which has nothing to do with commerce, issued an order[563] that landlords couldn't evict tenants who failed to pay rent during the Covid-19 "emergency." Biden claimed to have Executive authority[564] to force private businesses employing 100 or more people to make Covid vaccination a condition for employment.

The federal government's authoritarian orders are gradually being overturned by the courts. The public transportation mask mandate,[565] vaccine mandates,[566] and no-eviction rule[567] have all been ruled unlawful overreach[568]. However, during the months and years they were in place, the illegal mandates caused incalculable, and often permanent, harm to many people's lives and livelihoods.

Those who imposed the orders that caused so much harm continue to support them. For example, Los Angeles Unified School District, the second largest in the U.S. continued to mandate Covid-19 vaccines for all employees through September 2023, when the practice was successfully contested[569] in the Ninth Circuit Court of Appeals. In another example, New York Governor Kathy Hochul has appealed[570] the ruling that overturned the state's power grab[571] for authority to quarantine citizens for even *suspected* illness, for undetermined periods of time, in state-appointed locations, without due process. (For more about this case, see the Substack of Bobbie Anne Cox[572], the attorney who has taken on the state of New York in this matter.)

In June 2021, the Supreme Court let the CDC eviction ban[573] stand, because it was due to expire at the end of July, but expressed the opinion that congressional authorization would be necessary if the CDC wanted to extend the moratorium past July 31.

Then in July, a federal appeals court again upheld[574] the CDC eviction ban, noting the Supreme Court's ruling. When Congress declined to extend the eviction ban, the CDC issued a new 60-day eviction moratorium[575] order. Pres. Biden, in addressing the new ban said, "By the time it gets litigated, it will probably give some additional time, while we're getting $45 billion out to people who are in fact behind in the rent and don't have the money." The Supreme Court struck down[576] the eviction ban 20 days later on August 26, 2021, stating, [O]ur system does not permit agencies to act unlawfully even in pursuit of desirable ends."

The Supreme Court ruling stated[577]:

> It would be one thing if Congress had specifically authorized the action that the CDC has taken. But that has not happened. Instead, the CDC has imposed a nationwide moratorium on evictions in reliance on a decades-old statute that authorizes it to implement measures like fumigation and pest extermination. It strains credulity to believe that this statute grants the CDC the sweeping authority that it asserts.

The federal government and many governors and health departments seem to have knowingly entered into a pattern of setting and enforcing unconstitutional policies, using the time allowed by a slow judicial system to engage in unlawful agendas.

Missouri v Biden: government suppression of free speech through social media:

Legal discovery in the ongoing court case[578] *Missouri v Biden* has **revealed "that a dozen federal agencies pressured social media companies Google, Facebook, and Twitter to censor and suppress speech** that contradicts federal pandemic priorities. In the name of slowing the spread of harmful misinformation, the administration forced the censorship of scientific facts that didn't fit its narrative de jour."

With the NIH's control over billions of dollars in research and grant money throughout the world, along with its ties to Big Pharma, scientific and medical research has suffered. Following is an excerpt from an article highlighting the negative effect that NIH narrative control has had on science:

> The COVID-19 pandemic and lockdowns have not only been devastating for society, they have had a chilling effect on the scientific community. **For science to thrive, opposing ideas must be openly and vigorously discussed, supported, or countered based on scientific merit.**
>
> Instead, some politicians, journalists, and (alas) scientists have engaged in vicious slander of dissident scientists, spreading damaging conspiracy theories, even with open calls for censorship in place of debate. In many cases, eminent scientific voices have been effectively silenced, often with gutter tactics. People who oppose lockdowns have been accused of having blood on their hands, their university positions threatened, with many of our colleagues choosing to stay quiet rather than face the mob.

MARTIN KULLDORFF AND JAY BHATTACHARYA | March 18, 2021, The Federalist[579]

Missouri v Biden is working its way through the courts. The 5th Circuit Court ruling on October 3, 2021 gutted much of what the District Judge ruled previously. The U.S. Supreme Court has agreed[580] to take up the case. Judge Doughty, who heard the case in District Court, wrote:

> If the allegations made by plaintiffs are true, the present case arguably involves the most massive attack against free speech in United States' history.

JUDGE TERRY DOUGHTY | *Missouri v Biden* | July 4, 2023

28. COVID AUTHORITARIANS USE FEAR, IMPOSE MANDATES IN FALL 2023

On May 11, 2023, the Covid emergency officially ended. Heading into fall 2023, the official rhetoric began to heat up. Those in positions of power and influence **continually tried to scare people with the idea that the new Covid variants were going to lead to a surge in serious illness through the winter.**

The unfounded claims, in the winter of 2022-2023 that the **"triple threat" (Covid, flu, RSV) would overrun our hospitals, were rolled out again.** Some colleges have continued to require Covid shots[581] in order to enroll. Rumors of mask mandates and mandatory boosters have been circulating, and in some places it's actually happening again. (see here[582], here[583], here[584], and here[585])

Medical and public authorities, pharmaceutical companies, and the government agencies that regulate them have all ignored the multiple studies and databases that document the Covid vaccines do not prevent infection or spread, that face masks[586] do not prevent the spread of respiratory viruses (and in fact, are harmful[587]), and that lockdowns were a complete failure[588] at preventing the spread of Covid-19. Following are a few charts[589], compiled by Ian Miller[590], that illustrate the ineffectiveness of population-wide face masking and lockdowns, as well as showing the rise of Covid cases and deaths after the vaccine rollout.

A few charts that show the futility of mandates to prevent Covid spread:

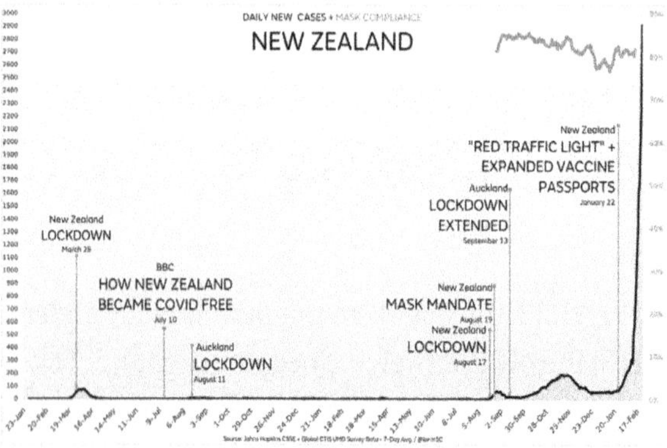

Because it's an island, New Zealand was able to keep out visitors and lock down residents. The lockdowns harmed the country in every way, but only delayed the arrival of Covid. In the above chart, the orange line shows mask compliance at almost 90% in September 2021. The black line shows daily new cases. Note the exponential rise of cases in February 2022, despite coerced/forced[591] Covid-19 vaccination.

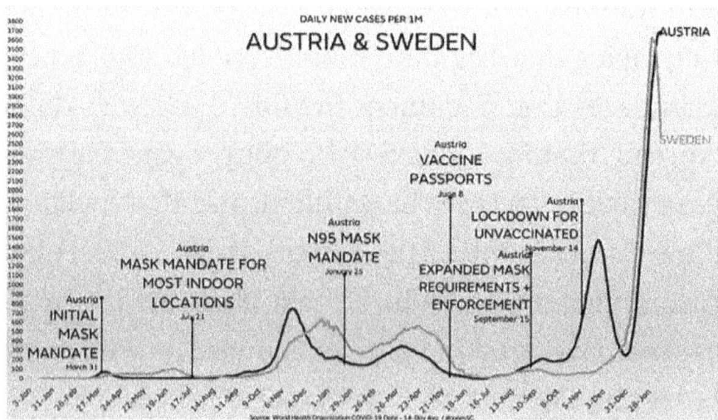

As the above chart shows, the rise and fall of cases in Austria and Sweden is almost identical. Sweden (orange line) did not implement mask mandates, did not lock down, and kept schools open for children up to age 14. Austria (black line) continually enforced mask mandates, then vaccine passports, and eventually locked down the unvaccinated[592].

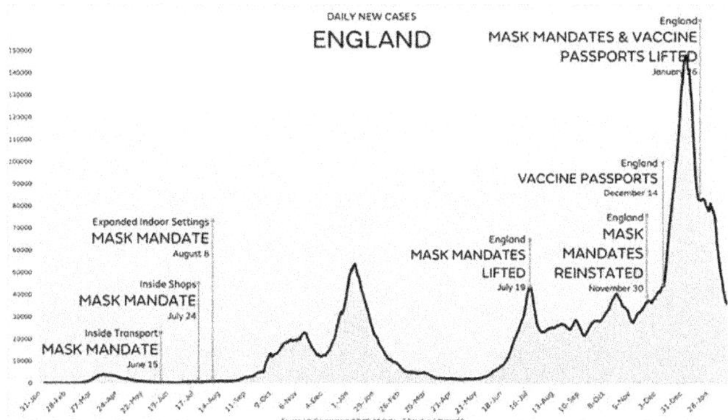

In the above chart, despite multiple strict mask mandates, and vaccine requirements, (and lockdowns, which are not tracked on this chart), cases soared in England. After mask mandates and vaccine passports were lifted, cases fell. There is no correlation between masking and the rise and fall of Covid cases.

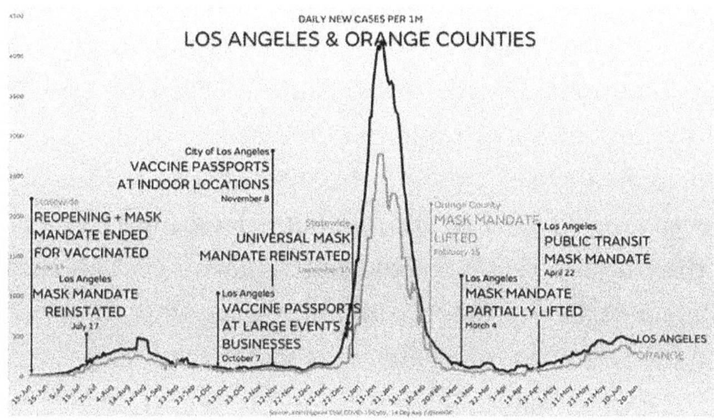

The up and down curve of Covid is almost exactly the same for Los Angeles and Orange Counties, despite different masking timelines, and L.A. requiring vaccine passports while Orange County did not.

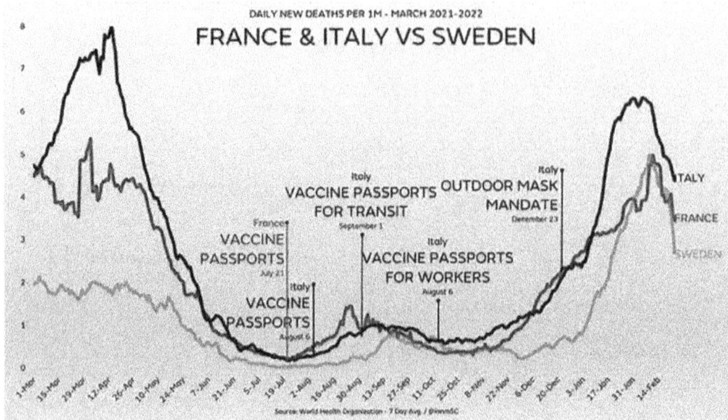

The above chart shows that despite wildly different masking and vaccine requirements (Sweden did not mandate either one), the total Daily New Deaths per million due to Covid-19 in Italy, France, and Sweden are almost the same. Italy is the black line, France is the red line, and Sweden is the orange line.

There is literally not one country, state, or country in the world where face masks and lockdowns led to overall reduction in cases, hospitalizations, or deaths. Some, as with New Zealand, delayed the arrival of Covid, but didn't prevent its spread, and that delay came at horrible economic, social, and societal costs. **Despite the data-based evidence showing the ineffectiveness of the pandemic measures, authorities continue to claim they handled things correctly.** Alarmingly, the disregard for long-established safety protocols, and the "forgetting" of over 200 years of medical and scientific knowledge continues to this day.

In interviewing Angus Dalgleish on October 11, 2023[593], Dr. John Campbell stated, **"I just hope we're not heading into an anti-scientific or controlled scientific Dark Age."** Dr. Dalgleish replied, "I'm afraid we're already there. We've already descended down the first tunnel. One of the reasons for writing this book is to wake everybody up as to what's happening." (Dagleish's book *The Death of Science*[594] was released in November 2023.)

29. DOMESTIC TERRORISM AND QUESTIONING THE OFFICIAL COVID NARRATIVE

The United States remains in a heightened threat environment fueled by several factors, including an online environment filled with false or misleading narratives and conspiracy theories, and other forms of mis- dis- and mal-information (MDM) introduced and/or amplified by foreign and domestic threat actors.

DEPARTMENT OF HOMELAND SECURITY BULLETIN

February 7, 2022

Homeland Security labels Covid- response questioners "domestic terrorists:"

As early treatment continued to be withheld from Covid-19 patients, **doctors who defied the consensus and continued to successfully treat patients with** off-label **drugs came under increasingly hostile attacks from federal authorities and hospital and medical boards.** The mainstream media vilified anyone who stepped out of the official narrative.

Social media platforms such as Facebook, Twitter, Instagram, and YouTube were instructed by various government officials to take down accounts that the government deemed misinformation. **The Department of Homeland Security labeled people who question the official Covid narrative, and the people who listen to them, domestic terrorists working hand in hand with people who claim there is election fraud.**

On February 7, 2022, Homeland Security issued a National Terrorism Advisory[595] System Bulletin titled "Summary of Terrorism Threat to the U.S. Homeland," identifying the primary terrorism-related threat to the

U.S. as people who spread "false or misleading narratives." Following are some excerpts from the Bulletin:

> **The primary terrorism-related threat to the United States** continues to stem from lone offenders or small cells of individuals who are motivated by a range of foreign and/or domestic grievances often cultivated through the consumption of certain online content. The convergence of violent extremist ideologies, **false or misleading narratives, and conspiracy theories have and will continue to contribute to a heightened threat of violence in the United States.**
>
> Key factors contributing to the current heightened threat environment include:
>
> 1) The proliferation of false or misleading narratives, which sow discord or undermine public trust in U.S. government institutions:
>
> - For example, **there is widespread online proliferation of false or misleading narratives regarding unsubstantiated widespread election fraud and COVID-19. Grievances associated with these themes inspired violent extremist attacks during 2021.**
> - Malign foreign powers have and continue to amplify these false or misleading narratives in efforts to damage the United States. (emphasis added)
>
> 2) Continued calls for violence directed at U.S. critical infrastructure; soft targets and mass gatherings; faith-based institutions, such as churches, synagogues, and mosques; institutions of higher education; racial and religious minorities; government facilities and personnel, including law enforcement and the military; the media; and perceived ideological opponents:

- As COVID-19 restrictions continue to decrease nationwide, increased access to commercial and government facilities and the rising number of mass gatherings could provide increased opportunities for individuals looking to commit acts of violence to do so, often with little or no warning. Meanwhile, COVID-19 mitigation measures—particularly COVID-19 vaccine and mask mandates—have been used by domestic violent extremists to justify violence since 2020 and could continue to inspire these extremists.

In issuing this bulletin, the Federal Government took First Amendment protected opinions, speech, and behaviors and labeled them as criminal acts.

Federal Judge Terry Doughty does not agree with Homeland Security: On July 4, 2023, US District Judge Terry Doughty issued a Preliminary Injunction[596] against the Biden Administration and various government agencies, including the FBI, CDC, and the U.S. Surgeon General, for their suppression of free speech on social media platforms. **Doughty pointed out that the government has conflated free speech with control over information, in order to deflect attention from its censorship.**

Judge Doughty concluded his 155-page Preliminary Injunction with these words:

> V. CONCLUSION
> Once a government is committed to the principle of silencing the voice of opposition, it has only one place to go, and that is down the path of increasingly repressive measures, until it becomes a source of terror to all its citizens and creates a country where everyone lives in fear.
>
> **HARRY S. TRUMAN | 33rd Pres.of the United States**
>
> The Plaintiffs are likely to succeed on the merits in establishing that the Government has used its power to silence the opposition.

> Opposition to COVID-19 vaccines; opposition to COVID-19 masking and lockdowns; opposition to the lab-leak theory of COVID-19; opposition to the validity of the 2020 election; opposition to President Biden's policies; statements that the Hunter Biden laptop story was true; and opposition to policies of the government officials in power. All were suppressed. It is quite telling that each example or category of suppressed speech was conservative in nature. This targeted suppression of conservative ideas is a perfect example of viewpoint discrimination of political speech. American citizens have the right to engage in free debate about the significant issues affecting the country.
>
> Although this case is still relatively young, and at this stage the Court is only examining it in terms of Plaintiffs' likelihood of success on the merits, the evidence produced thus far depicts an almost dystopian scenario. During the COVID-19 pandemic, a period perhaps best characterized by widespread doubt and uncertainty, the United States Government seems to have assumed a role similar to an Orwellian 'Ministry of Truth.'"

The Preliminary Injunction, restricting the Biden administration's communication with social media companies, was upheld in part[597] by the 5th Circuit Court of Appeals. The Supreme Court[598] put a stay on the 5th Circuit injunction and has agreed to hear the case in the next term, which is scheduled to end June 2024.

"Freedom of speech – not reach:"

Elon Musk's purchase of Twitter led to the **exposure of the government's interference with free speech via social media platforms.** For this opening of the Twitter Files[599] we are grateful. However, it should be noted that **Musk recently hired Linda Yaccarino[600] to be the CEO of X (Twitter).** Yaccarino is also the chairwoman of the World Economic Forum's task force on the Future of Work and is also a member of the WEF's Media, Entertainment and Culture Industry Governors Steering Committee.[601]

Yaccarino assures us that **reducing hateful content, pornography, conspiracy theories, and other similar tweets is part of how "X is committed to encouraging healthy behavior online**...It goes back to my point about our success with 'freedom of speech, not reach.' If it is 'lawful but it's awful,' it's extraordinarily difficult for you to see it." Yaccarino assures:[602]

> We have an extraordinary team of people who are overseeing, hands on keyboards, monitoring all day every day to make sure that 99.99% of [healthy] impressions remain at that number....But we also have to remember what's at the core of free expression. You might not agree with what everyone is saying. We want to make it a healthy debate and discourse, but free expression at its core will really, really only survive when someone you don't agree with, says something you don't agree with. And what a great place we would live in if we were able to return to a healthy, constructive discourse amongst people that we don't agree with.

Orwell's 'Ministry of Truth," referred to by Judge Doughty, is from the book *1984*:

George Orwell wrote *1984* to warn against a future totalitarian state in which government would be able to monitor and control its citizens through the use of technology. Manipulation of language is of central importance to the government in Orwell's fictional country of Oceania. The "Party" invents Newspeak, in which it changes the definition of words and eliminates others, with the goal that limited speech leads to limited ideas, including citizens being incapable of conceptualizing anything that questions the Party's power.

To that end, Party slogans are plastered everywhere across Oceania:

"War is peace."
"Freedom is slavery."
"Ignorance is strength."

Might we add "Freedom of speech, but not reach" and "Lawful but awful" to that list? Yaccarino's "extraordinary team of people," with "hands on keyboards monitoring all day every day to make sure everyone's tweets are "healthy communication" sounds menacing. Especially when the Biden Administration continues to claim in the Courts that it needs to control citizens' speech for their own safety. **In light of Yaccarino's Orwellian slogans, her little speech about the core of free expression being the ability for someone to say something you don't agree with sounds much more like *1984* than 1776.**

> We cannot yield to censorship, not because we want people to fill up the air with toxicity and hate, but because we know that if we try to control it, who has the right? Who are we granting the right to now? Have you investigated any of these organizations? Have you investigated their funding, their affiliations, their agenda, their imperative? Because to some degree I am sorry to report that I have, and I have found them wanting. They will not be getting my consensus for authority any time soon.
>
> **RUSSELL BRAND** | June 29, 2023[603]

30. COVID-19: HANDLED AS A BIOSECURITY THREAT, NOT A HEALTH EMERGENCY

In the councils of government, we must guard against the acquisition of unwarranted influence, whether sought or unsought, by the military-industrial complex. The potential for the disastrous rise of misplaced power exists and will persist.

PRESIDENT DWIGHT D. EISENHOWER

Farewell address | January 17, 1961[604]

Pres. Eisenhower's 1961 warning of a too powerful military-industrial complex:

In 1961, as his term of office was ending, President Dwight Eisenhower gave a farewell address to the American people. He stated, "Our military organization today bears little relation to that known by any of my predecessors in peace time, or indeed by the fighting men of World War II or Korea." The difference, Eisenhower explained, was that in the past the U.S. had no armaments industry. With the advent of the cold war, we could "no longer risk emergency improvisation of national defense." For this reason, **the U.S. had been "compelled to create a permanent armaments industry of vast proportions."**

Eisenhower saw the technological revolution as being both akin to, and largely responsible for, the sweeping changes in our industrial military posture. He said "the solitary inventor, tinkering in his shop, has been over shadowed by task forces of scientists in laboratories and testing fields." At the same time, because of the huge costs involved, government contracts had become "virtually a substitute for intellectual curiosity. For every old blackboard there are now hundreds of new electronic computers."

Although he was in favor of both the new technologies, and avoiding war through the strength of a permanent armaments industry, **Eisenhower understood the danger to liberty that a strong military-industrial complex could represent.** He warned that "an alert and knowledgeable citizenry" was necessary to ensure that "security and liberty may prosper together."

> So have they? Have security and liberty prospered together? Or are we now in the time that Eisenhower warned of when our respect for scientific research and discovery has caused public policy to "become the captive of a scientific-technological elite?" I think the fact that we heard "follow the science" invoked at every turn as our civil liberties were revoked answers that question.

Covid-19 vaccines were a military countermeasure:

> *Bioterrorism response plans – under the broader umbrella of counterterrorism – are not designed to incorporate the complicated nuances of public health principles, which balance the need to protect individuals from a pathogen with the need to keep society as functional as possible to maintain overall well-being.*
>
> *If counterterrorism measures are deployed against a public health threat, it is thus not surprising to witness massive disruptions to society, and harms to public health – as we have seen with the Covid-19 pandemic response.*
>
> **DEBBIE LERMAN** | Brownstone Fellow | January 18, 2023[605]

Brownstone Fellow Debbie Lerman has done extensive research and writing on the topic of Covid being handled as a military response to a bioweapon, rather than as a public health emergency. Lerman has documented[606] that Covid-19 pandemic policy was set by the National Security Council (NSC), not the Department of Health and Human Services (HHS). The NSC is "the President's principal forum for national security and foreign policy decision making," as described by the White House.[607]

Prior to Covid-19 this included all the defense department heads and the Chairman of the Joint Chiefs of Staff, among others. The White House now states that also included in the NSC meetings are "the Covid-19 Response Coordinator and the Special Presidential Envoy for Climate... when appropriate to address the cross-cutting nature of many critical national security issues, such as homeland security, global public health, international economics, climate, science and technology, cybersecurity, migration, and others."

As shown in the following chart, under the White House Task Force, we see that Covid Policy was set by the National Security Council, with the Emergency Support Function Leadership Group (ESFLG) led by FEMA managing interagency coordination, and Operations being carried out by the CDC and Health and Human Services.

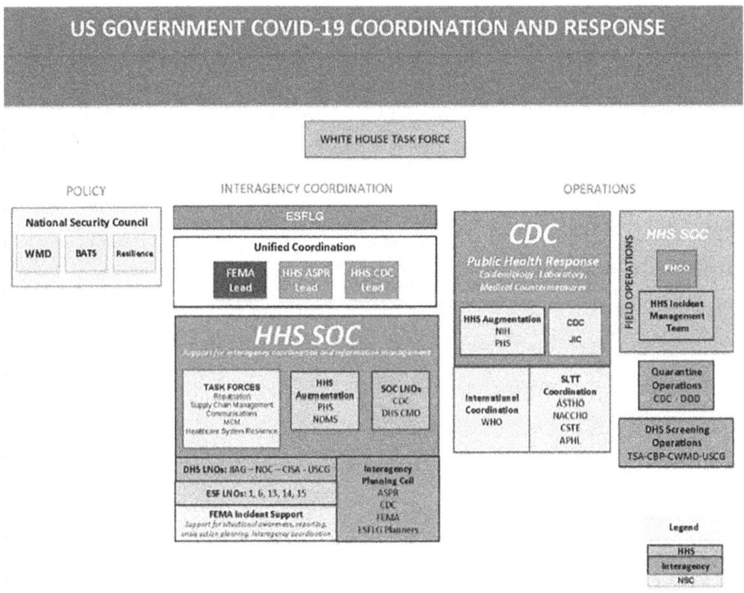

The Federal Emergency Management Agency (FEMA), which is under the Department of Homeland Security (DHS) was an interagency coordinator, along with Health and Human Services (HHS). **All pandemic plans prior to Covid-19 placed Health and Human Services at the helm of pandemic response, whereas FEMA, charged with handling response to natural disasters such as hurricanes and earthquakes, had never**

handled a public health response prior to Covid-19.

In an article[608] titled *Lockdowns Were Counterterrorism, Not Public Health*, Lerman explains that in the rare event of an actual bioweapons attack – **the biodefense strategy can be summarized as "quarantine-until-vaccine,"** in which individuals are kept isolated from the bioweapon, for as long as necessary, until an effective medical countermeasure (medicine/vaccine) is deployed.

Lerman points out that there was **no scientific evidence for universal mask mandates**, but masks instilled fear which promoted compliance with lockdowns and vaccine mandates. **The testing had no scientific basis once the virus was widespread**, but the test and isolate strategy effectively quarantined people and made them desperate for the exit strategy – the vaccine. Likewise, **the lockdowns were only instituted once the National Security Council took over Covid policy.** Lerman states:

> Arguing about whether mask mandates, testing and isolating, social distancing, lockdowns etc. are good public health policies or bad public health policies is a moot point. They are not public health policies at all. All of these measures were devised solely in the service of gaining compliance with the biodefense/counterterrorism plan of quarantine-until-vaccine.

Lerman believes that "national security authorities took control of the Covid pandemic response not just in the U.S. but in many of our allied countries (the UK, Australia, Germany, Israel and others) **because they knew SARS-CoV-2 was an engineered virus that leaked from a lab researching potential bioweapons."** (emphasis added)

> Our response to the Covid pandemic was led by groups and agencies that are in the business of responding to wars and terrorist threats, not public health crises or disease outbreaks.
>
> **DEBBIE LERMAN** | Brownstone Fellow

Also researching the Intelligence Community involvement in the pandemic response is former pharmaceutical research and development executive Sasha Latypova[609]. (Latypova is cited in Section 25 Dangerous Variations in Covid-19 vaccine batches.)

Latypova explains that the **Department of Defense (DoD) handled Operation Warp Speed. The DoD designated the Covid vaccines as "medical countermeasures,"** which meant they were designed to counter enemy activity – that is, a bioweapon.

> **Countermeasure**: That form of military science that, by the employment of devices and/or techniques, has as its objective the **impairment of the operational effectiveness of enemy activity**. (emphasis added)
>
> U.S. DEPARTMENT OF DEFENSE[610]

It was the Department of Defense that contracted with Moderna and Pfizer to create "countermeasures" against Covid-19. These **Covid shots** were going to be released to the public regardless of actual safety or efficacy; they were **ordered before any clinical trials were even started,** as shown in this excerpt Latypova provided from the contract with Moderna:

> **As OWS (Operation Warp Speed) products progress to clinical trials** to evaluate the safety and efficacy of vaccines and therapeutics, **it is critical that, in parallel, the USG (United States Government) supports large scale manufacturing so that vaccine doses or therapeutic treatment courses are immediately available** for nationwide access as soon as a positive efficacy signal is obtained and the **medical countermeasures** are authorized for **widespread use.**
>
> DEPARTMENT OF DEFENSE | Contract with Moderna

Latypova points out that the contract is for "manufacturing up to 500M doses." It's not for a product demonstration or prototype, but for the countermeasure - the Covid-19 vaccine. Latypova states, **"It is clear that the clinical trials are absolutely irrelevant to the approval of the injections by the FDA, as the large-scale manufacturing of these substances does**

not depend on them. It is performed in parallel with these fake exercises (clinical trials) intended to fool the public." (See Section 13 Clinical Trials Did NOT Show Covid Shots Were "Safe and Effective")

> Governments and the medical establishment approach infectious diseases and illnesses on a war footing, preparing for the next battle with the same aggressive approach that has been failing for decades. And just as in war, funds are allocated with a swiftness that benefits the chain of production.
>
> **DAVID MARKS** | Pandemonium[611] | November 29, 2023

Dr. Kory: "VAERS exploded...like seeing the battlefield strewn with soldiers:"

Dr. Pierre Kory was not keyed into these points made by Lerman and Latypova, coming at the pandemic from a medical angle, not from an investigative or national security perspective. But Dr. Kory was stunned during the pandemic by the violation of all the medical norms he had come to trust. His December 2020 testimony before the Department of Homeland Security was ignored, and steps were not taken to provide early, effective treatment for Covid.

In a May 2023 interview[612], Dr. Kory explains how he also came to the conclusion that the pandemic was not handled as a public health emergency, but more as a military response. In referring to the Covid vaccines, Kory states:

> There's never been a product, even a baby seat, a car seat, a car, a can of peaches, **nothing has had this level of adverse events reported, including deaths and the variation between lots. It was a manufacturing catastrophe. If the pharmaceutical industry was working correctly, those things would've been stopped and taken off the market immediately**.
>
> But it was just an unrelenting push through the media, even the government and the Department of Defense. I don't understand what this was. **It definitely came out of a lab, and that's already settled science, but was it leaked or was it an accident?**

> **Let's say it was an accident and our government had been preparing for this massive countermeasure**, and it really was the military that was employing a military countermeasure. **That actually makes some sense, because when you see all of the medical ethics that were violated, that's telling you it's not healthcare**. We still have a sense of ethics, and they forgot it overnight. **It sounded like a military exercise, where you have to sacrifice 100 people to take that hill**.
>
> The VAERS [Vaccine Adverse Event Reporting System] exploded, and with the amount of deaths that were reported within the first weeks, nobody looked at it. It was nothing but attacking VAERS as a source of data, "Nothing to see here." **It was like the military seeing the battlefield strewn with soldiers and just moving ahead, "You have to get your objective**." Do you understand what I'm saying? The corporations benefited and profited greatly, but **the conduct was like nothing we've ever seen before**. (emphasis added)

In August 2020[613], Peter M. Sandman, who has spent more than **40 years as a risk communication consultant**, expressed his puzzlement at the way the pandemic was handled. "Every pandemic plan I have worked on or looked at, starting in 2004," said Sandman, "emphasized the importance of responding quickly with what the CDC has called "targeted and layered" local non-pharmaceutical interventions to slow the spread of the pathogen. **I never saw a plan that contemplated telling everyone to stay home, locking down entire states and countries.** Even now, I am at a loss to explain how the US public health profession suddenly came to the conclusion that a nearly national lockdown was the right response to SARS-CoV-2."

Lockdowns weren't a form of "going medieval," they were brand new in 2020:

In September 2022 Dr. Aaron Kheriaty, psychiatrist and former director of Medical Ethics at the University of California at Irvine School of Medicine, addressed the concept of medical lockdowns (1:18:25[614]). Dr. Kheriaty said the Chinese Communist Party convinced Western nations that they had stamped out Covid in Wuhan through stringent lockdowns, but it was a lie. Dr. Kheriaty believes it was politically irresponsible to follow the example of an authoritarian communist state as a model for managing the pandemic.

> Lockdowns were never part of conventional public health measures...Lockdowns were not a medieval throwback. They were an entirely modern invention, wholly untested on actual human populations and based on a computer model.
>
> **DR. AARON KHERIATY**

Dr. Kheriaty says, "The biomedical security state has proven to be far more effective at controlling populations than anything previously attempted by Western nations...It would be hard to devise a better method than the widespread myth of asymptomatic spread, combined with the practice of quarantining healthy people, to destroy the fabric of society and divide us."

Dictionaries change the definition of lockdown:

Tellingly, "lockdown" is not a healthcare or medical term – **lockdown is a prison term.** Prior to 2020, the printed version of the Oxford English dictionary defines lockdown as "1) the confining of prisoners to their cells, typically after an escape or to regain control during a riot, 2) a state of isolation or restricted access instituted as a security measure."

Now that we've been through the Covid pandemic, the Oxford[615] Learner's online dictionary (on 12-04-23) defines lockdown as "an official order to control the movement of people or vehicles because of a dangerous situation." The first example of using the word in a sentence is this: "The government imposed a nationwide lockdown to contain the spread of the virus." The next example is "a three-day lockdown of American airspace." Then there are two usage-examples to prisons followed by, "The city schools were in lockdown."

The definition of lockdown in the Cambridge[616] online dictionary (on 12-04-23) includes this: "a period of time in which people are not allowed to leave their homes or travel freely, because of a dangerous disease."

Those of us with a few years in the rearview mirror, and a printed

copy of the dictionary on our bookshelves, remember that lockdown of entire populations is not something that is done in Western democracies. Even during World War II in England, when there were frequent enemy bombings on the country, people were not confined to their homes.

The lockdown of entire healthy populations for virus management was never previously considered a viable public health measure. Apparently the National Security Council was unconcerned about, or unaware of, the previous 200 years of Western democracy protocols when setting policies for handling the Covid-19 pandemic.

Lerman states[617], "We may not have a record of what the National Security Council's Covid-19 policy was, or what measures they came up with to implement that policy. **However, everything that was blatantly anti-public health, unscientific or downright insane in our lived Covid experience can be explained, if we assume the Covid response was based not on public health but on a counterterrorism, quarantine-until-vaccine policy."**

Cybersecurity Infrastructure Security Agency determined if you were "essential:"

Another example of the pandemic being handled by security agencies rather than public health agencies was the designation of workers as "essential" or "non-essential." **The list, made overnight and out of thin air, was compiled by Homeland Security's Cybersecurity and Infrastructure Security Agency (CISA),** without any input from Congress, from the Courts, or the public, and certainly without consulting the Constitution. CISA illegally decided who could keep working and providing for themselves and their families, and who could not. As Reverend John F. Naugle, M.A, S.T.B. wrote in April of 2020:[618]

> Under the guise of executive powers reserved for short-term disasters such as hurricanes, leaders across the West have done the previously unthinkable: they have FORBIDDEN entire segments of the population from working. Using a nonsensical distinction between essential and non-essential (as if providing for one's family is ever non-essential) our entire workforce has been divided into three groups: 1.) The upper class with jobs that can be performed in their

pajamas at home, 2.) Laborers lucky enough to still be able to go to work, and 3.) Those intentionally rendered unemployed.

Why would the Intelligence Community handle the Covid-19 pandemic response, instead of it being handled by the Department of Health and Human Services? Because SARS-CoV-2 was considered a bioweapon[619] that had escaped from a lab[620]. Whether that leak was purposeful, or accidental, the powers that be repeatedly did their best to hide the lab leak from the public, going so far as to call it a conspiracy theory. (see here[621], here[622], here[623] and here[624])

The so-called "lab-leak theory" was a taboo subject for over a year. To even suggest it as a possibility would get a person labeled as a conspiracy theorist[625] and booted off of social media, and out of polite society, so to speak. In fact on February 19, 2020 a group of scientists submitted a letter[626] to the *Lancet* praising China and declaring, "We stand together to strongly condemn conspiracy theories suggesting that COVID-19 does not have a natural origin...the coronavirus originated in wildlife."

The *Lancet* letter was coordinated by[627] Dr. Anthony Fauci, and signed by Peter Daszak, among others. Both Fauci and Daszak had a vested interest[628] in attempting to end the lab-leak theory. Fauci had provided funding[629] for Daszak's EcoHealth Alliance to conduct research in conjunction with the Wuhan Lab in China. In short, they didn't want their possible connection with the creation of the SARS-CoV-2 virus to be known. Especially because there was a moratorium[630] on gain-of-function research in the U.S. at the time.

Lerman summarizes[631], "even if the government considered Covid-19 to be a disease caused by a potential bioterror agent, how could the HHS Secretary justify an Emergency Use Authorization that required him to determine that 'there is a public health emergency that has a significant potential to affect national security' when it was known that Covid-19 was deadly almost exclusively in old and infirm populations?"

31. EVERY WAR INVOLVES PROPAGANDA

We are at war. Never has France had to take such decisions, albeit temporary, in time of peace. All our energy should be on one aim: to slow the progress of the virus.

PRES. EMMANUEL MACRON

France, March 15, 2020[632]

Adding these extra sources of uncertainty, reasonable estimates for the case fatality ratio in the general U.S. population vary from 0.05% to 1%. That huge range markedly affects how severe the pandemic is and what should be done. A population-wide case fatality rate of 0.05% is lower than seasonal influenza. If that is the true rate, locking down the world with potentially tremendous social and financial consequences may be totally irrational.

JOHN P.A. IOANNIDIS

Professor of Medicine, Epidemiology, Population Health & Biomedical Data Science March 17, 2020[633]

To this day, nobody has seen anything like what they were able to do during World War II. Now it's our time. We must sacrifice together because we are all in this together and we'll come through together... But we're going to defeat the invisible enemy.

PRES. DONALD TRUMP

March 19, 2020[634]

> *We are at war with a virus that threatens to tear us apart....Fight like your lives depend on it – because they do. The best and only way to protect life, livelihoods and economies is to stop the virus...Fight. Unite. Ignite. And let our singular resolve be: never again.*

W.H.O. DIRECTOR-GENERAL TEDROS
March 26, 2020[635]

Note in the above quotes the lone voice of the medical and biomedical data scientist. Prof. John Ioannidis[636] is known worldwide in scientific communities for development of meta-analysis protocols, and the quest for integrity[637] in medical research. In the middle of the pandemic war rhetoric, there were calmer voices asking questions, but they were silenced. A furor had already taken hold of world leaders and populations.

Propaganda techniques that have mainly been used on foreign enemies in the past were turned on citizens by their own governments during the Covid-19 pandemic, in order to control the narrative, and consequently to control the people. We've already discussed how propaganda was used to suppress effective treatments for Covid, paving the way for the vaccines (see Section 7 The War on Ivermectin and HCQ) But there was so much more.

> If you believe in democracy you must be suspicious of the use of psychology to manipulate you against your will. The use of fear is a sinister form of control.

LAURA DODSWORTH | *A State of Fear*[638]

In his book *A Plague Upon Our House*, Dr. Scott Atlas talks about serving from August to December 2020 as a Special Advisor to the President on the **White House Coronavirus Task Force. Dr. Atlas was struck by the lack of scientific and medical data discussed during Task Force meetings.** In one meeting, Drs. Fauci and Birx expressed that people should be warned even more strongly about the dangers of the virus spreading, about wearing masks and distancing, because people weren't "taking the virus seriously enough." Atlas writes:

> I was honestly surprised. I thought people were already panic-stricken. Normal life had virtually ceased to exist, even eliminating serious medical care or last visits with dying family. Meanwhile the media were on-message 24/7, instructing the public about masks and social distancing; there were signs and announcements demanding masks and diagrams about distancing everywhere; healthy young people were outside riding bicycles or driving their cars alone, wearing masks. Indeed, surveys showed that most adults perceived grossly exaggerated risks, particularly but not only younger people; and yes, a high percentage were obeying the edicts, distancing and wearing masks, according to virtually every published survey.

Atlas continues, **I challenged [Fauci] to clarify his point, because I couldn't believe my ears. "So you think people aren't frightened enough?" He said, "Yes, they need to be more afraid."** Dr. Atlas, the only member of the Task Force who was not a government bureaucrat, and the only one with experience in public health, disagreed. He writes, **"Instilling fear in the public is absolutely counter to what a leader in public health should do. To me, it is frankly immoral."** (See *A Plague Upon Our House*, p. 156)

Propaganda was key. Fear, lies, and "spin" were the order of the day. But many trusting, or non-discerning souls, didn't recognize the manipulations, and simply absorbed the messages and the fear. **Norman Baker, former Liberal Democrat MP in the U.K. emphasizes that "even in a democracy, we should never assume the government of the day is right or well-intentioned**...People have a very low opinion of politicians... But, for some reason, they believe the government unwaveringly about an issue of their security, such as an epidemic."[639]

> The systematic study of mass psychology revealed the potentialities of invisible government of society by manipulation...If we understand the mechanism and motives of the group mind, is it not possible to control and regiment the masses according to our will without their knowing it?
>
> **EDWARD BERNAYS** | From his book, *Propaganda*

The Propaganda of Policy Decisions based on Modeling:

> *In the (unlikely) absence of any control measures or spontaneous changes in individual behavior, we would expect a peak in mortality (daily deaths) to occur after approximately 3 months. In such scenarios, given an estimated R0 of 2.4, we predict 81% of the G.B. and U.S. populations would be infected over the course of the epidemic...* ***In total, in an unmitigated epidemic, we would predict approximately 510,000 deaths in G.B. and 2.2 million in the U.S.****, not accounting for the potential negative effects of health systems being overwhelmed on mortality.* (emphasis added)
>
> **NEIL FERGUSON** | Imperial College of London | March 16, 2020,[640] p. 6

Neal Ferguson's Imperial College model that predicted 510,000 deaths in Great Britain and 2.2 million in the U.S., was instrumental in implementing the devastating lockdowns in the U.K. and U.S. Once the U.S. locked down, the rest of the world largely followed.

In evaluating the Imperial College model in April 2020, Alan Reynolds of Cato Institute noted[641] that **in previous years, Ferguson and the Imperial College had been wrong, by orders of magnitude, in modeling the deaths that would arise from mad cow disease, bird flu, and swine flu.** Yet somehow Ferguson's modeling was taken as fact. Reynolds states:

> The trouble with being too easily led by models is we can too easily be misled by models. Epidemic models may seem entirely different from economic models or climate models, but they all make terrible forecasts if filled with wrong assumptions and parameters.

***The Telegraph*, in a retrospective on Ferguson's Imperial College Modeling of the pandemic, labeled it "the most devastating software mistake of all time."** Laura Dodsworth, author of *A State of Fear*, points out that the Imperial College model used **outdated code and contained multiple flaws**, but its "doom-laden" predictions grabbed headlines around the world.[642]

In August 2021, Dr. Sunetra Gupta of Oxford University explained[643] the drawbacks of modeling, in the following statement:

> As I've said over and over again since March of last year...models are absolutely crucial and invaluable as tools for generating testable hypotheses. They should never be used to predict, and while they should be used as a basis for formulating policy, they should not be treated as truth in any sense....**Only data can tell us where the truth lies**...There's a confusion between hypothesis and prediction. (emphasis added)

Modeling is not science. Modeling is a tool. When it's used to prove a preconceived point, modeling simply becomes a manipulation – enter in the data you need to get the result you want.

In 2020, researchers at Uppsala University in **Sweden adapted Ferguson's Imperial College model to their own country, deriving a projection of 96,000 coronavirus deaths** if Sweden did not lock down by mid-April 2020. Sweden never locked down[644], unlike most countries in Europe. **By April of 2021, there had been just over 13,000 fatalities attributed to Covid-19 in Sweden**, a smaller per capita loss than many European lockdown states. "The implications for Ferguson's work remains clear," concluded senior researcher Phillip W. Magness[645], PhD, **"the primary model used to justify lockdowns failed its first real-world test."**

Ferguson was defensive about negative conclusions involving the Imperial College of London modeling, claiming that Sweden developed their own model and did not use his. However, the Swedish researchers clarified[646] that they based their model "on the framework published by Ferguson and coworkers."

The Imperial College's original lockdown-justifying model was fatally flawed, as was its modeling study published[647] in June 2022 that claimed the vaccines prevented 20 million deaths[648] in the first year of the rollout. Manfred Horst, MD, PhD, MBA, has repeatedly pointed out that **the average age of death from Covid, is higher than the average age of death overall,** which is currently 76.4 years[649] in the U.S., but was closer to 78.8 at the start[650] of the pandemic. Horst reasons[651] that the **majority of those who were listed as Covid-19 deaths "would have left this world at the same time,** with Corona or from/with another virus or another disease." In response to the claims of "20 million lives saved" by Covid vaccines, Horst states:[652]

> Preventing the deaths of a cohort of people (the "Corona deaths"), who die at an average age equal to or higher than the average age of death in the general population, may be conceptually impossible. Be that as it may, **any serious epidemiological model would have to calculate and discuss the number of life years gained by "saving" all those lives.**
>
> Any medical intervention has side effects. The model from Imperial College does not take this into account at all. **There is also ever growing evidence that the Covid vaccines cause a number of very serious and severe side effects. We cannot exclude the possibility that their net effect on mortality – and/or above all on life years gained – is in fact negative.** (emphasis added)

Yet this modeling claim of "20 million lives saved through vaccination" is still unquestioningly and erroneously repeated, often, from mainstream media outlets to the halls of the US Congress. **Modeling used in this manner is propaganda on behalf of the official narrative.**

Following is a screenshot of top results from a search on October 26, 2023 of "Covid vaccines saved 20 million lives." Note the similarity of each headline and the fact that the search does not bring up links to any differing viewpoints or analysis of the Imperial College of London model, although it has been a year since its release. **Each outlet recites the "20-million lives saved" mantra as though modeling is data, and the Imperial College's modeling is fact.** There have been multiple articles, by credentialed, rational people that bring into question the Imperial College's modeling claims. None of those articles appear in a search. You have to know specifically where to look. Try entering "Imperial College" in a search at Brownstone Institute[653], and see what comes up.

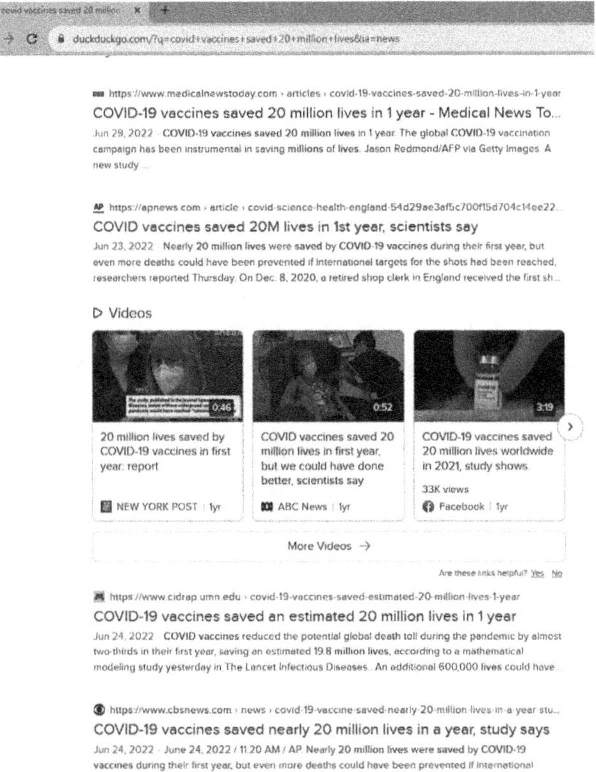

The Propaganda of Lockdowns and 15 Days to Slow the Spread:

> *No sooner had we convinced the Trump administration to implement our version of a two-week shutdown than I was trying to figure out how to extend it. Fifteen Days to Slow the Spread was a start, but I knew it would be just that. I didn't have the numbers in front of me yet to make the case for extending it longer, but I had two weeks to get them. However hard it had been to get the fifteen-day shutdown approved, getting another one would be more difficult by many orders of magnitude.*
>
> **DEBORAH BIRX**[654] | White House Coronavirus Response Coordinator

Dr. Deborah Birx, brought to the White House with dubious credentials[655] for managing a pandemic, never intended that "15 Days to Slow the Spread" would really only be for just 15 days, as evidenced by the above quote from her own memoir about the pandemic, *Silent Invasion*.

As mentioned in Section 1, Dr. Fauci spoke from the same playbook three days into "flatten the curve," on March 19, 2020[656]. Fauci said he thought the lockdown would go longer than 15 days because, "It doesn't just turn around over a week and a half or two." Not obvious to most of us at the time, but certainly obvious to the medical and science professionals who tried to speak up, a respiratory virus doesn't disappear because you stay in your house and give up your First Amendment rights.

The biggest lie of "two weeks to slow the spread," was that it would make a difference, when in fact it was pointless. Covid-19 had already been circulating among the world's population since 2019, as documented in Bill Rice Jr.'s[657] excellent work, and confirmed by others (see here[658] and here[659]). It's highly likely that it was Covid-19 that caused the flu-like illness that so many international participants contracted during the October 2019 Military World Games[660] in Wuhan, China. And it was Covid-19 that infected the crews of several U.S. Navy[661] ships during those same months.

May 16, 2020 Daily Mail, U.K. headline:
"Why DID so many athletes fall sick in Wuhan in October? More competitors reveal they were ill at the (2019) World Military Games months before China admitted coronavirus could be passed between humans."

On that fateful day[662], March 16, 2020, when the Trump Administration closed the country for 15 days, the horse, as they say, had already left the barn. Covid-19 had already spread to millions around the world, but quietly, because for most people it was almost not noticeable, or was just a bad flu from which they had recovered. With the stroke of a pen, and an announcement at a White House press conference, the biggest attack ever made on the U.S. Constitution and the human rights it enshrines was undertaken by our own government.

There was no scientific basis for shuttering businesses, schools, and churches, and no real plan for what to do after the 15 days were up. So the decision-makers pivoted from the claim that we were slowing the spread of the pandemic, to keep from overrunning our hospitals, to "We are going to crush Covid." Inexplicably the narrative moved from protecting our health system to the folly of attempting to eradicate a coronavirus.

Peter M. Sandman, who had worked on pandemic plans for over a decade before Covid-19, states:

> **So how could public health professionals make the near-universal lockdowns not seem like they were a horrible mistake, devastating millions of lives to no purpose? Offer a third rationale: The lockdowns prevented infections and thereby saved lives**...They sounded increasingly absolutist about the importance of minimizing the number of cases at all costs, and thus increasingly skeptical about the wisdom of moving from lockdown to some kind of New Normal. In doing so, they pointed the public away from the goal of living with the virus, toward the goal of beating the virus. That change from pandemic management narrative to a pandemic suppression narrative continues to have profound policy implications.(emphasis added)

An airborne respiratory virus cannot be contained and will run its course. Perhaps you heard during the pandemic: **"Virus is gonna virus." That's why we still have the common cold and influenza every year.** (see Section 16 There has Never Been an Effective Vaccine for a Respiratory Virus)

> As stated by Roger W. Koops, PhD, who has worked in the Pharmaceutical and Biotechnology Industry for over 25 years, "Humans have zero control over respiratory viruses, so 'allowing' them to propagate is the falsehood. If you want to talk about human hubris, there it is."

Contact tracing and testing for a respiratory virus that had already spread to millions of people throughout the world was "illogical and impossible to do," as explained by Dr. Scott Atlas on May 7, 2020[663]. By the time of such widespread infection, said Atlas, testing was basically only consequential for three groups – people entering nursing homes, people working in a hospital and taking care of patients/pandemic responders, and people with symptoms.

But somehow it all happened. The world closed down, and the contact tracing and PCR testing ramped up.

It was **Birx who traveled the country[664], meeting with each governor, pushing for statewide masking and lockdowns.** And Birx, who later found Dr. Scott Atlas, the only one on the Coronavirus Task Force with public health experience, to be full of "dangerous assertions," such as: schools should be open everywhere, children were not at risk from Covid and were not big spreaders of the virus, long Covid was being overplayed, masks were not preventing spread, the PCR testing was ineffective, and that asymptomatic people were not big spreaders of Covid.

Birx kept her White House position until December 23, 2020, when she resigned[665] over the **hypocrisy of having traveled to a multi-family Thanksgiving dinner, violating her own madly-pushed social distancing guidelines.** For his part, Atlas came under such heavy crossfire for his evidence-based recommendations on the pandemic response that he received death threats and had to hire security to protect his family and home. He also left the Task Force in December 2020.

Attorney Michael Senger credits Atlas with preventing Birx from implementing more of her totalitarian agenda. But she did enough. **Senger, who stated from the beginning that the U.S. was dangerously mimicking the China Communist Party response to the virus**, states:[666]

> I'm not saying Deborah Birx is a CCP agent. I'm just saying that if she was an agent for Xi Jinping's stated goal of gradually stripping the world of "independent judiciaries," "human rights," "western freedom," "civil society," and "freedom of the press," then every word of her book would read like that of *Silent Invasion*. If she did do it, this is how it would have happened.

In Section <u>30</u> we explored the idea that Covid-19 policy was set by the National Security Council – not by Health and Human Services. In his retrospective on the pandemic, Senger writes[667], "All over the world, across professions, **citizens have been depending on the idea that if there was anything corrupt or rotten about the response to Covid, the Western**

intelligence community would step in. Of course, that won't happen, because as the...record makes clear, **they're the ones who planned it.**

> But what if they were right and the SARS-CoV-2 virus was a bioweapon? And what if the Covid vaccines had actually prevented disease and spread of the virus? Would the lockdowns and mandates have been justified?

Bill Gates, the ever hopeful pandemic prognosticator, **warned[668] in January 2023 that a new pandemic is on the horizon** – possibly a man-made pandemic, and far more brutal than Covid-19. In October 2022 Gates attended **yet another wargame**, with 13 other participants made up of current and former health officials from around the world. This wargame, Catastrophic Contagion[669], **simulated a future pandemic that targets children and young people.**

Also in January 2023, author, neuroscientist, and philosopher **Sam Harris** commented[670] on how Covid-19 would have been viewed differently if it impacted children instead of the elderly. Harris stated:

> Leave Covid exactly as it is, but just make it preferentially dangerous to children rather than to old people...If kids were dying by the hundreds of thousand...we would have had a very different experience...**There would've been no f***ing patience for vaccine skepticism, right. And everyone would have recognized, this is not my body my choice. This is, 'You're not gonna kill my kids with your ignorance.'**

In looking at Gates and Harris we see examples of two different types of people who have embraced the militant and obsessive pandemic-to-vaccine view of the world. Gates is an example of **the moneyed elites** who want to reshape the world through virus patrol and control. Harris is an example of the **brilliant academics** who have been so thoroughly absorbed by the official narrative that he **cannot consider actual facts or nuances of thought.** Perhaps they sincerely believe what they're saying. Perhaps they mean well. But as Supreme Court Justice Louis Brandeis said[671], "The greatest dangers to liberty lurk in the insidious encroachment by men of zeal, well-meaning but without understanding."

> Or as C.S. Lewis put it, "Of all tyrannies, **a tyranny sincerely exercised for the good of its victims may be the most oppressive**... those who torment us for our own good will torment us without end for they do so with the approval of their own conscience...**To be 'cured' against one's will** and cured of states which we may not regard as disease **is to be put on a level** of those who have not yet reached the age of reason or those who never will; to be classed **with infants, imbeciles, and domestic animals**." (*From God in the Dock: Essays on Theology*, emphasis added)

Surely there will be another pandemic, but not because the world has become an unusually pathogenic place. It's because people like Gates will make it happen. After all, **Gates' entire existence is wrapped up in vaccines, pandemic simulations, and his push for constant detection and tracing of viruses throughout the world.** It's as if we never had viruses before Gates noticed them, never had immune systems that protected us before Gates got into vaccine development, and nothing is more important than pandemic policing.

> Respectfully, Mr. Gates, not everyone sees the world as you do. And just because you have money doesn't mean your worldview is correct.

There is really **something wrong** when we're **being told that the next pandemic will be much worse and will target children.** UK Health Security Agency Professor Dame Jenny Harries, has been leading an effort to develop a vaccine for the next pandemic, which she believes **will be caused by climate change**[672]. These obsessed bearers of bad fortune, and hopeful profits, certainly seem to have a lot of advanced knowledge about what's going to happen on the microbial level.

So what if the next pandemic targets children and young people? What if our government leaders can assure us in the face of a future pandemic, that in just 100 days they'll have the vaccine we need, and it will be safe this time because they'll correct the errors found in the first mRNA vaccines?

Would that justify another "limited" lockdown while we wait for the cure? What if governments say there's no need to lock down as long as we have vaccine passports that immediately certify that individual citizens are "safe" to be around (or not)? Will we comply, for our own "safety" and for the "good of society?"

The only answer to these questions, that comports with democratic principles of Western enlightenment, is a resounding "No!" The Nuremberg Code[673], developed as a result of the trial of Nazi doctors[674] who conducted cruel experiments on Jewish prisoners in concentration camps, states **that individuals are not to be coerced or forced to receive medical procedures or drugs. Whether or not the medical treatment would be effective or beneficial to the patient and/or to others is irrelevant.** "The voluntary consent of the human subject is absolutely essential."

An analysis of the Nuremberg Code, published in *JAMA* in 2017[675], states, Although "there has in the past been considerable debate among schools about the code's authorship, scope, and legal standing in both civilian and military science," **the Nuremberg Code is recognized as a milestone in the history of biomedical research ethics.** The Universal Declaration of Human Rights[676], proclaimed by the United Nations General Assembly in Paris on December 10, 1948, is considered a more binding international document, although it does not directly address medical freedom.

> Moving forward, are we going to model our societies and governments on the Chinese Communist Party model, or will we continue to pursue democracy? There is no real in-between.

The Propaganda of Face Masks:

> *I am, I believe, honest and humble enough to say...that perhaps what we didn't do well enough was, back then, was say, 'You know, we really don't have any idea whether or not something does or does not work, and therefore, maybe people should make up their own mind about wearing a mask.*
>
> **ANTHONY FAUCI** | October 26, 2022[677]

Dr. Anthony Fauci, of no-mask[678], yes-mask[679], actually wear two masks[680] fame, was in a friendly interview with Peter Staley at Harvard Institute[681] of Politics when he made that above statement. "Honesty" and "humility" are not something we've seen a lot when it comes to the pandemic response, nor has there been much respect for human rights and bodily autonomy.

In private, leaders acknowledged that the purpose of masks was not a medical intervention, but to keep people constantly reminded of the pandemic. For example, the Utah State Epidemiologist called for a mask mandate[682] for all citizens in June 2020. In an internal memo she said it was because, "This will **send the message** to Utahns that this outbreak continues to be a serious problem." (emphasis added)

Matt Hancock[683], who was the U.K. Health Secretary during the pandemic said the government should "frighten the pants" off its citizens in order to gain compliance with Covid measures. (Hancock later resigned due to being caught on camera engaged in an affair with his aide, violating[684] the masking and social distancing guidelines he was imposing on British citizens. Hancock was also part of then Prime Minister Boris Johnson's approved weekly wine parties for staff every Friday, held while U.K. citizens were cited[685] for violations of masking and lockdown orders.)

CDC Director Dr. Robert Redfield went so far as to testify[686] in a September 16, 2020 Senate Appropriations hearing, holding up a standard blue surgical mask, "Face masks are the most important powerful public health tool that we have." Redfield claimed that we could bring the pandemic "under control," if only we'd wear masks "responsibly" for "6, 8, 10, 12 weeks." Which is it, Dr. Redfield? That sounds a lot like two weeks to slow the spread, which we all know turned into two years plus.

Twirling the mask in his hand, Redfield stated, "We have clear scientific evidence that they work...I might even go so far as to say that this face mask is more guaranteed to protect me against Covid than when I take a Covid vaccine..." Redfield then appealed to all Americans, "particularly the 18-25 year-olds where we're seeing the outbreak in America continue like this (raises hand in a straight upward move) because we haven't got the acceptance, the personal responsibility that we need..."

It was all theater. Dr. Redfield was well aware that Covid-19 was spread through aerosols[687] that would not be blocked by a surgical mask. Redfield knew that surgical masks were introduced in surgery to keep spittle from getting in patient's open wounds – not to protect the surgeon. And he knew that three recognized clinical trials of surgical masks showed no difference[688] in wound infection rates whether the staff was wearing masks or not. Redfield knew that the CDC study[689] of May 2020, analyzing 10 randomized controlled trials from 1946 to July 2018 concluded, "Our systematic review found no significant effect of face masks on transmission of laboratory-confirmed influenza." Redfield knew that face masks were ineffective against respiratory virus spread, but he lied to prop up the narrative.

Face mask messaging was long on guilt[690] trips and short on data and facts, and hypocritically enforced. **The billions of face masks that have cluttered our landfills and littered our public spaces worldwide are testimony to science ignored, and illogical policies inappropriately enforced.**

> Perhaps nothing during the pandemic was granted stronger power of virtue-signaling than the wearing of a face mask. Immediately, almost subconsciously people judged others as good or bad, safe or unsafe, based on whether or not their face was covered.

The SARS-CoV-2 virus, which causes **Covid-19 is measured in nanometers**. A nanometer is 1,000 times smaller than a micrometer. A micrometer is 1,000 times smaller than a millimeter, which is 10 times smaller than a centimeter. Dr. Harvey Risch[691] has noted that wearing a face mask to prevent passage of the aerosol-borne SARS-CoV-2 virus is as effective as installing a chain-link fence to keep mosquitoes out of your yard.

As Dr. Anthony Fauci explained on February 5, 2020[692], "The typical mask you buy in the drug store is not really effective in keeping out virus, which is small enough to pass through material. It might, however, provide some slight benefit in keep(ing) out gross (large) droplets if someone coughs or sneezes on you."

However, **large droplets are not what spread Covid-19. SARS-CoV-2 is aerosol borne, meaning it's in the air.** The primary means of transmission of COVID-19 is not through someone coughing or sneezing, but through close contact with an infected person in an indoor space (especially with poor air circulation), for an extended period of time. This explains why the majority of cases were spread within households, which is another reason why lockdowns were medically counterintuitive.

> Then of course the masks come into play...If you believe spread is through droplets that are being spat out of somebody's mouth as they talk, then a mask is going to help...But if you realize that's not how it (Covid-19) spreads, then you've got two problems. One problem is the gaps all around the cloth mask where the air is going in and out. But the other problem is anything that you are spitting onto the mask, you're then breathing over it, so you're going to aerosolize what's on the mask. And if you look at the data in the real world of what happened when masking mandates were brought in, every single time there was a difference between one region and another, or one country and another, it was always the masked one that was doing worse.
>
> **DR. CLAIRE CRAIG** | November 26, 2023[693]

The SARS-CoV-2 virus passes right through the material of cloth masks, surgical masks, and N95 respirators, and out of every little opening and crack around the edges. Even if the material of the mask were impermeable, the air would escape around all the edges and cracks, and enter through those same openings.

OSHA[694] (Occupational and Safety Health Administration) requires full-on respirators for work that could damage the lungs, such as sanding. Military-grade respirators[695] are used when training for how to survive a toxic-gas (aerosol borne) attack. During the Oregon wildfires in the fall of 2020, the State issued a pamphlet[696] informing residents that "fine particles 2.5 micrometers or smaller can be inhaled into the deepest part of the lungs...dust masks and surgical masks...offer little protection." The pamphlet continued, "N95 respirator can filter 95% of smoke particles. However, N95 respirators do not filter toxic gases and vapors." The SARS-CoV-2 virus is more than 1,000 times smaller than a micrometer. It is carried in air vapor.

Yet what directive did CDC Director[697] Mandy Cohen provide when asked about protection during the 2023 holiday season? "Masks do work, so [wear] a mask if you're around a lot of other people." Also, wash your hands, stay home if you're sick, and get that all-important Covid vaccine that's formulated for a virus strain that is no longer circulating.

Then there's always this from Canada:

3:19 PM · Oct 19, 2023 · **29K** Views

All About the Air - Oct. 2023 (1:52 min)[698]

There are entire books[699] and hundreds of articles[700] addressing the illogic, the ineffectiveness[701], and the historic symbolism[702] of oppression represented in mandated face masking. Others address the psychological[703] and physical[704] harms of mask wearing. Still others, the damage to young children[705] who could not see faces during the pandemic and have reduced cognitive abilities and need speech therapy, as a result.

> Facts do not matter to totalitarians and cult members. What matters is loyalty to the cult or the party.
>
> **CJ HOPKINS** | Playwright | Political satirist[706]

Heading into the pandemic, the most respected[707] and up-to-date[708] studies of the day had already established that face masks were not effective at preventing infection or spread of aerosol borne diseases, but they were ignored. During April and May 2020, a large study was conducted in **Denmark[709]** to assess if surgical mask use by the general public, when going about daily life, reduced wearers' risk for SARS-CoV-2 infection. **Roughly 6,000 people participated. Half wore face masks; half did not. The difference in Covid-19 infection rates between the two groups was -0.3 percentage points,** which is obviously almost nothing.

The Journal of the ***American Medical Association*** clarified[710] on **March 4, 2020** the differences between surgical masks and respirators. The absence of any reference to cloth face masks is noticeable, as is the statement that "there is no evidence to suggest that face masks worn by healthy individuals are effective in preventing people from becoming ill:"

> **There are 2 main types of masks used to prevent respiratory infection: surgical masks, sometimes referred to as face masks** (the typical blue masks), and respirators (N95). These masks differ by the type and size of infectious particles they are able to filter. **Face masks are used more commonly for respiratory viruses that spread via droplets**, which travel short distances and are transmitted by cough or sneeze. **Face masks often fit loosely, and prevent the wearer from spreading large sprays and droplets, as well as preventing hand-to-face contact**.
>
> **N95 respirators block 95% of airborne particles. They are tight fitting and prevent inhalation of smaller infectious particles that can spread through the air over long distances after an infected person coughs or sneezes**. Diseases that require use of an N95 respirator include tuberculosis, chickenpox, and measles. **N95 respirators cannot be used by individuals with facial hair or by children because it is difficult to achieve a proper fit**. In those cases, a special respirator called a powered air-purifying respirator may be used instead.
>
> **Face masks should be used only by individuals who have symptoms of respiratory infection such as coughing, sneezing, or, in some cases, fever**. Face masks should also be worn by health care workers, by individuals who are taking care of or are in close contact with people who have respiratory infections, or otherwise as directed by a doctor. **Face masks should not be worn by healthy individuals to protect themselves from acquiring respiratory infection because there is no evidence to suggest that face masks worn by healthy individuals are effective in preventing people from becoming ill**. Face masks should be reserved for those who

need them because masks can be in short supply during periods of widespread respiratory infection. **Because N95 respirators require special fit testing, they are not recommended for use by the general public.** (emphasis added)

It was known in early 2020 that face masks do not prevent the spread of aerosol borne respiratory diseases, as confirmed at the time by Dr. Fauci, the American Medical Association, and numerous scientists and doctors. The study conducted in Denmark early in the pandemic confirmed this fact. Masks don't prevent the spread of air borne diseases, but during the pandemic masks were very effective at separating people, and determining who was complying with government edicts, and who was not. They were also great training for the idea that there should be a visible sign to "prove" you are safe to be around in public, paving the way for mandated vaccines and vaccine passports.

Sadly, the inappropriate and ineffective face-masking impacted our young people the most. While European countries did not require face masks for children, U.S. children were forced into masks for hours a day at school, and during their extracurricular activities. Children under the age of 5, who were the least at risk for Covid-19 disease, were the last age group to be freed of face masks in the U.S. (see here [711], here[712], and here[713]) This video of a little boy in daycare (location unidentified) being forced into a useless cloth mask repeatedly, should make any normal person angry:

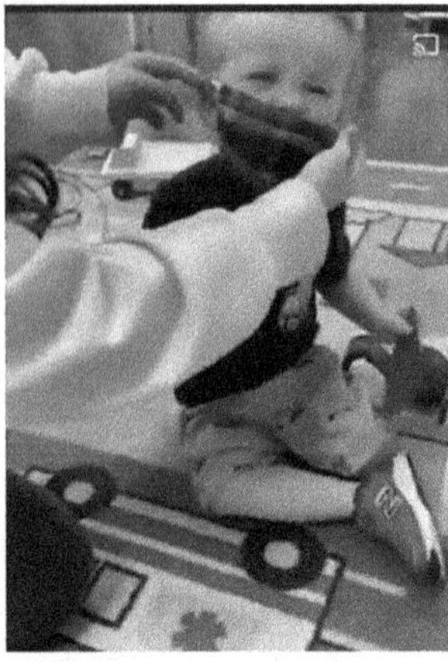

Toddler cries as he is forced into a mask (1:19 min)[714]

Despite the fact that we knew aerosolized respiratory viruses are not stopped by cloth face coverings or surgical masks, and we knew that the young were not vulnerable to Covid-19, nor big spreaders of the virus, children and young people around the world bore the brunt of our Covid policies. As explained[715] by Dr. David Bell and former Headteacher Hugh McCarthy:

> [S]ociety "closed schools and forced children to cover their faces, reducing their educational potential and impairing their development. Knowing that school closures would disproportionately harm low-income children with poorer computer access and home study environments, we ensured that the children of the wealthy would widen their advantage for the next generation. In low-income countries, these school closures worked as expected, increasing child labour and condemning up to 10 million additional girls to child to child marriage and nightly rape."

> [W]hat could be more dangerous to a democratic nation's health than forcing upon its children an extended tutorial on the need for absolute and unquestioning compliance to authority, even when the commands make no logical or ethical sense?

NAOMI WOLF | *The Bodies of Others*, p. 200

Dr. Mark McDonald, a child psychologist in Los Angeles stated in an April 2022[716] interview:

> **"Well, it's just a mask." How many times have we heard that in the last two years? "It's just a mask." Well, actually, it isn't. It's not just a burqa either**. We would not allow American women to be covered head-to-toe in black cloth like they do in the Islamic world, largely speaking, because we find it dehumanizing. **Whether we're told that it's done for good reason or not is irrelevant**. We would say no. Why? Because we're not scared to say no. But we have been incredibly, incredibly frightened and intimidated in the last couple years from just thinking for ourselves and using our rational faculties. We've turned them off. We don't use them anymore. **Now, we make our decisions based upon fear. Therefore, we are compliant**. (emphasis added)

The purpose of face mask mandates in the U.K. was to instill fear and compliance in the population. **David Halpern**, head of the Behavioural Insights Team in the U.K., also known as the "Nudge Unit," (because its purpose is to give citizens subliminal nudges toward desired behavior), **referred to the useful "signal" masks give**. Notably, he didn't say anything about them preventing the spread of disease.[717]

The Cochrane[718] review is considered the gold standard of systematic review of research in health care and health policy. **By eliminating studies that are observational, cross-sectional, or that use modeling, and relying on randomized controlled trials, the Cochrane review is able to minimize bias from confounding elements.**

On January 30, 2023, the Cochrane's latest review[719] of quality studies of face masks for prevention of disease-spread confirmed what their 2020 review had found:

> There is uncertainty about the effects of face masks...**The pooled results of RCTs (randomized controlled trials) did not show a clear reduction in respiratory viral infection with the use of medical/surgical masks.** There were no clear differences between the use of medical/surgical masks compared with N95/P2 respirators

> in healthcare workers when used in routine care to reduce respiratory viral infection.

As explained by Tom Jefferson[720], one of the lead authors of the study, **to continue to rely upon, promote, and even mandate face masks to prevent the spread of disease,** in light of no conclusive evidence that they are effective, **is "a complete subversion of the 'precautionary principle' which states that you should do nothing unless you have reasonable evidence that benefits outweigh the harms."**

In today's climate where The Science™ is touted, but actual evidence-based data that doesn't fit the narrative is not, the Cochrane review on masks immediately came under fire[721]. Dr. Fauci, when asked about the Cochrane review, said[722] there were other studies that proved masks DID make a difference.

> Jefferson counters that Fauci didn't name the studies. **Since Cochrane reviews all available studies and chooses the most solid and properly conducted, Fauci must be relying on "trash studies."** "Many of them are observational, some are cross-sectional, and some actually use modeling," said Jefferson. "That is not solid evidence. Once we excluded such low-quality studies from the review, we concluded there was no evidence that masks reduced transmission.

It has been scientifically established, through evidence-based studies, that face masks do not prevent the spread of respiratory illnesses. One can only conclude that the face masks were used for purposes of virtue signaling, to give a false sense of protection, and to control the population, rather than for health benefits.

Despite mountains of evidence, including the gold standard Cochrane[723] review, that find no evidence for the effectiveness of face masks in preventing spread of respiratory disease, the propaganda about face masks continues. For example, this is what the Mayo Clinic[724] website stated on December 14, 2023, in response to the question "Can face masks help slow the spread of the virus that causes coronavirus disease 2019 (Covid-19)?":

> Yes. When used with measures such as getting vaccinated, hand-washing and physical distancing, wearing a face mask slows how quickly the virus that causes COVID-19 spreads.

The Mayo Clinic then goes on to show different kinds of masks, ranging from cloth to K95 and details how to wear them. Moving into 2024, the Mayo Clinic is still talking about face masks, physical distancing, and vaccines slowing the spread of Covid-19, despite all three claims having been scientifically and observationally disproven.

> Masks do not prevent the spread of aerosolized respiratory viruses, but even if they did, forcing people to wear masks goes against principles of agency and autonomy. Mask mandates have no place in Western democracies.

The Propaganda of Asymptomatic Spread:

> *We have a number of reports from countries who are doing very detailed contact tracing. They're following asymptomatic cases, they're following contacts and they're not finding secondary transmission onward. It's very rare and much of that is not published in the literature...We are constantly looking at this data and we're trying to get more information from countries to truly answer this question. It still appears to be rare that an asymptomatic individual actually transmits onward.*
>
> **DR. MARIA VAN KERKHOVE** | World Health Organization | June 8, 2020

Dr. Kerkhove made that statement in a WHO virtual press conference. The very next day she was told to walk it back[725]. But nonetheless, in an unguarded moment she had said it. **All the decision-makers and the media seemed to run with the idea that asymptomatic spread was a primary cause of Covid-19 disease.** Constantly reinforcing this idea contributed to making people fearful of each other, and viewing each other as germ-spreaders, rather than as fellow human beings on the planet.

Early data from China[726] and Germany[727], and the data being reported to the **World Health Organization** by June of 2020, **all concluded that asymptomatic spread was "very rare." Yet this false rumor persisted throughout the pandemic** and is still widespread today.

> This switch from the idea that it's fine to be out in public if you don't look or feel ill, to the idea that you're a dangerous infective agent just for existing, has had a damaging influence on the social fabric of our families, friendships, and communities. It also led to unnecessary and invasive quarantining of healthy individuals, simply because they had been within six feet of someone who tested positive for Covid-19.

The Propaganda of PCR Tests, Tests, Tests, Tests....

> *[The PCR test] tells you something about nature and about what's there. But it allows you to take a very miniscule amount of anything and make it measurable...It doesn't tell you that you're sick and it doesn't tell you that the thing you ended up with really was going to hurt you or anything like that.*
>
> **KARY MULLIS** | Inventor of the PCR Test[728]

There used to be jokes about the hypochondriac who was constantly taking his temperature so he would know if he was sick. That was then, this is now. **Although the inventor of the PCR test explained that it could not diagnose active illness, but only the presence of the thing being tested, the PCR test became ubiquitous during the pandemic.**

Yuri Biondi, Senior Tenured Research Fellow at the French National Centre for Scientific Research explains[729] that constant PCR testing was a flawed approach:

> While contact tracing and isolation may be important for some infectious diseases, it is futile and counterproductive for common infections such as influenza and Covid-19. A case is only a case if a

person is sick. Mass testing asymptomatic and non-vulnerable individuals is harmful to public health, useless and expensive.

Not only was the constant testing **"harmful to public health, useless, and expensive,"** but the accuracy of PCR testing is erratic; results will vary widely depending on how and when the sample is collected, and how it is tested in the lab. PCR stands for polymerase chain reaction, and is a process[730] in which the sample is obtained from a nose swab, throat swab, or from saliva. The sample is then run through a thermocycler machine. The cycle threshold (Ct) is the number of cycles after which the virus can be detected. If the cycle threshold is low and it detects the virus, that indicates the person is carrying a significant viral load. The more cycles a specimen is run through in the thermocycler machine, the more likely it is that non-infectious old fragments of virus will be detected.

Prof. Carl Heneghan[731] of the University of Oxford conducted research on the optimal cycle threshold. The study[732] found that a positive result for presence of SARS-CoV-2 in the sample did not necessarily mean infectious virus was present." Infectious virus was associated with lower cycle thresholds. At higher thresholds, non-infectious amounts of virus still triggered a positive PCR test result. **Heneghan called for a standardized number of cycles, to prevent positive results being triggered by an insignificant (non-infectious) amount of virus, but the number of cycles was never standardized.** The often too-high cycle threshold led to numerous "positive" results in people who looked and felt fine, but were told to quarantine, even as their test went into the "casedemic" totals for the day.

The constant, daily tallies of Covid-19 cases were almost unavoidable. They were broadcast in every news medium, put up on billboards, and exhaustively highlighted[733] in press conferences, and news alerts. With people being encouraged to get tested whether or not they were symptomatic, and the PCR tests being run at too-high a number of cycles, positive cases soared. **The number of cases reported daily became highly influential in establishing public policy about the pandemic, while often being largely separated from the number of hospitalizations and deaths.**

For example, in the fall of September 2020, a *Wall Street Journal* article[734] pointed out that although 29 large universities had reported some 26,000 cases by September 9, there had been no hospitalizations. Cases were mild or asymptomatic in most college students, but that didn't prevent the universities closing campuses to in-person learning. Common sense went out the window.

This chart by Yinon Weiss that tracked cases and deaths per million in the U.S. from March 5, 2020 to November 17, 2020 gives a clear visual of the **lack of correlation between PCR-confirmed cases and Covid deaths:**

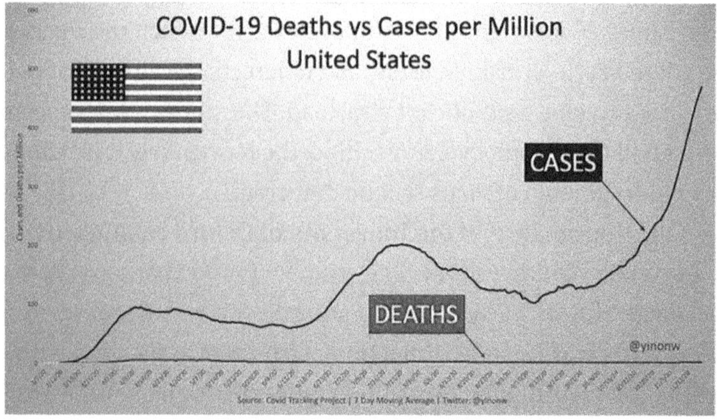

As shown in the above graph, when considering deaths per million vs cases per million, there is no correlation between the two. The red line, Deaths per million, is almost flat, compared to the rising blue line showing positive PCR-confirmed Cases per million. Therefore, positive PCR cases are not a good metric for determining the seriousness of the disease, or for setting public policy.

For many months only PCR tests were available, meaning there was a lapse between testing and the results from the lab. Eventually lateral flow antigen tests were developed and widely distributed as at-home test kits that returned results in minutes. Unlike PCR, which tests for the presence of genetic material, lateral flow[735] detects the presence of a particular protein target. Prior to Covid-19, the best known lateral flow tests were over-the-counter pregnancy tests.

The at-home lateral flow tests were designed to detect proteins found in the Covid-19 virus, and were distributed for "free" in many countries for quite some time. ("Free" being a definite misnomer, as taxpayer dollars are all that a government has to work with.) The accuracy of the lateral flow test was in question from the start, further compounded by the tests being manufactured by multiple companies[736] under Emergency Use Authorization, and self-administered by untrained medical professionals.

A CDC article[737] last updated September 25, 2023 states, "Every home in the U.S. is eligible to order 4 free at-home tests beginning November 20, 2023. If you did not order 4 tests earlier in the fall, you can place two orders for a total of 8 tests." The CDC reminds us that, "Although self-tests are usually faster, they are not as good at detecting the virus as PCR tests, meaning you might get a false negative result." The article states that the FDA recommends a second test 48 hours after a negative result, and a third test after another 48 hours, if the second test was negative.

A May 2021 article[738] in the Pharmaceutical Journal in England reported that the Healthcare products Regulatory Agency (MHRA) "emphasised that lateral flow tests are only authorized to be used as a 'red light' test to find infectious people and ensure they self-isolate quickly, and not as a 'green light' for people who test negative to enjoy greater freedoms." That's certainly not how testing was presented by governments, or viewed by the majority of people who were just trying to get their lives back.

Throughout the pandemic, the number of Covid cases became the carrot on the stick for many jurisdictions. If cases were high, people were supposed to mask up and stay home. If cases started to drop, people were told they could venture out again, but that they should keep masking – just in case. Dodsworth wrote, [A]n MP told me **masks were introduced to give people confidence to go shopping.** People were scared and the economy didn't bounce back hard enough after the first lockdown. Thus, masks were mandated." (*A State of Fear*, p. 238)

> The use of fear intrudes on our private lives, our minds and our physiological health. The messaging of fear exposes us, against our will, to harmful and offensive messages and creates unnecessary anxiety.
>
> **LAURA DODSWORTH** | *A State of Fear*, p. 240

In commenting on the obsession with case numbers, Professor Sunetra Gupta said[739], "We have to stop being afraid of infection itself. That's just a condition of life." She stated that the focus on keeping cases low, with the intent to eliminate them altogether, was "national vanity."

Propaganda about Hospitals Being Overrun with Covid Patients:

Case numbers were constantly combined with statements about hospital capacity. In the U.S., footage of the hospital situation in New York City in April 2020 was called up over and over as a warning of what would happen if citizens didn't comply with masking, social distancing, and lockdowns. **Yet what happened in New York City was unique, and was never once repeated in another part of the country.**

Much of what we heard and saw about the overflowing hospitals and morgues in NYC was misconstrued[740] and propagandized[741]. That statement is not meant to discredit the hardworking doctors and other medical professionals who were taking care of patients in April 2020, but is rather a condemnation of how the government and the media handled the situation.

It seemed every newscast ran footage of the freezer trucks outside NY hospitals, brought in to accommodate the bodies piling up. What was neglected in reporting is that all the funeral homes, morgues, and cemeteries were closed by government order, and not receiving bodies. Also not reported was that during this dramatic period of lockdowns and rising cases and deaths in New York, NYC hospitals saw an overall 50 percent drop[742] in admissions.

PANDA research[743] explains that we knew in February 2020 that SARS-CoV-2 "was not a particularly dangerous virus to the broad population," based on the data from the Diamond Princess cruise ship. PANDA notes that "something else beyond Covid or the spread of a respiratory virus was responsible for a significant majority of the 'first-wave' excess deaths in NYC," and continues:

> [M]ore likely causes of the unprecedented excess mortality in NYC need to be considered, including iatrogenic harms, psychological effects, neglect, panic stoked by constant media propaganda, ill-advised use of ventilators and sedatives, and policies relating to

residents in care homes. **If these were indeed the causes behind the excess in deaths in NYC in early 2020, a tragic proportion of the deaths were avoidable.**

Hospital ICUs routinely operate at about 70% full, as shown in a 3-year study[744] by the National Center for Biotechnology Information. The study examined 97 U.S. ICUs and found that total ICU occupancy ranged from 57.4% to 82.1%, even during influenza season. A 2018 *Time* article[745] about the worse-than-usual flu season highlighted maxed-out ICUs in California, where some patients were treated in the hallways, triage tents were set up outside to accommodate the large number of patients, and medical personnel were putting in long hours. In interviews, doctors acknowledged it was an intense time, but not unprecedented and not of long duration. Few of us remember the influenza surge in 2017-2018 because no one in government, public health, or the media presented it as anything other than a bad flu season.

Yet during the pandemic it was **almost as if we forgot that the purpose of hospitals is to treat ill patients.** Media and public leaders continually engaged in hand wringing about hospital capacity without providing perspective. Also, something that was never highlighted by the media is that the overflow field hospitals[746] that were set up to handle the expected massive numbers of Covid-19 patients were hardly utilized, or never used[747], not even in New York City[748]. The navy hospital ship that was sent to New York City by the federal government, to handle overflow, treated less than 200 patients[749], while Andrew Cuomo[750] sent elderly Covid patients back to nursing homes, leading to thousands of preventable infections and deaths.

Propaganda through Inflated Covid-19 Death Counts:

At the time, it seemed impressive and helpful that Johns Hopkins almost immediately made publicly available its amazingly detailed Covid-19 Map of the World[751]. Now we might wonder how the map was ready to go from Day 1. It was extremely dramatic watching the red dots spread across the map and reading the daily increases in numbers of cases and deaths.

Constantly focusing on the number of Covid-19 deaths was a powerful fear tactic during the pandemic. No attempt was made to put Covid-19 deaths in context with overall deaths and normal statistical

trends. For example, on July 6, 2021 former U.S. Surgeon General Jerome Adams stated that lack of full FDA licensure for the Covid-19 vaccines was interfering with the "quickest path to 70% [herd immunity] threshold." Adams tweeted, "US citizens are dying (~1 in every 5 minutes) and variants are rising while we await an update from the FDA and @WhiteHouse on #COVID19 vaccine licensure." That "1 every 5" stat sounded alarming. But at the same time, cancer had been killing 1,600 people a day[752] in the U.S. for several years, which breaks down to losing one person to cancer every minute of every day. The U.S. Population clock says that there is one birth every 9 seconds, and one death every 10 seconds in the U.S. **We can all agree that births and deaths go on around us all the time, as part of the ebb and flow of mortality, but the normal process of death was highlighted as an unusual and constant tragedy.**

The CDC's Covid-19 death data[753] in August 2020 showed that **out of 161,392 Covid deaths to date, just six percent, about 9,700 deaths, were attributed to the coronavirus alone. The other 94% had an average of 2.6 additional conditions such as heart disease, diabetes, and sepsis.** Fact-checkers went into overdrive to try to suppress that statistic, but it speaks for itself no matter how you twist it.

In addition, over 80% of Covid deaths were in the 65 and older age group. Observationally, and statistically, it was clear that Covid was not a high risk disease for young people. Many of the elderly and those with chronic health issues who contracted Covid-19 recovered. **Many of the people who died of Covid were victims, not of the disease, but of the suppression of early effective treatments, and of the deathly hospital Covid protocols.**

There were large financial incentives for hospitals to admit people as Covid-19 patients, as was discussed in Section 2 under Riches from Remdesivir: In addition, there was pressure to list Covid-19 as the cause of death, even if it wasn't. **Dr. Scott Jensen went public with his concerns about the improper attribution of Covid-19 on death certificates, and the financial motives[754] for doing so.** He also created videos on social media putting Covid in perspective[755] with the annual flu. Until that moment, Dr. Jensen was a respected member of his community, having been a family physician for 40 years, and serving as a Minnesota State Senator. **For pointing out that Covid deaths were being inflated, and bringing**

up other questions about the official narrative, Jensen became the target[756] of multiple investigations by the Minnesota Board of Medicine for alleged "unprofessional conduct" relating to the "spread of misinformation." All the Board actions were eventually dropped[757], but not until Jensen had gone through several years of stress and expensive efforts to defend himself.

We were **manipulated with propaganda about Covid-19 deaths as part of the unethical efforts to achieve ends largely unrelated to public health**. Medical professionals like Dr. Jensen were trying to educate people. But the government and Big Pharma spread propaganda. The public suffered as a result.

Propaganda of Blaming the Unvaccinated for the Continued Pandemic:

> *Cruelty became as contagious as any disease.*
>
> NAOMI WOLF | *The Bodies of Others*, p. 274

In his September 9, 2021 speech[758], Biden said the unvaccinated were overcrowding hospitals and "overrunning ER units, hospitals and intensive care units, leaving no room for someone with a heart attack, or pancritis (sic), or cancer."

Biden's statement was not true. **It wasn't the unvaccinated filling the hospitals. It was the vaccinated,** as shown in the early data from Israel (see Sec 17 The vaccinated become the bulk of Covid hospitalizations:), and later confirmed by data from other studies such as that of the Cleveland Clinic (see Section 14 The More Covid Shots You Take, the More Likely You Are to Get Covid)

Nurse Gail Macrae[759] was working in a Los Angeles hospital at the time of the vaccine rollout in her community in March 2021. **Within two weeks of the rollout, the hospital, which was never overrun by Covid patients in 2020, filled up with the vaccine-injured.** One of Macrae's direct managers told her he was evaluating the records, and the hospital was experiencing three times higher admissions than it ever had since it opened. This continued into June, although respiratory illness season is during the winter months.

On one day when she worked as the patient care coordinator during a 16-hour shift, Macrae noted that every patient was there for a **heart attack, stroke, clotting, or Guillain-Barré syndrome, among other heart or neurological issues.** Macrae also noticed that during her shifts after the vaccine rollout began, there was a **significant increase of "code blue" calls (life-threatening emergency) to the first floor where the vaccination clinic was operating.** Two of her own colleagues went into anaphylactic shock immediately after receiving the Covid shot.

Macrae said it was understood that they were not allowed to report any of these events as vaccine injuries. "If we did, we would get a head shake and they would say, 'No, no we can't prove that this is from the injection.' It was very hush hush." Macrae's colleague who was a Covid-injection clinic nurse at the hospital witnessed 3-10 episodes of anaphylaxis during her shifts. She asked her manager about reporting those events and the manager said if she reported a single adverse event, she would be fired.

> We have a very hard time imagining that terrible things are happening, and we really do go to lengths to trust these 3-letter agencies that are setting protocols, like the CDC and the WHO and the AMA...There are so many practitioners that are genuinely good people and they just can't imagine that these terrible things are happening.
>
> **NURSE GAIL MACRAE**

Clayton Cobb, who has closely examined the data[760], points out that the CDC defined the status of a person this way: "An unvaccinated person had SARS-CoV-2 RNA or antigen detected on a respiratory specimen and has not been verified to have received Covid-19 vaccine." **In other words, they tested positive for Covid and it wasn't confirmed they were vaccinated.**

Macrae's hospital charting system used the CDC definition[761] of "vaccinated," as outlined in the following paragraph from CDC guidelines:

> Vaccination status: A person vaccinated with at least a primary series had SARS-CoV-2 RNA or antigen detected on a respiratory specimen collected ≥14 days after verifiably completing the primary series of an FDA-authorized or approved COVID-19 vaccine. An **unvaccinated**

> **person had SARS-CoV-2 RNA or antigen detected on a respiratory specimen and has not been verified to have received COVID-19 vaccine. Excluded were partially vaccinated people** who received at least one FDA-authorized vaccine dose but did not complete a primary series ≥14 days before collection of a specimen where SARS-CoV-2 RNA or antigen was detected. (emphasis added)

Macrae notes that the CDC definition of "vaccinated" led to these outcomes:

- Patients who had **just received a Covid shot** and were having a bad reaction were listed as **unvaccinated**, because it hadn't yet been two weeks. For example, the people who went into anaphylactic shock in the vaccination lab were counted as unvaccinated.
- Patients who had **received a first Covid shot, but** who had **not received their second shot**, were counted as **unvaccinated**. (J&J would be the exception, which was a one-shot vaccine, in which case, two weeks out from the J&J would be considered fully vaccinated.)
- Patients who had received their second Covid shot, but it had been less than two weeks, were counted as unvaccinated.

Macrae said they were not allowed to report the cases of fully vaccinated patients and staff who came into the hospital with Covid. Macrae said there were "glitches" in the charting for the first few months where **the only boxes that could be checked were "unvaccinated" or "vaccinated status unknown."** She believes that was remedied after some months, "But when the media was reporting that all of the patients who were admitted to the hospital were unvaccinated, that was never true. I was working on those Covid floors taking care of those patients, and that came down **to the charting system… [that] intentionally excluded 'fully vaccinated' status."**

Macrae's experience has been confirmed by others, including Dr. Pierre Kory[762], who worked as an ICU critical care doctor in multiple hospitals during the pandemic. Dr.Kory states that the popular EPIC electronic medical record (EMR) system only allowed for two categories under the Covid-19 vaccine status section, "Vaccinated" or "Unknown." It wasn't until August or September of 2021 that he admitted a patient whose EMR indicated "Vaccinated." Dr. Kory had suspected "something was off" with the U.S. statistics as early as February 2021, when the data coming out of the U.K. showed that the majority of hospitalizations and deaths were in the vaccinated. (Data from multiple countries[763] including Israel, Germany, and New South Wales[764] also showed "breakthrough" infections occurring in the fully vaccinated.) Dr. Kory recalls, "I knew it could simply not be true that in a 9-month period, only one patient that I took care [of] in the ICU was 'fully' vaccinated."

Dr. Kory learned that when patients presented their vaccination card from an outside source, hospitals were categorizing their vaccination status as "Unknown," because the patient data didn't appear in the hospital EMR system. A July 2023 study[765] published in *JAMIA Open* performed a manual evaluation of local Immunization Information Systems (IIS) for 4,114 adult patients whose vaccination status was listed as "Unknown." The results were that 44% of the patients listed with vaccination status "unknown" were actually vaccinated. The study concluded:

> When the interface between the patient chart and the local Immunization Information System depends on a manual query for the transfer of data, the COVID-19 vaccination status for a panel of patients is often inaccurate.

Clayton Cobb notes that another factor leading to inaccurate vaccination status is the CDC's method of calculating the estimated unvaccinated population. The CDC subtracts the total known number of fully vaccinated people from the estimated total population count of the whole country. **The CDC used the population numbers from 2019 (which were from the 2016 census) for their calculations, even after the 2020 U.S. census data was available.** This **skewed** both the **total population** and **unvaccinated population** tallies.

Propaganda of The Science™

Follow the science is a phrase that should make any serious scientist blush. And anyone who says he represents The Science™ should really be embarrassed, but that doesn't seem to be in Fauci's repertoire of emotions. (see 56:56 to 58:07[766])

Certainly science is an evolving process, but government officials arbitrarily changed the parameters and rules on Covid policy so fast it gave the public metaphorical whiplash[767]. Stay six feet apart, but three feet is okay at school. You can gather, but only with 50 people, no 20 people, no 10 people, make that just six people and no one from outside your household. You have to wear a mask to walk to your table at the restaurant, but it's safe to take it off while you eat. If you're vaccinated, you don't need to wear a mask – actually everyone still needs to wear one to keep each other safe. Only 20 people are allowed in person at church, but there can be 50 at other places. Exercise outdoors in this plastic bubble. Wear a mask alone in your car. Follow the arrows in the grocery store aisles. You can't open your business until case numbers fall below this certain level. Don't gather for Thanksgiving. Don't sing. Wear a mask with a hole in it to play your musical instrument. Shut off all the drinking fountains. You can go to the theater, but only if you wear a mask. You can only lower your mask to take a bite, and then you have to pull it back up while you chew. And on, and on, and on.

All of this and so much more because, you know, we were following The Science ™. No wonder Professor Dalgleish said we've entered into an anti-scientific Dark Age. The introduction to Professor Dalgleish's book, *The Death of Science*[768], co-written with Paul R. Goddard, contains this observation from oncologist Karol Sikora:

> **When public opinion and policy decisions are driven by emotions rather than facts, the death of science becomes an imminent threat to societal progress.** When overwhelming scientific evidence is rejected or undermined, policymaking becomes susceptible to biases and vested interests. Issues such as climate change, vaccination, and genetically modified organisms (GMOs) have become **battlegrounds for ideological debates rather than reasoned discussions based on scientific consensus.** This erosion of trust in scientific consensus hampers our ability to make informed decisions and address critical challenges effectively. (emphasis added)

Until the people who foisted the anti-science decisions of the pandemic upon us acknowledge that they made a mistake, and/or until enough of the public refuses to go along, we are at risk. Until we say "No," the bureaucrats, pseudo-intellectuals, corporatists, and globalists who controlled our daily lives and jerked our long-term prospects out from under us will use The Science™ in service of the next "crisis" to get what they want at the expense of everyone else.

Propaganda of "The Experts:"

> *[S]o many of these people, who by dint of their educational backgrounds should have found it more easy than most to go to the primary sources of scientific information on the virus and the measures taken to lessen its impact, chose in large numbers – with doctors being very prominent among them – to instead "educate" themselves on these important matters with curt summaries derived from the mainstream press, social media or Pharma-captured agencies like the CDC and the FDA. This, paradoxically, while millions of intrepid and less credentialed people with a great desire to know the truth, often became quite knowledgeable about the actual state of "the science."*

THOMAS S. HARRINGTON | *The Treason of the Experts*[769]

We hear this phrase a lot these days: **"Stay in your lane!"** Merriam-Webster says the phrase is "used as a term of admonishment or advice against those who express thoughts or opinions on a subject about which they are viewed as having insufficient knowledge or ability."

Although there are multiple-credentialed doctors and scientists who have spoken out against the official Covid pandemic response, there have also been many people without medical or science degrees, who have researched, questioned, and contributed to the pushback against medical tyranny. Most do not consider themselves experts, but rather they are self-educated, informed people, who have taken up the cause of truth and freedom in a world where "the experts" have largely failed us. These "intrepid and less credentialed people," as Harrington described them in the above quote, have become the voice for millions who do not want the world that is being fashioned for us by elites and globalists.

> You don't have to be a medical scientist to understand the data. You just have to be a critical thinker.
>
> **DR. SCOTT ATLAS** | Former advisor to White House Coronavirus Task Force

Others who were somewhat obscure before the pandemic sprang to prominence with less beneficial results for society. Combined with bureaucrats and self-serving professionals, these individuals caused a lot of harm by taking upon themselves the mantle of "Expert," while they promoted ideas that led to pandemic-response madness. **They can't be entirely blamed for the harm they caused; they weren't kept in check, by medical, scientific, government, media, and other leaders who should have questioned them and course-corrected.**

An **obscure epidemiologist, Eric Feigl-Ding, quickly became an "expert"** at the beginning of the pandemic. His first viral tweet in all cap bold letters and emojis claimed an infected individual would in turn spread the illness to 3.8 people – an astounding (and totally inaccurate) infection rate. His "alarmist misinformation," labeled as such by Professor Steve Templeton[770], led to many Twitter followers, which in turn led to interviews with CNN, MSNBC, and major newspapers. He was recommended by Twitter as a "Covid expert." Feigl-Ding's pre-pandemic expertise was in the health effects of diet and exercise. He had **no experience in pandemic or respiratory virus epidemiology.**

De Kai[771], a computer science expert with no experience in aerosol science or respiratory diseases, **claimed** on April 21, 2020 that **his mathematical model projected 80% of Covid-19 infections would be prevented if the general public wore face masks.** In a June 2020 interview [772], epidemiologist Dr. Michael Osterholm explained that the **De Kai model had "serious flaws [773] in its assumptions and exposure parameters. For example, it does not take into account dose or time contact,** supporting the conclusion that the authors have a very poor understanding of the critical issue of aerosol versus droplets exposure, and what a cloth mask can do to limit such exposure." **The model was not yet published before being widely promoted by media around the world.**

On May 8, 2020, *Vanity Fair* ran an article[774] covering **De Kai's "compelling new study and computer model"** that provided "fresh evidence for a simple solution to help us emerge from this nightmarish lockdown." The formula was, "Always social distance in public, and, most importantly, *wear a mask*." (sic) The article stated, De Kai is "the chief architect of an in-depth study, set to be released in the coming days, that **suggests that every one of us should be wearing a mask—whether surgical or homemade, scarf or bandana—like they do in Japan and other countries, mostly in East Asia,**" and stated, "If 80% of a closed population were to don a mask, COVID-19 infection rates would statistically drop to approximately one twelfth the number of infections – compared to a live-virus population in which no one wore masks." (emphasis added)

On May 14, 2020 Jeremy Howard, a data scientist at the University of San Francisco, published an article[775] in *The Conversation*. Howard cited De Kai's model, and claimed mandatory masking with cloth masks could be among the most powerful tools to stop the community spread of Covid-19.

Howard had no expertise in respiratory illnesses or viral epidemiology, but stated with certainty it was "clear and simple: COVID-19 is spread by droplets. We can see directly that a piece of cloth blocks those droplets and the virus those droplets contain. People without symptoms who don't even know they are sick are responsible for around half of the transmission of the virus."

Howard formed a group called MASKS4ALL and **pushed for universal face masking. He and "100 of the world's top academics" released an open letter to all U.S. governors** urging them to require mask-wearing in public. As explained[776] by Dr. Michael Osterholm in a June 2, 2020 podcast, aerosol science is a very technical discipline:

> Please know that the vast majority of information you're hearing every day in the popular literature or even in the news about cloth masks is not coming from anyone with any expertise in aerosol science…**An MD or PhD in disciplines other than aerosol science or respiratory protection does not automatically make one an expert in these areas.** (emphasis added)

Howard and his colleague, **Zeynep Tufekci,** a professor at the University of North Carolina's School of Information and Library Science, and a contributor to the *New York Times*, worked together to push face masking. Attorney Michael Senger posits in in an article[777] how Tufekci and Howard, two **unqualified "experts" came to have so much influence on public policy:**

> The first is that their advocacy was essentially scripted theater, a pretext for seemingly-spontaneous actions that a network of institutional leaders was actually planning to take anyway, unbeknownst to the public...
>
> The second possibility is that it really was this easy for ambitious activists with no relevant expertise to convince institutional leaders to reverse longstanding public health guidance in the early days of COVID—these being the same leaders who then spent years closing their eyes and ears to any evidence that their interventions weren't working, even from some of the world's most-qualified scientists.

To be fair, as Dr. Scott Atlas was quoted previously, "You don't have to be a medical scientist to understand the data. You just have to be a critical thinker." And **therein lies the difference between an "ambitious activist," and the grassroots effort by hundreds of people around the world** who are trying to understand, and shed light on, the wrongs that have been perpetuated under the auspices of the pandemic response.

"Ambitious activists with no relevant expertise," should never have had so much influence on public policy. In a sense, Howard and Tufekci are like Bill Gates, albeit on a very small scale. Just because someone has money, or a platform, doesn't make that person an expert, nor give him or her the right to govern the lives of others. In democracies, we elect our leaders who in turn govern within the framework of the Constitution and the will of the people. This established democratic model is in danger of being overthrown by an elite, moneyed class that has desires to rule the world.

Psychiatrist and philosopher Jordan Peterson captures well the concern felt by those who are pushing against the elitists' grab for global top down rule in his article[778] and video[779] (starts at min 2:17) *"Back off, Oh Masters of the Universe."*

The Propaganda of Endless Sanitizing to "Stop the Spread"

> *The whole point of a virus is to invade a living host...viruses like Covid-19 don't survive well on their own. ..It's like a fish out of water. Sure, your goldfish might be OK to flop around for a few seconds while you're transporting them into a new bowl, but you better act fast. Viruses need humans like fish need water..., "[J]ust because the virus is detectable on a surface doesn't necessarily mean that there's enough there to infect you...viruses start to die when they're not in the body..."*
>
> **CLEVELAND CLINIC**[780] | November 13, 2023

As previously discussed, in early 2020 it was known that Covid-19 was an airborne disease likely spread through aerosols[781] (not surfaces), that infections were most commonly spread within households[782], and that asymptomatic spread[783] was not a strong driver of infection. It was common knowledge that it was good hygiene to wash our hands, important to cover our mouths when coughing, use a tissue when sneezing, and to stay home if we were ill. Cold and flu season came and went each year without significantly impacting society at large.

Yet despite over 100 years of experience studying and dealing with coronaviruses, the narrative around Covid-19 was that it was a completely unknown and unpredictable deadly virus. "Experts" presented Covid-19 as if it could jump on you like a flea if you touched an infected surface or passed too close to someone, even if they seemed perfectly healthy.

Historical medical memory of the few was drowned out by a non-scientific official narrative that turned the Covid-19 response into a germaphobe's dream, or nightmare, convincing a majority of people that fellow human beings are basically bags of germs. Society was suddenly riddled with people behaving like Adrian Monk[784] or Bob Wiley[785], worried about touching the gas pump, or an elevator button, or standing in line at the grocery store, turning away from each other outdoors lest we inhale another's germs. They had us sanitizing our keys and credit cards, and wiping down groceries and Amazon packages. They told us to avoid human touch. Pre-schoolers were taught, "Don't share." Who can forget the images of people hugging through plastic sheeting, and the heartbreak of the elderly in care homes putting a hand to the glass that separated them from loved ones?

Hand sanitizer by the gallon appeared everywhere, and "hygiene theater" became the norm in all public spaces. From restaurants to entertainment venues, from gyms to schools, from offices to churches, the official message was that thorough sanitizing would keep us safe from Covid. In fact, "protecting others" was called our civic duty, and questioning the Covid policies was not allowed. We were offered paper menus or QR codes and plastic utensils, instead of fine dining. We couldn't get refills on popcorn because we'd touched the carton. Custodians' workloads increased dramatically, as they obsessively wiped down walls, floors, restrooms, and surfaces throughout the day. Drinking fountains and drink dispensers were shut off. Public restrooms were closed. Caution tape and chain link fences were installed around playgrounds and swing sets. Skate parts were filled in with sand. Retail hours were reduced in order to meet sanitizing protocols to "keep you safe." People covered keypads with plastic, wore disposable gloves and masks, and forgot they had immune systems.

Although completely ineffective at preventing the spread of Covid-19, the persistent sanitizing was very effective at keeping the idea of "There's a pandemic" on everyone's mind. As with masks, there will be some who say, when confronted with the ineffectiveness of sanitizing surfaces to prevent Covid-19, "Well, if it helps even a little bit, it's worth it." That's not how public health works. Because every intervention has unintended consequences[786], there must be evidence that benefits outweigh harms before imposing measures on an entire population.

For example, the requirement to sanitize everything cost[787] millions of dollars in cleaning products, paid work hours, lost productivity, and misdirected resources. The costs were often passed on to customers[788]. The CDC estimated the cost for increased cleaning and sanitizing of K-12 schools at anywhere between $55 and $442 per student. It was simply a lie that sanitizing surfaces would prevent the spread of Covid-19, but the consequences of the requirements were costly, and even devastating. Small businesses[789] that often operate on a tight profit margin were especially hard hit by the added expenses.

In January 2021 *Nature* published an article[790] titled "Covid-19 rarely spreads through surfaces. So why are we still deep cleaning?" The subheading read, "The coronavirus behind the pandemic can linger on doorknobs and other surfaces, but these aren't a major source of infection." The article used to be available to the general public, but is now behind a paywall. It was not until May 7, 2021 that the CDC finally acknowledged[791] aerosol transmission through infected people as the primary driver of Covid-19, yet the sanitizing compulsion[792] continued. It wasn't just that the emphasis on sanitizing everything was silly and ineffective; the practice had real impacts on everything in society, including the supply chain.

For example, in Los Angeles where 40% of imports arrive in the U.S., shipping containers backed up at the ports, largely due to illogical and onerous Covid-19 policies. The American Association of Port Authorities protocol[793] in October 2021 included limited staffing on shifts due to social distancing, and thorough sanitizing of all surfaces between each shipment being unloaded. Where normally there would be one cargo ship waiting at each of the Los Angeles ports, in October 2021 there were 65 cargo ships waiting[794] to unload goods. In the fall of 2021 the maritime, road, and aviation industries issued an open letter[795] to world leaders with an "urgent plea...to remove restrictions hampering the free movement of transport workers." The illogical Covid-19 policies had led to employee shortages, crews unable to leave their ships, flights being restricted, and truck drivers "forced to wait, sometimes weeks, before being able to complete their journeys and return home."

The Cleveland Clinic article[796] quoted at the beginning of this section ends on this obligatory official Covid-narrative note, "[C]leaning, sanitizing and taking care not to needlessly touch public surfaces is still important to staying healthy. So is keeping up with your COVID-19 vaccines, washing your hands properly and following other steps to stop virus transmission." A January 2024 search of "How does Covid-19 spread?" brings up a CDC article last updated in August 2022. As if in a time warp[797] the CDC is still talking about droplets, contaminated surfaces, and the danger of asymptomatic spread:

> COVID-19 spreads when an infected person breathes out droplets... In some circumstances, these droplets may contaminate surfaces they touch...Anyone infected with COVID-19 can spread it, even if they do NOT have symptoms.

The Propaganda of Super-Spreader Events:

> *Why develop policies that punish, encumber and restrict human contact in humane/analog (unsurveilled, unmediated) spaces? Because human contact is the great revolutionary force underlying human freedom.*
>
> **NAOMI WOLF** | *The Bodies of Others*, p. 200

Were there lots of super-spreader events during the pandemic? Not really, but if asked, most people will tell you there were because that's what they heard over and over. The super-spreader narrative was inflamed early on by the account of the unfortunate church choir[798] that held practice and spread Covid infection throughout the group, leading to the deaths of a couple members.

When South Dakota Governor Kristi Noem welcomed the Sturgis Motorcycle Rally to her state in August 2020, public health professionals and the press were beside themselves with dire predictions of how it would spread Covid-19 all over the country. Despite 450,000 people gathering in Sturgis, South Dakota during a period of 10 days, no big outbreaks of COVID-19 were traced to the event.

There was a study[799] touted by the media, that used modeling to claim Sturgis led to 260,000 Covid cases, but it was found to be grossly flawed. Researchers at Johns Hopkins[800] questioned the study's accuracy, and found it to be incorrect. In reality[801] South Dakota identified 124 cases tied to the rally, with another 290 people in 12 states testing positive after attending. The numbers resulted in an infection rate of between 0.02% and 0.09%. Sturgis was *not* a super-spreader.

There were dire warnings[802] from government officials about holiday travel in 2020. People traveled anyway and cases dropped dramatically in January. Dr. Fauci and CDC Director Walensky gave solemn statements[803] of the devastation that would happen if people held Super Bowl parties on February 7, 2021. Thousands of people traveled to Florida for the event, resulting in just 57 cases[804].

When Texas and Mississippi lifted mask mandates and restrictions on gathering in March 2021, Pres. Biden accused[805] them of "Neanderthal thinking" for dropping Covid restrictions. Cases dropped. A crowd of over 38,000 people, mostly without masks, sat together during the Texas Rangers game on April 5, 2021. As with the Super Bowl, there were no outbreaks[806], but there also weren't any retractions from Dr. Fauci, Dr. Walensky, or Pres. Biden.

It was the same with the Alvarez vs Saunders boxing match at the AT&T Stadium in Arlington, Texas on May 7, 2021. The match set a record for the largest attendance in history at an indoor boxing match. It was also the largest indoor event since the pandemic began, and "doctors and other experts" feared[807] the "lax rules" (masks and social distancing encouraged, but not required, no temperature or vaccine checks) could lead to an upswing in Covid-19 cases. But there was no upswing in cases. This great news was met with silence from the media, CDC, Dr. Fauci, and all the others who were squawking that the match would be a super-spreader.

The 2021 Indy 500 held on May 30 was the largest sporting event held since the beginning of the pandemic. Despite "experts"[808] expressing the danger[809] of the gathering, no temperature checks were taken and proof of vaccination was not required to enter, but the venue *was* kept at 40% capacity. A scan of people in the crowd of about 135,000 showed many maskless spectators next to each other in the stands. There were no subsequent outbreaks from attendees returning to other states, and Indiana's cases continued to decline.

The myth of super-spreaders was an effective propaganda tool, causing many people to be afraid to gather with loved ones and friends, and participate in community events. The arrows in the grocery store aisles, the little circles that kept us six-feet apart, the restrictions on how many people you could host in your own home, the burden on businesses to limit capacity and restrict hours, the dividing of people into essential and non-essential, the prohibition on worship services and singing, the cancelation of performing arts and entertainment – all of it was paranoid, unscientific, unconstitutional, anti-human, soul-crushing evil.

> **Restrictions in human history always preceded theft: theft of the lands, assets, and opportunities of the restricted. It was so this time too. For the first time in our species' history, those restricted were not one or even several distinct**

subgroups, but humans as a whole. Those who restricted them had human form, but they were traitors to humanity.

NAOMI WOLF | *The Bodies of Others*, p. 140

Propaganda of Equating Covid-19 with the Spanish Flu:

We will have a better chance of suppressing infectious diseases only if we adopt what the WHO calls a One Health approach and integrate predictive modelling and surveillance used in both infectious disease control and climate change.

WORLD ECONOMIC FORUM | August 18, 2022[810]

According to Bill Gates[811], the United Nations World Health Organization (see here[812], here[813], and here[814]), and the World Economic Forum [815], the next big pandemic is just around the corner. We've been told many times that the Covid-19 pandemic was the worst to hit the globe since the Spanish flu of 1918. However, the two illnesses are not comparable. **The Spanish flu killed 50 million people worldwide. Adjusted for today's population, that would be more than 216 million people. Covid-19 killed around 7 million[816] worldwide.**

The Spanish flu[817] targeted mostly those under age 5, those in the 20-40 year old range, and those 65 years and older. Covid-19 targeted the elderly and those with compromised health. **In 1918 it was often the bacterial pneumonia that developed in flu patients that led to loss of life.**

No antibiotics had been developed yet in 1918, and treatments for respiratory ailments were primitive; modern medicine was largely in its infancy. **In fact, it is now believed that aspirin[818], a new drug at the time, caused many of the deaths during the Spanish flu** because it was unknowingly administered in toxic levels.

Medicine aside, **in 1918 water and sanitation systems, and sewage disposal systems, were also in developmental stages to their modern counterparts.** All of these drawbacks have largely been rectified or improved, yet to hear the decision-makers and influencers talk, you would think humankind is more at risk from viruses than it ever has been.

Viruses and bacteria are in the environment all around us and have been since the dawn of time, as Steve Templeton reminds us in his book *Fear of a Microbial Planet*[819]. Templeton explains that the biomass of bacterial cells on Earth is second only to plants, and exceeds that of all animals by more than 30-fold. Microbes make up to 90% of the ocean's biomass. Templeton states, "The very air you breathe contains a significant amount of organic particulate matter that includes over 1,800 species of bacteria and hundreds of species of fungi airborne in the form of spores and hyphal fragments." We are an integral part of our microbial planet. **Viruses and bacteria are not a threat to life for the most part, but just a part of life.**

The world didn't shut down during the Spanish flu, and certainly should not have closed during the Covid-19 pandemic. Most definitely we should not be doing research to create more virulent viruses than those that occur in nature; all the gain-of-function experiments should be discontinued.

What is needed now is not the obsessive model of Virus Detection and Eradication, but **a calm voice of reason to take a step back from pandemic panic**. We must never again respond to a virus in the manic manner with which we handled Covid-19.

> There is no such thing as a medical emergency nowadays because modern medicine has come such a long way and the doctors around the world are good enough to take care of anything, if it comes up. Let it come up first and then let us see whether there is an emergency or not.
>
> **DR. SUCHARIT BHAKDI**

Messaging about Covid-19 – a Master Class in Propaganda:

> *Some things are believed because they are demonstrably true, but many other things are believed simply because they have been asserted repeatedly.*
>
> **THOMAS SOWELL**, senior fellow | Stanford University, Hoover Institution Documentary from Jan 25, 2021[820]

The above statement by economist and social theorist Thomas Sowell was certainly demonstrated during the pandemic. The phrases about social distancing, lockdowns, face masks, and other so-called public health safety measures were constantly repeated, similar in feel to the type of propaganda that is regularly plastered over walls and pamphlets throughout communist countries:

AT THE BEGINNING OF THE PANDEMIC:

Be considerate; wear a mask
Two weeks to flatten the curve
Do the right thing; stay home
We're all in this together
Prevent the spread; wear a mask
Show you care; social distance
Stay home - Stay safe
Six-feet apart
Stop the spread
Mask up to stay open
Together we can win the fight against COVID-19
Studies show that masks work

LATER IN THE PANDEMIC:

Do your part – get vaccinated
Safe and Effective
Take the shot to stop the spread
Vaccination is better than natural immunity
The only way out of this pandemic is through vaccination

By attacking the mind with a constant stream of never-ending facts the ability to use logic and reason is further shut down. A campaign of endless slogans and phrases, including wartime-loaded language, has been pumped into the zeitgeist to guide our opinions and fill any cracks in the narrative.

From the video short *Covidian Cult*[821]

These mantras were repeated as fact without scientific proof, both to prove a point and silence dissent. For example, the "six-feet apart" rule was chosen on a whim. **Former FDA Commissioner Scott Gottlieb[822] stated on December 10, 2021 that the 6-foot rule was "arbitrary," and had no impact on the aerosol-borne spread of Covid-19.** Gottlieb's claim was confirmed by Dr. Fauci in January 2024 when he testified[823] before the House Select Subcommittee on the Coronavirus Pandemic. Fauci was unable to provide scientific data for the 6-foot rule and said it "sort of just appeared."

When the CDC was providing guidance for opening schools in March 2021, desks were allowed to be 3-feet apart [824], because 6-feet apart wasn't functionally practical, not because Covid was now contained in a smaller space. At the same time, students and staff were to maintain 6-feet of distance in the lunchroom, during sports, exercise, and choir and band rehearsals, because "increased exhalation occurs" during such activities[825]. The continuous stream of CDC pandemic propaganda created logistical difficulties that interfered with education and hampered normal social interactions.

Efforts to enforce the "six-feet apart" rule led to the use of Plexiglas barriers, which were just a psychological rabbit's foot – another way to separate people, without providing any protection from a virus that was in the air flowing over and around the pieces of upright plastic.

When the CDC lifted masking for a time, Director Rochelle Walensky said[826], "We wanna give people a break from things like mask wearing when our levels are low and then have the ability to reach for them again, should things get worse in the future." Think about that for a minute. **Mask wearing was all about "the levels," but never about the efficacy.**

In January 2022[827] three economists released[828] a "Literature Review and Meta-Analysis of the Effects of Lockdown on Covid-19 Mortality." The three researchers, who were with the Johns Hopkins Institute for Applied Economics, Global Health and the Study of Business Enterprise, **found that Covid-19 lockdowns in Europe and the U.S. reduced Covid-19 deaths by 0.2 percent.** Shelter-in-place orders were found to have reduced Covid deaths by 2.9 percent. The report stated, "We find no evidence that lockdowns, school closures, border closures, and limiting gatherings have had a noticeable effect on Covid-19 mortality."

Neil Ferguson[829] of the catastrophic Imperial College of London modeling, questioned the economists' methods, and said the report "does not significantly advance" understanding of how effective lockdowns are. Other critics accused the researchers of cherry-picking data to prove their point. However, Professor Robert Dingwall, a sociologist and former Government Covid adviser said, **"What is really troubling is the lack of interest by governments in properly evaluating these drastic social interventions, in comparison with the investments in research on vaccines and drugs."**

Our pandemic response should have centered on protecting the vulnerable, while the rest of society went about its business. The government should have emphasized early treatment for Covid, and the importance of vitamin D[830], exercise[831], and weight loss[832] as highly impactful preventative measures. Instead, governments around the world unleashed propaganda campaigns against their own citizens, preparing them to accept the vaccines – either voluntarily or by force.

> Within a deluded and constrained view of disease, it is no surprise that in responding to the threat of a new strain of flu, **natural immunity continues to be mocked relative to glorifying vaccination**. An unreported statistic is the number of unvaccinated people with a nutritious diet and clean bill of health who were hospitalized or died from COVID-19.
>
> It is because there are no such cases.
>
> **DAVID MARKS** | Pandemonium[833] | November 29, 2023

32. THE PROPAGANDA OF PRIZES AND AWARDS FOR PANDEMIC PLAYERS

Historian: It's been said that I am the greatest historian in history.
Little Prince: That's wonderful sir! Who said it?
Historian: I said it. I wrote it. I read it. It's printed. Consequently, it's fact.

From the movie *The Little Prince*
Book by Antoine de Saint-Exupéry

Nobel Prize for technology that led to "two highly effective" mRNA Covid vaccines:

On October 2, 2023, it was announced that Katalin Karikó, PhD, and Drew Weissman, MD, PhD have been named winners[834] of the Nobel Prize in Physiology or Medicine for their work on the mRNA platform. The award is **for their early 2000s work that showed "it was possible to dampen the body's inflammatory responses to lab made mRNAs** by making specific chemical changes to the component bases of the molecules." The way the body's inflammatory response was suppressed was to replace uridine with pseudouridine.

Essentially, the uridine in natural messenger RNA allows the mRNA to be quickly broken down and disposed of by the immune system after delivering its message to the cells. But the pseudouridine used in modified mRNA - used in the Covid shots – prevents the body from breaking down the modified mRNA. That's great for getting injected mRNA to the cells, but is disastrous for the body, which apparently can't eliminate the injected mRNA, nor turn off the message it sends to create spike protein.

Rickard Sandberg, of the Nobel Assembly said that Karikó and Weissman's research, was added to others' work on stabilized spike protein and mRNA delivery using lipid nanoparticles, which led to "two highly effective mRNA vaccines against Covid-19" being developed and approved "in record time."

Dr. Anthony Fauci pronounced it a "wonderful choice of Nobel Prize." Olle Kampe, vice chair of the committee that chose the winners of this year's award stated, **"Giving a Nobel Prize for this Covid-19 vaccine can make hesitant people take the vaccine and be sure that it's very efficient and it's safe."**

In awarding the Nobel to Kariko and Weissman, the Committee ignored the mounting evidence that the genetic-based vaccines are not efficient or safe. A study[835] out of Germany, published November 17, 2023, used mass spectrometry to examine biological samples from people who have received the Covid shots. The study found vaccine-induced spike protein was still circulating in the body up to six months after injection, in 50% of the samples analyzed.

The idea that indestructible messenger RNA is circulating for months after injection, instructing the body to make toxic spike protein, is bad enough. But it is now known that the fragments of DNA that contaminate the vaccines, are causing the body to make other, unintended proteins in a process called frameshifting.

A study published[836] in the prestigious journal *Nature* on December 6, 2023, found that these unintended proteins occurred in 25-30 percent of people studied.

Dr. Paul Marik explains[837], "What the study shows is that when you put a pseudouridine in where a uridine should be, the ribosome jumps or misreads messenger RNA. And as a consequence of this, it results in a bogus protein being made. So instead of making spike protein, it makes a nonsense protein that is possibly toxic."

As there is no way to predict what kind of protein the frameshift cells make, and as each mRNA shot contains trillions of mRNA packages, the potential for harm is astounding.

> In fact the Nobel Prize was awarded to two physicians for this discovery. So in a way they've kind of shot themselves in the foot because the technology on which the platform is based has now been shown to be functionally defective because it doesn't do what it's meant to. And it applies to the entire messenger RNA platform

because this entire technology is based on the pseudouridine. Which means whatever vaccine, or whatever protein they want to make, is going to be defective based on this problem.

DR. PAUL MARIK | December 18, 2023

Both Weissman and Karikó are listed as "University of Pennsylvania messengerRNA pioneers." The University of Pennsylvania, in its glowing report[838] of the recent Nobel Prize announcement, mentions this at the end:

> The Pfizer/BioNTech and Moderna COVID-19 mRNA vaccines both use licensed University of Pennsylvania technology. As a result of these licensing relationships, Penn, Karikó and Weissman have received and may continue to receive significant financial benefits in the future based on the sale of these products. BioNTech provides funding for Weissman's research into the development of additional infectious disease vaccines.

There won't be as much financial benefit as the university might hope, as Pfizer and Moderna's stocks are tumbling [839] , due to lack of consumer uptake of the latest Covid shots.

As a side note, **Dr. Weissman co-authored a paper[840] in 2018 "MRNA vaccines – a new era in vaccinology" in which both advances in, and safety concerns about, mRNA vaccines were outlined**. The safety concerns included **possible blood clots, systemic inflammation, lack of control over where the vaccine goes in the body and how long it stays there, autoimmune diseases, and the potential toxicity of lipid nanoparticles.** As has been outlined throughout this paper, **none of these safety concerns were addressed and resolved before the Pfizer and Moderna Covid vaccines were manufactured and distributed to the world.**

Although the Nobel Prize is a prestigious award and is often well-deserved, it is also known to be quite political. Dr. McCullough notes[841], **"Ironically, this invention has altered synthetic mRNA turning it deadly since there is no way of shutting off production of the lethal Spike protein produced from Covid-19 genetic vaccines."**

McCullough believes as more people become aware of vaccine harms,

Weissman and Karikó's work will join Nobel Laureates remembered for inventions that later were proven to have injured and killed untold number of humans, including Nobel (dynamite), Haber (chlorine gas used in gas chambers), Muller (DDT), and Moniz (prefrontal lobotomy).

There is a movement to protest the awarding of the Nobel Prize to the scientists who helped develop the flawed and deadly mRNA vaccines. The petition[842] calls for "signatures to honor the memory of the people worldwide who have died tragically as a result of these improperly tested products."

> There's a love affair with messenger RNA - our US government, governments all over the world are so deep into this - tens of billions of dollars of investments since 1985... And now we've learned that Moderna just reported a messenger RNA RSV trial in adults. RSV is not a threat. Way less than 1 % of people ever get RSV as adults. It's just an unnecessary vaccine. But they are advancing messenger RNA for RSV, for Epstein-Barr, for influenza. The vaccine syndicate as we chronicle in our book *The Courage to Face Covid-19*, this biopharmaceutical complex appears to be hellbent on messenger RNA technology no matter how unsafe it is.
>
> **DR. PETER MCCULLOUGH**

Anthony Fauci Announced as the 2024 Inamori Ethics Prize Winner

The Inamori Center[843] at Case Western Reserve University is devoted to "ethical inquiry in both its practical and theoretical aspects, and to facilitating the development of future leaders who will, in the words of Dr. Inamori, "serve humankind through ethical deeds rather than actions based on self-interest and selfish desires."

Humble public servant Dr. Fauci, whose household net worth during the pandemic increased[844] from $7.6 million to $12.6 million, was announced[845] on October 25, 2023, as the 2024 Inamori Ethics Prize winner.

"Dr. Fauci has cared not only for the nation's health but also the health of the world," said Case Western Reserve President Eric W. Kaler[846]. "As a scientist, research leader and public health advisor, his contributions to scientific discovery have truly improved lives. His leadership through one of the most challenging times in history – the Covid-19 pandemic – serves as a model for us all."

Fauci, who has declared that attacks on him are attacks on science[847], will be conferred the prize and monetary award of $35,000 at a September 2024 ceremony where he will "deliver a free public lecture about his work." No doubt Dr. Fauci will take time off, to receive the Inamori award, from his work at Georgetown University where he has been awarded multiple lucrative professorships, but is reported to not teach any classes.

One could hazard a guess that Case Western has not spent much time reading *The Real Anthony Fauci* by Robert F. Kennedy, Jr.

Peter Hotez receives Inaugural IDSA Anthony Fauci Courage in Leadership Award

On October 11, 2023, it was announced[848] that "Dr. Peter Hotez, co-director of the Texas Children's Hospital Center for Vaccine Development (CVD) and professor and dean of the National School of Tropical Medicine at Baylor College of Medicine, has been awarded the inaugural Infectious Diseases Society of America (IDSA) Anthony Fauci Courage in Leadership Award **for his efforts to uphold and speak to scientific truths.**"

The Anthony Fauci Courage in Leadership Award[849] "celebrates individuals who inspire and encourage others to make a difference and is awarded to a person who has demonstrated courage in leadership, a commitment to promoting **scientific integrity, advocating for sound science, and advancing the field of infectious disease** at their institutions or in their local, national, or global communities."

Hotez, author of the newly released book *The Deadly Rise of Anti-Science: A Scientist's Warning*, states,

> I'm thrilled to be honored by the IDSA, and it's especially meaningful to have the name of my longstanding colleague and mentor, Dr. Anthony Fauci, attached to this award. Beyond being a vaccine scientist, over the past decades I have also sought to counter anti-vaccine activism.

The summary of ***The Deadly Rise of Anti-Science*** reads like a propaganda love note, full of self-praise and misinformation:

> During the height of the COVID-19 pandemic, one renowned scientist, in his famous bowtie, appeared daily on major news networks such as MSNBC, NPR, the BBC, and others. **Dr. Peter J. Hotez often went without sleep, working around the clock to develop a nonprofit COVID-19 vaccine** and to keep the public informed. During that time, he was **one of the most trusted voices on the pandemic and was even nominated for a Nobel Peace Prize for his selfless work.** He also became one of the main targets of anti-science rhetoric that gained traction through conservative news media...
>
> [Hotez] explains how anti-science became a major societal and lethal force: in the first years of the pandemic, **more than 200,000 unvaccinated Americans needlessly died despite the widespread availability of COVID-19 vaccines.** Even as he paints a picture of the world under a shadow of aggressive ignorance, **Hotez demonstrates his innate optimism, offering solutions for how to combat science denial and save lives in the process.** (emphasis added)

In another show of selflessness, Dr. Hotez asked that on the recent "Dr. Peter J. Hotez Day," October 25, 2023 (so-declared by Houston Mayor Sylvester Turner), he hoped "more people will recognize the accessibility of adult immunizations, including a new annual Covid immunization and get their flu vaccinations." Hotez also expressed the "stress and heartache" he feels as he combats "growing anti-vaccine activism."

A Russian immigrant who could enlighten Hotez, if the doctor would listen:

Dr. Hotez seems mystified that anyone would question the Covid-19 vaccines, or any vaccine, for that matter. He appears to think that people who question the official narrative are all uneducated conspiracy theorists who must be stopped. Hotez and others would do well to listen, rather than to blame and vilify those who have a differing viewpoint from theirs.

For example Hotez could learn something from **Konstantin Kisin, a Russian immigrant in the U.K.** Kisin's grandmother was born in a Russian Gulag (essentially a concentration camp for citizens who didn't support the

Soviet regime). His parents worked and raised a family under the harsh control of the Communist Party in the Soviet Union. **Because of his family history, Kisin, who is a comedian and social commentator, is finely attuned to the signs of totalitarianism and the eroding of democratic values.** With regard to the Covid-19 vaccine mandates, Kisin states:[850]

> We went way too far in this country (the U.K.)...we had ...a genuine conversation about **forcing people, individual free citizens to inject something into their body because the governments decided that this is what they must do**. I don't agree with the government being able to force you to inject something into your body...**you've gotta be able to decide what goes in and outta your body...We are free citizens** and we're free to make decisions for ourselves, even if they have negative consequences for us.
>
> I was out in Parliament Square with a tiny number of people who felt the same as I did, **not because I was some wacko who thought the vaccines were, you know, some globalist conspiracy** to, to whatever. Whatever the nonsense was that was being spread around. **I felt the lines about our freedoms and our rights as individuals were...being crossed.** And the reason I love the West...is I think this is one of the few societies that respect your right as an individual to make your own choices.

...We reacted (to the pandemic) in a way that was disproportionate, people and government, and I hope we learned some lessons from that. I really, really do. Because to me, I think it's a very real threat that in order to buy more safety, we give away more and more of our freedoms. And that is not a trade-off that we should be making in the West because the West is built on very different principles.

KONSTANTIN KISIN | Russian immigrant | April 29, 2022[851]

The White Hats versus the Black Hats:

Brownstone founder and president Jeffrey Tucker, notes that the fearful, embattled views shared by Hotez reflect those of a very large group. Tucker writes[852], "Life today feels ever more like an old Western film with white hats and black hats in a forever battle for the control of the town." He invites us to consider the situation and speculate on which side is going to win:

> Let's begin with a gratifying scene, a gathering of public health officials, large foundations, tech companies, operatives of the Democratic National Committee, and big media muckety mucks. You will hear alarming tales of loss, sadness, and near defeat.
>
> They speak as victims, surrounded on all sides, overwhelmed by opposition. **They describe a world awash in disinformation, misinformation, and malinformation.** And they don't have any idea of what to do about it.
>
> You can get a sense of this view by looking at Peter Hotez's new book "*The Deadly Rise of Anti-science: A Scientist's Warning*" (2023). Talk about a tale of woe! **You would think that all that stands between them and marauding masses with pitchforks and torches is a handful of truth-telling scientists, forever underfunded and embattled, and vastly numbered and outspent by mobs bent on the destruction of modernity.**

Tucker states, "Yes, I read and hear these things and just laugh. After all, **these are the very people who were powerful enough to lock down nearly the entire world for a virus with a 99.8 percent survival rate** in which the median mortality extended beyond the average lifespan. **Then they had the wherewithal to force shots on billions of people who didn't want them or need them**, and harm so many people...These are very powerful people, in fact, and funded by the world's largest governments, foundations, non-government organizations, and global agencies. **They are the establishment and rather well-to-do, thriving as never before. But they don't see it this way. They imagine themselves as an embattled minority, fighting for their lives and careers.**" Tucker continues:

> [M]edia opinion on the topic was nearly uniform. Daily and hourly, we were badgered and bullied. They harangued us to mask up, lock down, and be afraid of getting sick to the point that we gave up the freedom of assembly and worship. If you had a different view and posted it on social media, you risked being throttled, struck down, and even fully canceled.

Contrary to the black hat view, what Tucker sees as he attends various conferences of the "**white hats" are "small business owners, churchgoers, parents with school-age kids, and just average young professionals struggling to pay the bills, keep healthy, and get by as best they can.** They have had the living daylights beat out of them for nearly four years straight." Tucker finds:

> They are still in shock at what happened. Life seemed more-or-less normal and then suddenly it was not. The businesses, schools, and churches were closed. They couldn't see their parents in retirement homes. They couldn't travel. They were told not to go out in public without covering their faces. The only consolation was the time and opportunity to watch endless movies on streaming services for which they paid with stimulus checks.
>
> **Later the inflation arrived and hasn't gone away.** It keeps eating away at the standard of living, with the dollar having lost between 17 and 20 cents of value in a mere three years. Household income is down, and the cost of borrowing is sky-high. They are trapped in their homes with large mortgage bills. They dare not sell because they cannot afford to buy another home of the same size.
>
> Meanwhile, they are surveilled in their online activities, monitored in their movements, and censored in their communications. Government keeps getting bigger and more intrusive, preaching inanities that are inconsistent with the good life. Drive less. Give up your gas stove. Get a solar panel. Have fewer kids. Eat less meat. Instead eat bugs.

> So yes, it's true, there are masses of people out there in the United States and around the world that have a strange feeling that very powerful people run the world but do not care about the well-being of average people. (emphasis added)

Tucker summarizes, "The last four years have taught us much about the workings of the world that we did not know. **The real rulers of the political social order who once hid in the shadows came out into the open.** To use a metaphor drawn from poker, **they tipped their hand.** So we see who is involved and what is up, perhaps for the first time in our lives." Tucker concludes:

> We won't forget these lessons, at least not soon. The passion, knowledge, and energy is on the side of the white hats. What we lack in money we make up for in creativity and determination. This is why we see so many trends going in the right direction. We aren't winning yet but you can feel it coming. More and more people are discovering what's up and what needs to be done to fix it.

These are good points and encouraging words, which is so important, because we white hats have a long way to go, and everyone who sees things clearly is needed.

33. THE COVID-19 VACCINE MERRY-GO-ROUND

I signed off this morning on...a request for additional funding for a new [Covid] vaccine that is necessary, that works...Tentatively...it will likely be recommended that everybody get it, no matter what they got before.

PRES. JOE BIDEN

August 28, 2023

We are now entering one of those more extreme periods, where the hierarchy really becomes clear. Those pulling the public health strings have gained enormous power and profit from Covid-19 and are focused on getting more. Their chosen enforcers did their job during Covid-19, turning a virus outbreak that kills near an average age of 80 years and at a rate globally perhaps slightly higher than influenza into a vehicle to drive poverty and inequality. They continue to do this, pushing 'boosters' associated with rising rates of the infection they are aimed against, and with unusual evidence of harm, ignoring prior understanding of immunology and basic common sense.

DAVID BELL[853]

Former medical officer World Health Organization

30-year Director of FDA vaccine research resigned in protest of Covid boosters:

In September 2021 Dr. Marion Gruber, the highly-respected director of the FDA's Office of Vaccines Research and Review, resigned[854] after 32 years with the FDA, over the way the **Covid boosters were being pushed without**

normal FDA safety protocols. Gruber's Deputy Director, Dr. Philip Krause, also resigned from the FDA for the same reason at this same time. In a September 2023 opinion piece[855] for the *New York Post*, Dr. Marty Makary and Dr. Tracy Beth Hoeg note:

> Ever since the loss of these two vaccine experts, the agency's vaccine authorizations have been consistent with an overly cozy relationship between pharma and the White House. Pushing a new COVID vaccine without human-outcomes data makes a mockery of the scientific method and our regulatory process.

None of the concerns with the original Covid mRNA shots have been resolved, and **the mounting reports of vaccine injury are being ignored** by those who are busily promoting the latest Covid shots[856]. The fall 2023 "Covid booster" was rebranded the "updated vaccine." An FDA spokesperson told CBS News[857] in an email, "To clarify, **these vaccines would not be considered 'boosters' per se.** These vaccines, as previously announced, would be updated with a new formulation for the 2023-2024 fall and winter seasons."

> **Bye bye, booster. We are no longer giving boosters, and it's going to be very difficult to stop using that word because that word has become pervasive...We are beginning to think of COVID like influenza. Influenza changes each year, and we give a new vaccine for each year. We don't 'boost' each year.** (emphasis added)
>
> **DR. KEIPP TALBOT** | Member CDC committee of vaccine advisers September 2023[858]

Note Talbot's comparison of Covid-19 to the flu, a comparison that was taboo[859] until this year, and could get you censored and canceled, as happened in the case of Stanford Professor of medicine, epidemiology, and statistics John Ioannidis[860] in 2020.

Pfizer, Moderna, and Novavax were all granted Emergency Use Authorization for their fall 2023 "updated vaccines." The taxpayer funded money will keep flowing. Unfortunately, much of that money compromises the very government agencies that should be looking out for the health and best interests of American citizens.

Might we ask why a slew of expensive new Emergency Use Authorized vaccines are needed for an illness that the government is now comparing to the annual flu, and for which the national emergency ended on May 11, 2023? That answer won't come from the government, as evidenced by their calling the end of the Covid-19 emergency not "The End," but a "Public Health Emergency *Transition*." (see here[861], here[862], and here[863])

The Illusion of Consensus for Covid-19 Vaccines:

> *Covid jabs may not be good for much in the way of protection. But they sure have immunized a huge swath of reporters from anything resembling reality.*
>
> **ALEX BERENSON** | Former *New York Times* Reporter | November 10, 2023 [864]

Throughout the entire pandemic, there has been a narrative from government and public health officials, and the pharmaceutical industry, backed by the legacy media (traditional media sources), that vaccines are good, and the Covid-19 vaccines are especially good. End of discussion. But the "consensus" that that has been carefully crafted by those with the money, power, and influence is only an illusion, based largely on propaganda.

Covid shots post-National Health Emergency: Team Reality vs The Narrative

Following are some quotes extending from the official end of the Covid-19 Public Health Emergency on May 11, 2023, to the release of the Covid booster shots in the fall.

Quotes from proponents of the official Covid narrative are in regular type, while questions and concerns that are largely silenced and kept from the general public are in italics. While scrolling through, it is interesting to ask the question: How might our pandemic response, and more specifically the vaccine rollout, have been different if people had been allowed to hear both sides of the debate and had informed consent?

* * *

"As the COVID-19 Public Health Emergency (PHE) ends on May 11, 2023, the Administration has taken significant steps to ensure all individuals have continued access to lifesaving protections such as vaccines, treatments and tests, and that the nation is well prepared to manage the risks of COVID-19 going forward... The Biden-Harris Administration is also continuing its strong support of efforts to enhance global genomic surveillance and associated surveillance networks, to rapidly identify emerging variants, their spread and impact around the world."

THE WHITE HOUSE | May 9, 2023 [865]

"We have this huge industry being set up to surveil for variants. And the model that they are pushing for public health is that they will find variants. They will lock people down. They will mandate or allow them to have a vaccine to get their freedom back. And then the vaccine will make a vast profit. And then the cycle will go around, they'll find another variant. And this is a future that is sort of mapped out."

DR. DAVID BELL | Public health physician
Former WHO medical officer | September 3, 2023 [866]

"There's a new vaccine that's coming around the corner, a new mRNA COVID-19 vaccine, and there's essentially no evidence for it...There's been no clinical trial showing that it is a safe product for people — and not only that, but...there are a lot of red flags... [these vaccines] actually cause cardiac injury in many people."

DR. JOSEPH LADAPO | Florida Surgeon General | September 9, 2023[867]

"We have experience with this type of vaccine in billions of people. It's a safe vaccine."

DR. ANTHONY FAUCI | Former Director NIAID | September 10, 2023[868]

"Today, the U.S. Food and Drug Administration took action approving and authorizing for emergency use updated COVID-19 vaccines formulated to more closely target currently circulating variants and to provide better protection against serious consequences of COVID-19, including hospitalization and death...these vaccines have been updated to include a monovalent (single) component that corresponds to the Omicron variant XBB.1.5...The FDA is confident in the safety and effectiveness of these updated vaccines...the benefits of these vaccines for individuals 6 months of age and older outweigh their risks."

FDA PRESS RELEASE | September 11, 2023[869]

"The Food and Drug Administration has authorized and approved an updated mRNA vaccine for use in adults and children aged 6 months and up. This new shot targets the XBB.1.5 Omicron subvariant, which currently makes up only 3% of cases in the United States... The government's ill-advised effort to encourage another vaccine booster lacks medical justification and erodes the public's already waning trust in our public health authorities."

FLCCC PRESS RELEASE | September 12, 2023[870]

"The vaccine booster rollout is deeply flawed. We do not know enough about the COVID vaccines to make the broad recommendations that the FDA is making. There is so little data available on the safety of this latest booster, the FDA's actions create an unnecessary risk to the public's health, especially children and those who might still be suffering the life-altering side effects from previous COVID vaccinations and boosters."

DR. PIERRE KORY, MD, MPA | September 12, 2023[871]

"Vaccination against COVID-19 remains the most important protection in avoiding hospitalization, long-term health complications, and death. I encourage all Americans to stay up-to-date on their vaccines."

THE WHITE HOUSE | September 12, 2023[872]

"We have more tools than ever to prevent the worst outcomes from COVID-19. CDC is now recommending updated COVID-19 vaccination for everyone 6 months and older to better protect you and your loved ones."

MANDY COHEN, MD, M.P.H | Director Centers for Disease Control
September 12, 2023[873]

"[T]he Covid-19 vaccines and all of their progeny and future boosters are not safe for human use."

DR. PETER MCCULLOUGH | Cardiologist
Testimony before European Parliament | September 13, 2023[874]

"Those who have received the product are more likely to develop Covid than those who have not been injected. And there are suggestions in the data that the period of time between injection and "negative effectiveness" is getting shorter. Even if the "vaccine" products have zero rather than negative effectiveness, they certainly have toxicity risks, so why would anyone be willing to receive these products if they knew this?"

DR. ROBERT MALONE | September 15, 2023[875]

"[W]hen I think about the fall ahead, I'm really looking at a period where there could be a lot of serious illness, a lot of suffering – but so much of it is preventable...[T]he data so far suggests that the new Covid vaccine should be very effective against the circulating strains. We're getting into a new cadence now where we're going to have annual COVID vaccines, and it becomes in that way very much like flu – very manageable."

DR. ASHISH JHA | Former White House Covid-19 Response coordinator | September 19, 2023[876]

"These signals (deaths and disabilities) are being seen by all the health authorities that are issuing and following health edicts, and they're ignoring them. These signals are so large that there has to be a reason why. My thesis... is it's the Pfizer and Moderna vaccines that they mandated on us, and this is a ginormous cover-up."

ED DOWD, ANALYST | Author of Cause Unknown | September 19, 2023[877]

A coronavirus disease 2019 (COVID-19) vaccine can prevent you from getting COVID-19 or from becoming seriously ill or dying due to COVID-19... Each COVID-19 vaccine causes the immune system to create antibodies to fight COVID-19.

MAYO CLINIC STAFF | November 4, 2023[878]

Our World In Data – Covid-19 in 2023:

Was Covid-19 a big threat in the fall of 2023? Was it in 2022? The data says no. Almost everyone in the world has been exposed to Covid, had Covid-19 infection, and/or been vaccinated. At this point in time, the endemic virus continues to circulate, causing little hospitalizations and almost no deaths, as evidenced by a sampling of countries in these charts from Our World In Data:

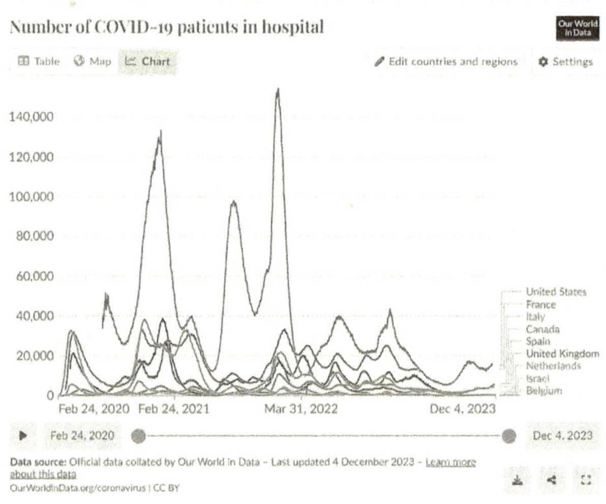

https://ourworldindata.org/covid-hospitalizations[879]

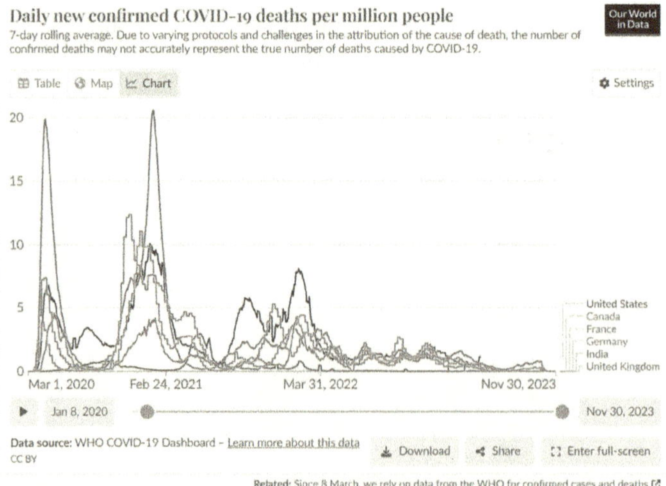

https://ourworldindata.org/covid-deaths[880]

34. PANDEMIC RESPONSE: A SUCCESS OR FAILURE? IT DEPENDS...

That's the iron law of bureaucracy. They serve their own internal conveniences and interests, and they rarely get smaller.

MATTHEW B. CRAWFORD

American Philosopher | Jan 17, 2023[881]

What's that old definition of insanity? Doing the same thing over and over and expecting different results. But then again, it all depends on what you were trying to accomplish.

If the goal of the official pandemic response was to handle a virus in the least intrusive way possible, protecting the vulnerable, while helping society continue on with normal life, then **the pandemic response was an abject failure.**

But.....

If the overall goal of the pandemic response was to gain power, money, and control over the global population through fear of a virus, and compliance with visible government mandated behaviors, then **the official pandemic response was quite successful.**

Christine Anderson of the European Parliament said at the International Covid Summit 3 (18:15[882]) on May 7, 2023, "Covid was just sort of a test balloon and the lessons they learned from this, namely... where did they fail to get the people to just do what they were told, to...comply. **And they will take this to the next level...digital I.D. It's in preparation**...The next thing we will be seeing is digital state currency that will be the ultimate blow to depriving us of all our freedom and privacy..."

The WHO declared an end to Covid-19 as a global health emergency on May 5, 2023. Its umbrella organization, the U.N., noted[883] with crocodile-tears concern, "the enormous damage inflicted on all aspects of global life by the virus, including enormous economic upheaval, 'erasing trillions from GDP, disrupting travel and trade, shuttering businesses, and plunging millions into poverty.'"David Bell, public health physician and former medical officer at the WHO counters:[884]

> To restate the obvious, this does not happen due to a virus targeting sick elderly people. It occurs when children and productive adults are barred from school, work, healthcare, and participation in markets for goods and services. Economic, social, and health catastrophes inevitably result, disproportionately harming poorer people and low-income countries, conveniently far indeed from the halls of Geneva and New York.

Concerned individuals worldwide need to better inform ourselves and look at our governments, and the globalists, with a critical eye. American philosopher Matthew B. Crawford notes[885] the **use of emergencies, and the propaganda surrounding them, has led to increasing top-down control of our day-to-day lives.** Crawford references political theorist in Italy, Giorgio Agamben, pointing out how in the liberal democracies of the West, the 'state of exception' has almost become the norm. Crawford states:

> So the language of war is invoked to pursue ordinary domestic politics. In the U.S., in the last 60 years we've had the war on poverty, the war on drugs, the war on terror, the war on Covid, and now the war on disinformation, the war on domestic extremism. So it sort of becomes normalized to have a state of emergency.

Interviewer Freddie Sayers asks Crawford, "Do you now feel like we are more **vulnerable to...danger of having our fundamental way of life being shifted through this never-ending emergency?"** Crawford answers:

> I think it has shifted. And I also think that the further normalization of emergency as the basic sort of idiom of government, you see it all around. Of course, climate provides the ultimate emergency...So you can be fully convinced of climate change, and the dire consequences of it, while also noticing that it has this feature of being ideal for purposes of centralizing power.

Note the rhetoric from the 2023 U.S. Fifth National Climate Assessment[886] (NCA 5) which connects climate change and pandemics:

> Climate changes can also speed the transmission of pathogens and promote the establishment of new diseases. Coupled with global travel networks and dense urbanization, novel pathogens can rapidly spread to areas far from their origins. Our understanding of COVID-19 is evolving, but the pandemic demonstrates the global threat of emerging infectious diseases and has raised awareness of linkages between climate change and zoonotic diseases.

The NCA 5 falsely claims that Covid-19 was a zoonotic disease, meaning it was the result of virus transfer from animals to humans. (There has not been one shred of scientific evidence[887] that Covid-19 was a zoonotic disease.) Aside from that falsehood, the **NCA 5 calls out elements of our modern way of life** – urbanization, daily living with sufficient fuel and food, global travel – **and turns them into negatives that foster both climate change and disease spread.** The Assessment calls people "vectors," and is loaded with garbled nonsense, such as this statement:

> Climate-related disasters have interacted with COVID-19 throughout the course of the pandemic in multiple ways. Certain communities—including essential workers, older adults, low-wealth communities, and communities of color—are disproportionately impacted by these compounded exposures.

When Charlie Chester, Technical Director of CNN told an undercover reporter in April 2021 that **climate change was going to be the next big**

thing in the news cycle, once the pandemic lost its newsworthy status, he wasn't exaggerating. (start at 5:30[888] on the video):

> I think there's, like, Covid fatigue, so whenever a new story comes up they're going to latch onto it. They've already announced in our office that once the public is, will be open to it, we're going to start focusing mainly on climate. Climate, like global warming…it's going to be our focus…[the next] pandemic-like story, that we'll beat to death, but that one's got longevity. You know what I mean?... "[The pandemic will] taper off to a point that it's not a problem anymore. Climate thing is gonna take years. So they'll probably be able to milk that for quite a bit.

Matthew Crawford points out that, "Climate catastrophism is overblown… but it has to be sort of catastrophized maximally in order to scare people into giving up not just freedoms, but a whole lot of activities that are woven into life at every level. Because…**in order to actually reduce carbon, you'd have to radically transform life.** I think the radicalness of it is precisely what's attractive about it to a certain mindset of social engineering…

"This isn't simply a libertarian point – 'Government off my back,' it's not that but there's a kind of militancy to this entity. It is hard to identify this blob, this governmental-like thing – often corporations are a big part of it – that **seems determined to reach into every aspect of life, including the most intimate, and sort of police it and reorder it.** And I think people are recoiling against that. So it isn't even a question of should I try to save the world or just hunker down? It's more like, how do I defend just the most basic way of life because it seems to be very much under threat."

Witness this **assault on our freedom of speech** coming from the Department of Defense in July 2023, in which every facet of life including the "linguistic," is viewed as part of a unified "whole-of-government" control of "information as an instrument of national power."

> The Information Environment is the aggregate of social, cultural, linguistic, psychological, technical, and physical factors that affect how humans and automated systems derive meaning from, act upon, and are impacted by information, including the individuals, organizations, and systems that collect process, and disseminate, or use information.

To truly make Operations in the Information Environment an effective national capability it must be applied by the Department of Defense in a deliberate and unified manner...**The United States must embrace a whole-of-government approach to the development and employment of Information as an instrument of national power.**

35. HOW DO YOU DECIDE WHAT IS TRUE?

History shows us some reasons why the administration may be so intent on sending lies into the press stream and accusing those who tell the truth of lying. Perhaps the barrage of lies serves a more substantial purpose than simply advancing a certain position. Sending a current of lies into the information stream is part of classic psychological operations to generate a larger shift—a new reality in which the truth can no longer be ascertained and no longer counts.

NAOMI WOLF

December 14, 2023

These are hard times. Many people I have spoken with feel there is no way they can know who is right, or what is true, because there are data and experts on both sides of the Covid-19 pandemic equation.

The CDC released a multi-page document quoting studies and experts, claiming success in handling the pandemic. In July 2023 the International Coalition of Medicines Regulatory Authorities (ICMRA[889]) prepared a similar document. To go page-by-page, and chart by chart, to analyze or refute what they said is almost impossible. But it's also not necessary.

> Some have decided it's too hard to figure it out for themselves, so they'll just "follow the science," or "listen to the experts," and do as they're told. But which science will they follow? And which experts? And what if the ones who have been choosing the narrative and setting the policies and rules for the rest of us don't have our best interests in mind? For the sake of our children and grandchildren, we have to ask these questions.

One of the strongest ways to determine truth-telling is to follow the money and the motivations. Ed Dowd states, "There's a bunch of people that still are trapped in the old mindset that...our institutions are okay." Dr. David Bell says people should stop complying[890] with authorities and agencies that take away freedom and tell people how to live, in the name of public health and safety. Also, states Bell, we "have to get away from this left-right thing," because this is a choice between freedom or fascism.

Robert F. Kennedy, Jr. states, "Today many of my liberal chums are still crouched in a knee jerk posture defending 'our' agencies against Republican slanders and budget cuts, never quite realizing how thoroughly the decades of attacks succeeded in transforming those agencies into subsidiaries of Big Pharma." (*TRAF*, Intro)

Journey of a "lefty liberal" from being Asleep to being Awake:

In a May 2023 interview[891], Dr. Pierre Kory, who identifies himself as a "lefty liberal," explained his own journey of being awakened to the fact that our institutions are NOT okay, in the following excerpts:

> **At the time when Covid started, I was a regular *New York Times* reader, and I believed it to be the paper of record and a symbol of top journalism.** If you really wanted to know what's going on, you read the *New York Times*.
>
> I felt similarly about medical journals, in particular the high-impact medical journals, like the *New England Journal of Medicine* and *JAMA* (*Journal of the American Medical Association*). I thought the best scientists were published there. **I believed that the health agencies had the primary focus of protecting the public health of our population, and they were the most expert at doing that.**
>
> I believed very strongly in those things. I was also a believer in the concept of good government...**I really thought the government could fix things and that government would probably be the best way to provide the solution.** I thought it was supposed to be there to level the playing field. I had that general political orientation. Obviously, politics is more complex than that, but that's where I started.

Now, I've been subjected to immense amounts of propaganda and censorship for three years in almost every august media outlet of journalism. I got transformed by becoming an expert on a few very important things around the science of Covid. I became deeply knowledgeable.

Being exposed to the control of mass media has been really disorienting. It's frightening to think that I've been relying on them as a source for accurate information for most of my life.

It's not just the *New York Times*, it goes across the mediascape. That's what I've come to today. I know how to spot narratives, or at least I think I do. I'm sure we're all still at risk of falling for narratives, but I have a sense of what propaganda is, and how it's deployed.

I've learned to be much, much more skeptical, if not just plain ignoring, of what the media says. I have to do my own deep dives and do my own research. I'm not going to believe anything just because it's printed in some newspaper.

I always knew pharma was bad. I didn't understand that they are literally a criminal syndicate, who have been committing crimes for decades. They pay fines, then move on and continue their standard operating business.

When it came to government, **I wasn't aware of how corporations have literally taken over almost all the agencies of government.** The response to **Covid was controlled and conducted by the pharmaceutical industry, with probably even bigger powers behind them.** But to understand the pharmaceutical industry, I looked at three years of every policy issued by those agencies.

All you had to ask yourself was, "What would a pharmaceutical company want?" Voila, there was your policy. Every single policy was in line with serving the interests of a pharmaceutical company. **Guess what that brought us? It brought us multiple**

> **humanitarian catastrophes**, millions of lives lost from the suppression of early treatment, millions of people dead around the world from the vaccines, and now epidemics of vaccine injury and long Covid with very little treatment.
>
> To say I've been transformed is an understatement. With a lot of folks that I know, **the adage is about going from blue pill to red pill. The journey has been very disturbing with what I've had to see and had to learn. It has been quite frightening and even dystopian, but I wouldn't trade it for the world.** That's number one.
>
> Number two, **some of the beauty has been in connecting with people who didn't fall for those lies, and who could spot the propaganda.** They taught me things and made me understand things that maybe I didn't want to understand, but that I've come to understand now.
>
> **Ultimately, it was a war of information. All of the destruction was about information and how it was controlled**...It was a war where the voices of truth and sanity were getting drowned out by lies that were told for different objectives. The CEO of Moderna has $4 billion of wealth. **The pharmaceutical companies made tens to hundreds of billions, with this massive transfer of wealth**. It has been a really difficult three years.
>
> Most of the population thinks it was a rough time and we're just going to move on now. That's not what my experience was. (emphasis added)

Dr. David Bell states, "A lot of people still trust the authorities and haven't figured out the basic [fact] that in the end it's about profit and money... **We've got to start seeing this as an industry and not as some altruistic arm of government that's just there for the greater good.** It's there to make money for the investors who have invested in the pharma companies. This includes the large investment houses, you know BlackRock, Vanguard, etc. **So people have really got to start to be skeptical about the world we're in. It's not the world we thought we had in Western countries."**

36. THE CULPABLE ARE ATTEMPTING TO REWRITE HISTORY

I didn't shut down anything.

ANTHONY FAUCI

August 23, 2022[892]

Former NYC Governor Andrew Cuomo Lies About His Pandemic Role: The lies and manipulation surrounding the Covid-19 pandemic are unprecedented, as are the attempts to rewrite history and the refusal to take responsibility. Former New York **Governor Andrew Cuomo**, whose administration was ground zero for the Covid-19 panic in the U.S. embodies the general attitude of all the authoritarians who rescinded civil liberties during the pandemic. In a September 20, 2023 interview[893] with Dr. Leana Wen, who was a media darling "professional expert" throughout the pandemic, Cuomo expressed concern about the loss of public confidence in the government:

> "Government had no capacity to enforce any of this," said Cuomo, referring to his rose-colored retro-pandemic world. If New York citizens had said "No!" to face masks or business closures, there was nothing he could have done about it, Cuomo stated, because, **"It was really all voluntary...society acted with that uniformity voluntarily, because I had no enforcement capacity."**

Cuomo reflects, "We still, if Covid happened today, we would be in worse shape than we were because now we have the governmental distrust factor... and we really haven't increased the operational capacity...The Federal government turned to the states...There is no public health workforce of any scale...None of that has advanced... since Covid."

He apparently claims no responsibility or memory for the very NOT voluntary Excelsior Pass[894] (vaccine passport required[895] to participate in public life), mask mandates[896], business closures[897], loss of jobs[898], and loss of life[899] that were direct results of his and Hochul's overreaching[900], unconstitutional[901], and outright creepy[902] mandates throughout the pandemic. (Kathy Hochul was Cuomo's Lieutenant Governor, and stepped into the role of Governor after Cuomo resigned in disgrace in August 2021, due to sexual misconduct.)

Cuomo and Wen lament the loss of public trust. "Public trust has been eroded by people who are promoting misinformation and disinformation," said Wen, as Cuomo nodded sagely. "Science is evolving," Wen continued. Based on the interview, both Cuomo and Wen think the answer to future pandemics is More – more central coordination on a national and global scale, more control, more enforcement.

California Governor Newsom tries to reframe his pandemic totalitarianism:

Others claim that we just didn't know much about Covid-19 in early 2020, so we did the best we could. But we did know[903], and the pandemic was not handled in accordance with years of established pandemic plan protocols.

Gavin Newsom, Governor of California and inflictor of some of the most harsh, long-term Covid policies in the U.S., stated in a September 10, 2023 interview[904], "I think we would've done everything differently...I think all of us in terms of our collective wisdom, we've evolved. **We didn't know what we didn't know. We're experts in hindsight. We're all geniuses now.**" This is the same **Gavin Newsom** who was unconcerned enough about Covid in November 2020 that he was partying it up, unmasked and indoors at an obviously *not* socially distanced dinner at the French Laundry[905] restaurant, while imposing strict masking[906] and social distancing[907] rules on California citizens.

Although Newsom claimed "we would've done everything differently," he was **quick to defend his abysmal pandemic policies when confronted with the opposite approach that was taken by Gov. Ron DeSantis in Florida.** Florida indeed did "everything differently"[908] from California.

With the exception[909] of the almost nationwide closures at the beginning of 2020, Florida kept schools, businesses, and churches open, rejected mask mandates[910] starting in September 2020[911], and passed laws[912] to prohibit vaccine passports.

According to the official narrative, Florida should have done much worse than California, because it didn't mandate masks, closures, or vaccines. **But CDC mortality data[913] shows similar Covid stats for both California and Florida, despite Florida's older median age.** Meanwhile in Florida, kids were in school, small business owners thrived, churches were open, and the Super Bowl brought big crowds to Florida in 2021, which despite critics' doom-saying, was not a super-spreader event.

During the pandemic, people from around the country moved to Florida, and vacationed in Florida (see here[914], here[915], and here[916]), where it still felt like the U.S. was a free country. California saw an exodus of residents[917] from the state, can't account for 152,000 school-age children[918] not returning to school after a year-and-a-half of closures, and caused thousands of small businesses[919] to close their doors permanently, due to Covid regulations.

The substantial difference between California and Florida during the pandemic, is California was run by a governor who violated human rights and freedoms during the pandemic, while Florida's Ron Desantis did not. Gavin Newsom should be held accountable[920] for his horrific pandemic response, instead of being allowed to smugly sit unquestioned while he says, "We didn't know what we didn't know."

AFT President Randi Weingarten Lies: "I fought to reopen schools in 2020":

American Federation of Teachers President Randi Weingarten had significant influence[921] on the failure[922] of public schools to be open for regular in-person teaching in the fall of 2020. Weingarten was also instrumental in requesting that U.S. Attorney General Merrick Garland investigate parents[923] who were expressing displeasure at school board meetings over Covid policies, and curriculum.

Weingarten's insistence on strict "safety measures" led to many schools being closed, or partially closed, to in-person learning for many months and even years. But when she testified[924] on April 26, 2023 before the House Select Subcommittee on the Coronavirus Pandemic, Weingarten said, "We spent every day from February on trying to get schools open." Apparently she forgot how she encouraged the CDC to put such strict and unrealistic pandemic rules in place for opening schools, that it made it almost impossible for many to do so.

Mindful members of the public noted that "areas with high union influence remained closed much longer" than areas where unions have lower influence. The *New York Post* reported[925] that Weingarten "consulted with [then] CDC Director Rochelle Walensky and White House officials prior to a Feb. 12, 2021, announcement that slowed a full reopening of in-person instruction for students." The *NY Post* also shared a string of texts[926] exchanged wherein Walensky calls Weingarten "friend," and appears to take counsel on how to word the February 12, 2021 CDC guidance.

Yet Weingarten tweeted[927] on April 28, 2023, "Covid scared the nation, but we tried to do everything we could to reopen schools safely."

When asked to provide scientific data backing the claim that school was an unsafe place for students to return to during the pandemic, Weingarten sidestepped[928]. The AFT, which has about 1.7 million members, donated[929] more than $2.3 million to Democratic candidates during the 2022 election cycle. Six of the eight Democratic members of the House Select Subcommittee on the Coronavirus Pandemic received thousands of dollars in campaign contributions from AFT, according to a *New York Post* article[930].

Jennifer Sey advocated for the opening of public schools in San Francisco, and lost a lucrative and prestigious job at Levi's[931] because of it. Sey explains[932] that Weingarten did **not** go to bat for children during the pandemic:

> **Weingarten and the CDC ignored all actual evidence that open schools did not increase risk and spread in communities, regardless of community spread levels.** Evidence in red states, in Sweden, in Denmark and all across Europe abounded, as early as spring and summer 2020. Often schools served as brakes on transmission, and were the safest places for teachers and kids to be.
>
> Yet **Weingarten persisted in vilifying children. So, while bars and strip clubs opened, schools remained closed.**

The fact is, no one fought harder to keep kids out of the classroom than teachers' unions.

A Fairfax County Virginia Parents Association spokeswoman said[933] on June 2, 2023, "As parents who very closely watched the political maneuvering over school closures at the local, state, and federal level, **it was apparent to us from the start that the CDC was deferring to teachers unions and not science when it came to reopening schools**...Since June 2020, parents in Fairfax County were sounding the alarm about the **emotional, mental, academic, and social harms happening to children during the school closures.** As we have always told our parent advocates, **teachers unions exist to represent the interests of their dues-paying members. They do not exist to advocate for the interest of children.**"

Sey writes that schools are about our children, "But **Weingarten** didn't care. She made it all about her. And she's doing it again now in her attempt to rewrite history. She **wants to be remembered as a hero in the open schools debate, not the villain responsible for generational harm.**"

The moment calls for truth, not for self-justification:

The severe-pandemic-restriction players now call for amnesty[934], for understanding, for a little bit of grace, because "Nobody knew, and we were just doing our best."

For another example beyond Fauci, Cuomo, Newsom, and Weingarten, observe New York University Professor Scott Galloway on Bill Maher, October 27, 2023:

1:15 AM · Oct 29, 2023 · **970.1K** Views

Prof. Scott Galloway calls for "grace" Oct. 29, 2023 (41 seconds)[935]

Now here is Prof. Scott Galloway in October 2021, telling the unvaccinated that they had their "head up their a**," and are incompetent, ignorant, and endangering their own and other people's lives. It was well known by October 2021, for anyone willing to listen, that the vaccine was not stopping infection, transmission, hospitalization, or death. People were losing their jobs for refusing the shot. The vaccine-injured were crying for help, and many people had already lost loved ones to the Covid-19 shots. There was ample evidence for concern, caution, and an end to the vaccine rollout, but Prof. Galloway chose to tweet this:

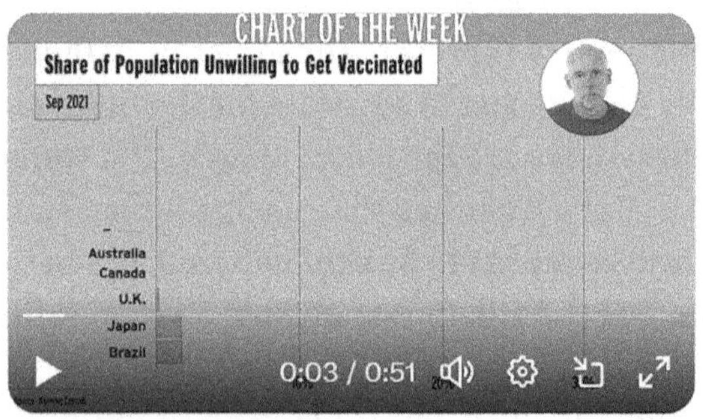

Prof. Galloway dumps on unvaccinated Oct. 2021 (51 seconds)[936]

Galloway admits he was a proponent of stringent lockdowns, including at his own child's school, and says "I was wrong. The damage to kids of keeping them out of school longer was greater than the risk. But here' s the bottom line...**we were all operating with imperfect information and we were doing our best. So let's learn from it. Let's hold each other accountable. But let's bring a little bit of grace and forgiveness** to the s*** show that was Covid."

Biologist and scientist Alex Washburne[937], who was **bullied and canceled for questioning lockdowns and other pandemic policies,** responds to Galloway and to all those who claim we should just forgive each other and move on:

> While I'm grateful to see people apologizing for lockdowns, apologizing for harming kids, and so on, there's still some unsettled dust we need to discuss before the balm of mercy can be applied...

For all their talk about diversity, equity, inclusion, and justice, many white, liberal, and privileged academics have a lot to learn about tolerance. The response to my personal advocacy was not tolerance, curiosity, understanding, and compassion, but rather call-outs from people who grew up in private schools, and a persistent blocking and bullying...

When I hear these people call for Covid amnesty, while I remain blocked and shunned by people with immense power in our academic institutions, while my reputation is dragged through the mud with lies and mischaracterizations about my truths and my character, forgive me but I have a difficult time being merciful...

Until we have meaningful reconciliation, amnesty will merely cement the incumbents' hold on academic, media, and narrative power, all but ensuring we repeat the failures of pandemic public health policy.

ALEX WASHBURNE | Biologist/Scientist

These hypocritical government and public health officials, and other public leaders, these 3-letter organizations, these individuals and organizations who are now backpedaling about their policies and trying to rewrite pandemic history – they are an offense and a threat. **They make token apologies to the applauding herd, explaining that they were working from "imperfect knowledge." But as Jeffrey Tucker points out, "Information is always everywhere 'imperfect' on all topics in all times and places."** Hence the need for freedom of speech and thought, so as to make the best decisions possible in this world where imperfect knowledge is par for the course.

The continuous manipulation and lies from these people are an insult to every dissident doctor, scientist, journalist, and citizen who was censored and canceled for shouting into the void of officialdom from Day 1, "Something is wrong with how this is being handled!" Those who implemented the devastating pandemic response, and now continue to support the official narrative, are tone-deaf and disrespectful of the fact that their choices and actions harmed millions of people.

Don't Let Them Get Away With It:

The poem "Mistakes Were NOT Made," by Margaret Anna Alice, captures the gravity and scope of what was inflicted on humanity by self-serving and scheming forces during the pandemic:

> The Armenian Genocide was not a mistake.
> Holodomor was not a mistake.
> The Final Solution was not a mistake.
> The Great Leap Forward was not a mistake.
> The Killing Fields were not a mistake.
>
> Name your genocide—it was not a mistake.
> That includes the Great Democide of the 2020s.
> To imply otherwise is to give Them the out they are seeking.
>
> It was not botched.
> It was not bungled.
> It was not a blunder.
>
> It was not incompetence.
> It was not lack of knowledge.
> It was not spontaneous mass hysteria.
>
> The planning occurred in plain sight.
> The planning is still occurring in plain sight.
>
> The philanthropaths bought The $cience™.
> The modelers projected the lies.
> The testers concocted the crisis.
> The NGOs leased the academics.
> The $cientists fabricated the findings.
> The mouthpieces spewed the talking points.

The organizations declared the emergency.
The governments erected the walls.
The departments rewrote the rules.
The governors quashed the rights.
The politicians passed the laws.
The bankers installed the control grid.

The stooges laundered the money.
The DoD placed the orders.
The corporations fulfilled the contracts.
The regulators approved the solution.
The laws shielded the contractors.
The agencies ignored the signals.

The behemoths consolidated the media.
The psychologists crafted the messaging.
The propagandists chanted the slogans.
The fact-chokers smeared the dissidents.
The censors silenced the questioners.
The jackboots stomped the dissenters.

The tyrants summoned.
The puppeteers jerked.
The puppets danced.
The colluders implemented.
The doctors ordered.
The hospitals administered.

The menticiders scripted.
The bamboozled bleated.
The totalitarianized bullied.
The Covidians tattled.
The parents surrendered.
The good citizens believed ... and forgot.

This was calculated.
This was formulated.
This was focus-grouped.
This was articulated.
This was manufactured.
This was falsified.
This was coerced.
This was inflicted.
This was denied.

We were terrorized.
We were isolated.
We were gaslit.

We were dehumanized.
We were wounded.
We were killed.

Don't let Them get away with it.
Don't let Them get away with it.
Don't let Them get away with it.

An annotated version of the poem, with hyperlinks to each of the referenced events, is found here[938]. Following is a link to Dr. Tess Lawrie, of the World Council for Health, reading Mistakes Were Not Made:

Mistakes Were NOT Made (4 min.)[939]

There were some standout people in positions of leadership and influence during the pandemic, who pushed back against the erosion of human rights. We need more like them, and they deserve our support. Anyone who values freedom and the enlightened principles that have shaped Western thought and democracies has a responsibility to stand up at this time and in the future.

Some are doing that in the form of Declarations, such as the Westminster Declaration[940] for freedom of speech, and the Student Declaration 2023[941], "A statement by students against past, present, and future mandates in order that the grievous harms and losses of the past three years of Covid-19 mandates are not repeated." The Student Declaration is an outgrowth of No College Mandates[942], a group of concerned parents, doctors, nurses, professors, students and other college stakeholders. Others have joined the Alliance for Responsible Citizenship[943] (ARC) bringing people together who "do not believe that humanity is necessarily and inevitably teetering on the brink of apocalyptic disaster." ARC sees beauty and potential in each of us fellow travelers on this planet Earth.

> We do not believe that we are beings primarily motivated by lust for power and desire to dominate. We do not regard ourselves or our fellow citizens as destructive forces, living in an alien relationship to the pristine and pure natural world... We hope to encourage the development of an alternative pathway uphill, out of both tyranny and the desert, stabilizing, unifying and compelling to men and women of sound judgement and free will.
>
> **ALLIANCE FOR RESPONSIBLE CITIZENSHIP**

Doesn't that above quote sound better than the constant catastrophizing, divisiveness, and dehumanizing that so many in positions of influence invoke today?

Each of us has a responsibility to ask the hard questions, look for answers beyond the official narrative, and take action in our own circle of influence. We are in a battle of ideologies. I choose freedom of thought, speech, movement, religion, association, health, and commerce. What do you choose?

MEDICAL TERMS GLOSSARY (LAYPERSON'S VERSION)

Definitions gleaned from all over the internet, without attribution, meaning the language in each definition is largely not my own.

ACE2 receptor – angiotensin-converting enzyme 2 (ACE2) is the protein that provides the entry point for the coronavirus spike protein to hook into and infect human cells.

adenovirus – common viruses typically causing mild cold- or flu-like illness. Adenoviruses can cause illness in people of all ages any time of year.

adenoviral vector – an engineered virus, in which a gene of interest is added to a common adenovirus. Adenoviral vector vaccines are called DNA vaccines. In the case of the Covid shots, a modified adenovirus carries the gene that instructs cells to make spike protein. Spike protein then surfaces on the cells, causing the immune system to detect foreign protein and begin to make antibodies.

adjuvant – a substance incorporated into some vaccines to enhance the immune response of the vaccinated individual. The Novavax Covid-19 vaccine uses a Matrix-M adjuvant which contains saponin extracts from the bark of the Soapbark tree that is native to Chile.

anergy – a lack of response by the body's immune system to an antigen or antibody.

antibiotic – a substance produced by or derived from certain microorganisms, that can destroy or inhibit the growth of other microorganisms, especially bacteria. Antibiotics do not work against viruses because there is no target organism to attack in a virus.

antibody – Any of a large number of proteins that are produced by specialized B cells after stimulation by an antigen and act specifically against the antigen. Antibodies consist of four subunits including two heavy chains and two light chains, and usually do not react with self-antigens due to negative selection of T cells in the thymus and B cells in the bone marrow.

antibody dependent enhancement (ADE) – Also known as pathogenic priming. The process whereby instead of providing a robust immune response, pathogen exposure acts like a Trojan horse, allowing the pathogen to get into cells more easily, creating an overreaction.

antigen – a substance that when introduced into the body stimulates an immune response, either alone or after forming a complex with a larger molecule (such as a protein), and is capable of binding with a product (such as an antibody or T cell) of the immune response. Antigens can originate from the body (self-antigens) or from the external environment (non-self). Antigens exist on viruses, bacteria, allergens, parasites, proteins, tumor cells and normal cells in your own body.

antigen escape – also known as immune escape or viral escape. A condition in which the body's immune system is no longer able to recognize and eliminate a pathogen, such as a virus.

antigen-specific – can only react to and bind one specific antigen.

antigenic imprinting – also known as original antigenic sin or immunological imprinting. Antigenic imprinting leaves the immune system less able to mount an effective response during subsequent infections because the body mounts a response to the prior pathogen, not the current mutation of that pathogen.

antiviral – a substance that fights against viruses and inhibits their growth.

attenuated vaccine – a vaccine that uses a living, but weakened, version of the virus or one that's very similar. Because these vaccines are so similar to the natural infection they help prevent, they create a strong and long-lasting immune response.

autoimmune diseases – Diseases in which antibodies react to self-antigens and damage the body's own cells. In other words, the body perceives itself as a threat and the immune system attacks the body instead of a pathogen.

B cell – also called B lymphocyte, one of the two types of lymphocytes (the other is the T cells). All lymphocytes begin their development in the bone marrow. B cells are involved in so-called humoral immunity. On encountering a foreign substance (antigen), the B cell differentiates into a plasma cell, which secretes antibodies. Following their release into the blood and lymph, they bind to the target antigen and initiate its neutralization or destruction.

bacteria – organisms, often consisting of one biological cell, that are not dependent on a host cell to reproduce. Humans and most other animals carry vast numbers of bacteria, most of which are in the gut, and many are on the skin. Most bacteria are harmless, and even helpful, but some are pathogenic and cause infectious diseases.

Bifidobacteria – a species of "good bacteria." Bifidobacteria are one of the major types of bacteria that make up the gastrointestinal tract microbiome in mammals, and comprise an important aspect of the immune system.

biologic – a biologic is manufactured in a living system such as a microorganism, or plant or animal cells. Most biologics are very large, complex molecules or mixtures of molecules. Many biologics are produced using recombinant DNA technology. In contrast to most drugs that are chemically synthesized and their structure is known, most biologics are complex mixtures that are not easily identified or characterized.

biopharmaceutical – also known as a biological medical product, or biologic, is any pharmaceutical drug product manufactured in, extracted from, or semi-synthesized from biological sources.

coronaviruses – a group of related RNA viruses that cause diseases in mammals and birds. Prior to Covid-19 there were four recirculating (endemic) coronaviruses, including one which is a cause of the common cold. SARS-CoV-2 is the coronavirus that causes Covid-19.

Covid-19 – stands for coronavirus disease 2019. Covid-19 is the disease caused by the SARS-CoV-2 virus.

Creutzfeldt-Jakob disease – also known as CJD, is a rare brain disorder that leads to dementia. It belongs to a group of human and animal diseases known as prion disorders. Early symptoms include memory problems, behavioral changes, poor coordination, and visual disturbances. Death usually occurs within a year, usually due to medical issues associated with the disease such as having trouble swallowing, falls, heart issues, lung failure, or pneumonia or other infections.

cytokine – any of various small proteins that regulate the cells of the immune system.

cytokine storm – a sometimes life-threatening systemic inflammatory condition in which excessive cytokines are released in the body in an uncontrolled fashion. Cytokine storm occurs when the immune system responds too aggressively to infection.

dirty placebo – a "placebo" that is in fact not a saline or pure placebo, but an altered version of the drug being tested, or another drug that is designed to treat the same condition. For example, new vaccines are often tested, not against a placebo, but against a vaccine already in use.

DNA – or deoxyribonucleic acid, is the hereditary material in humans and almost all other organisms. DNA is located in the cell nucleus.

DNA vaccine – a vaccine that contains DNA that codes for specific antigens from a pathogen. A DNA vaccine induces an immunologic response in the host agent against bacteria, parasites, viruses, and potentially cancer. The Johnson & Johnson (Janssen) and AstraZeneca Covid-19 vaccines are DNA vaccines, using an adenoviral vector to carry the messenger RNA to the cells.

DNA virus – a virus that has a genome made of DNA.

effector cell – Any of various types of cell that actively responds to a stimulus and effects some change.

endemic – prevalent in a particular locality, region, or population. A disease outbreak is endemic when it is consistently present but limited to a particular locality, region, or population. Disease spread and rates are predictable in the endemic stage, such as with the expected reoccurrence of the flu or common cold each year.

endothelium – endothelial cells that line the interior surface of blood vessels and lymphatic vessels.

epidemic – an unexpected increase in the number of disease cases in a specific geographical area. The difference between a pandemic and an epidemic isn't in the severity of the disease, but the degree to which it has spread.

epistemology – the branch of philosophy concerned with knowledge.

epitope – The specific piece on the surface of an antigen molecule to which an antibody attaches itself; the part of an antigen that is recognized by the immune system, specifically by antibodies, B cells, or T cells; that part of a biomolecule (such as a protein) that is the target of the immune response.

EUA – Emergency Use Authorized.

excess deaths – also called "excess mortality," is the difference between the total number of deaths for a specific time period and the number that would have been expected.

gain-of-function – research that involves experimentation to increase the transmissibility and/or virulence of pathogens.

gene therapy – medical technology which alters the biological properties of living cells.

genome – the genome is the entire set of DNA instructions found in a cell. A genome contains all the information needed for an individual to develop and function, and in humans, the genome consists of 23 pairs of chromosomes located in the cell's nucleus.

glycans – chain-like structures composed of single sugar molecules linked together by chemical bonds. There are plant, animal, and microbial glycans. The glycans coating some viruses, such as SARS-CoV-2, help the viruses to evade the immune system.

heterogeneity – a dissimilarity of structure in different parts of an organism.

histopathology – using a microscope to look at human tissue to see if it has signs of diseases, damage, or other abnormalities.

homologous/homology – similarity often attributable to common origin.

iatrogenic – induced unintentionally by a physician or surgeon or by medical treatment or diagnostic procedures.

IgG4 antibodies – One of four subunits of antibodies in the immune system. IgG4 antibodies do not mount an attack, but rather dampen inflammation and instruct the body to live with, rather than eliminate, the IgG4 specific antigen. IgG4 has a blocking effect either on the immune response or on the target protein of IgG4.

immune – *prior to Covid-19:* resistant to a particular infection or toxin owing to the presence of specific antibodies or sensitized white blood cells.

immune – *definition post Covid-19:* having a high degree of resistance to a disease.

immune escape – occurs when the immune system of a host, especially of a human being, is unable to respond to an infectious agent: the host's immune system is no longer able to recognize and eliminate a pathogen, such as a virus.

immunization – *prior to Covid-19:* make immune to infection, typically by inoculation.

immunization – changed by CDC in October 2021**:** A process by which a person becomes protected against a disease through vaccination.

immune imprinting – Also known as antigenic priming or original antigenic sin. A process in which the immune system deploys the response to a previous infection when a subsequent mutation of that foreign pathogen is encountered. Immune imprinting leaves the immune system less able to mount an effective response during subsequent infections.

immunogen – a substance that produces an immune response.

immunogenic – relating to or producing an immune response.

immunological escape – Also known as viral escape or antigen escape. The condition in which the body's immune system is no longer able to recognize and eliminate a pathogen, such as a virus.

inactivated vaccine – a vaccine that uses a killed virus or bacteria to stimulate the immune system to protect the body against infection. Because the bacteria or virus is dead, it cannot replicate or cause disease.

inoculate – *prior to Covid-19: tr*eat with a vaccine to produce immunity against a disease.

inoculate – definition post Covid-19: to introduce immunologically active material (such as an antibody or antigen) into the body in order to treat or prevent a disease in vitro – experimentation conducted in an artificial environment outside a living organism, colloquially called "test-tube experiments."

in vivo – experimentation conducted in a whole, living organism – animals, humans, plants

Johnson & Johnson Covid-19 vaccine – a vaccine that delivers the virus' DNA to your cells to make the spike protein. An adenovirus acts as a delivery vehicle used to carry the coronavirus genetic material (DNA). The adenovirus delivers the little piece of DNA to the cell that will then make the spike protein.

lipid nanoparticle – lipid nanoparticles are spherical vesicles composed of ionizable lipids, which are positively charged at low pH. LNP's with an outer shell of lipid molecules have the ability to encapsulate and transport complex active ingredients, such as mRNA, to cells in the body.

lymphocyte – a type of white blood cell. Lymphocytes are the cells that determine how to respond to infectious microorganisms and other foreign substances. T cells and B cells are lymphocytes.

macrophage – a type of white blood cell that helps eliminate pathogens.

memory B and T cells – Part of the immune system. B and T cells replicate during the primary immune response and produce effector cells and long-lived memory cells. Memory B and T cells are antigen-specific and, on encountering the antigen again, can mount a more rapid and effective immune response, known as the secondary immune response.

microbiome – the collection of all microbes, such as bacteria, fungi, viruses, and their genes, that naturally live on our bodies and inside us.

mRNA – or messenger RNA – is a type of single-stranded RNA involved in protein synthesis. mRNA molecules in cells carry codes from the DNA in the nucleus to the cell machinery that makes proteins.

mRNA vaccine – an injection that contains messenger RNA, encased in a lipid nanoparticle envelope. The lipid nanoparticle envelope prevents the body from breaking down the mRNA, which arrives at the cells and instructs them to make the Spike protein of the SARS-CoV-2 virus. The idea is the body will then recognize the foreign Spike protein and mount an immune response against it, thus preparing the immune system to fight SARS-CoV-2 before encountering it. The vaccines produced by BioNTEch/Pfizer and Moderna are the first mRNA vaccines to be used as vaccines in humans. mRNA shots are biopharmaceutical products, also known as biologics.

myocarditis – inflammation of the heart, also known as inflammatory cardiomyopathy. Symptoms can include shortness of breath, chest pain, decreased ability to exercise, and an irregular heartbeat (arrhythmias), although often someone with myocarditis experiences no symptoms. This is dangerous as untreated myocarditis can lead to heart attack, stroke, or sudden cardiac death.

nanotechnology – the understanding and control of materials on the molecular, atomic, or even subatomic scale; manipulation of materials on an atomic or molecular scale especially to build microscopic devices. Nanotechnology encompasses nanoscale science, engineering, and technology.

nosocomial spread – also called health-care associated or hospital-acquired infection, meaning the infectious disease is acquired in a health-care facility. To be considered nosocomial, the infection cannot be present at admission; rather, it must develop at least 48 hours after admission.

off-label use – also known as "repurposed use" of a drug for conditions not specifically listed on the FDA-approved label.

original antigenic sin – also known as antigenic imprinting or immunological imprinting. A process in which the immune system deploys the response to a previous infection when a second slightly different version of that foreign pathogen is encountered. Original antigenic sin leaves the immune system less able to mount an effective response during subsequent infections.

PEG – polyethylene glycol – an active ingredient used in the pharmaceutical industry as a solvent, plasticizer, surfactant, ointments, and suppository based, and tablet and capsule lubricant. PEG is an additive in both the Moderna and Pfizer-BioNTech Covid-19 vaccines. It protects the mRNA as it is transferred into human cells.

pandemic – a pandemic is when a virus covers a wide area, affecting several countries and populations, characterized by rapid growth from day to day. The difference between an epidemic and pandemic isn't in the severity of the disease, but the degree to which it has spread.

pathogenic – capable of causing disease.

pathogenic priming – Also known as antibody dependent enhancement (ADE). A priming process whereby instead of providing a robust immune response, pathogen exposure acts like a Trojan horse, allowing the pathogen to get into cells more easily, creating an overreaction.

pericarditis – swelling and irritation of the thin, saclike tissue surrounding the heart (pericardium). Pericarditis often causes sharp chest pain as the irritated tissue layers rub against each other. Pericarditis is usually mild and goes away without treatment, however long-term pericarditis can lead to thickening and scarring of the heart lining. Damaged heart lining prevents the heart from filling and emptying properly, which in turn can lead to swelling of the legs and abdomen, shortness of breath, and dangerously low blood pressure.

pharmacokinetics – sometimes described as what the body does to a drug, refers to the movement of drug into, through, and out of the body; the study of how the body interacts with administered substances for the entire duration of exposure.

pharmacovigilance – is the detection, monitoring, understanding, and prevention of adverse events (AEs) for a medicine.

plasmid – A circular, double-stranded unit of DNA that replicates within a cell independently of the chromosomal DNA. Plasmids are most often found in bacteria and are used in recombinant DNA research to transfer genes between cells.

primary immune response – The immune system's response upon first contact with an antigen.

prophylactic – a medicine or device that guards against or prevents the spread of disease.

reactogenicity – the capacity of a vaccine to produce common, "expected" adverse reactions to vaccination, such as redness at the injection-site and pain and/or swelling, fever, tiredness, and headache.

recombinant DNA – Recombinant DNA molecules are DNA molecules formed by laboratory methods of genetic recombination that bring together genetic material from multiple sources, creating sequences that would not otherwise be found in the genome. Recombinant DNA is the general name for a piece of DNA that has been created by combining two or more fragments from different sources.

RNA – or ribonucleic acid, is essential for most biological functions, either by performing the function itself (Non-coding RNA), or by forming a template for production of proteins (messenger RNA).

RNA vaccine – a vaccine that introduces a piece of mRNA that corresponds to a viral protein, usually a small piece of a protein found on the virus's outer membrane.

RNA viruses – known as retroviruses, have RNA as their genetic material.

repurposed drug – also known as off-label use of a drug for conditions not specifically listed on the FDA-approved label.

SARS-CoV-2 – the virus that causes Covid-19 disease. The acronym stands for Severe Acute Respiratory Syndrome Coronavirus 2 (to distinguish from SARS-CoV-1 in 2002.)

secondary immune response – The response of the immune system to a known antigen, due to prior exposure, whereby memory B and T cells are deployed for a rapid and effective response against the specific antigen.

seropositivity – having a positive serum reaction, especially in a test for the presence of an antibody.

seroprevalence – the rate of seropositivity in a population; the proportion of a population whose blood serum tests positive for a given pathogen.

subunit vaccine – subunit vaccines include only parts of a virus that best stimulate your immune system. The Novavax Covid-19 subunit vaccine contains "harmless" spike protein, which causes the body to create antibodies and defensive white blood cells.

T cell – also called T lymphocyte, a type of leukocyte (white blood cell) that is an essential part of the immune system. T cells are one of two primary types of lymphocytes—B cells being the second type—that determine the specificity of immune response to antigens (foreign substances) in the body.

therapeutic – a medicine or device that treats disease.

thrombosis – the formation of a blood clot inside a blood vessel.

transfection – the insertion of foreign nucleic acid (DNA or RNA) into a cell, typically with the intention of altering the properties of the cell.

turbo cancer – particularly aggressive and fast-growing cancers, occurring especially in young patients.

vaccine – definition pre-Covid-19: "A substance used to stimulate the production of antibodies and provide immunity against one or several diseases.

vaccine – definition changed by the CDC in October 2021: A preparation that is used to stimulate the body's immune response against diseases.

vector – a carrier of disease or medication. For example, in malaria the mosquito is the vector that carries and transfers the infectious agent. With regard to the Johnson & Johnson and AstraZeneca Covid shots, a modified adenovirus (common cold virus) is the vector used to carry the mRNA message to the cell.

viral escape – Also known as antigen escape or immune escape. A condition in which the body's immune system is no longer able to recognize and eliminate a pathogen, such as a virus.

virus – a virus is an infectious particle that reproduces by "commandeering" a host cell and using its machinery to make more viruses. A virus is made up of a DNA or RNA genome. There are millions of viruses in our environment at all times, but the human immune system is designed to fight them at all times. Viruses do not have cells and are basically just packages of nucleic acid and protein. Tens of millions of viruses could fit on the head of a pin. (See Khan Academy: Viruses)

ACKNOWLEDGMENTS

Thank you to Chris, my husband and supporter extraordinaire! The many hundreds of hours it took to research and write this book seeped in and around all the other moments of our lives. Thanks for your love and patience, for being on the same page, for preparing food and reminding me to eat, for all the walks (while I kept writing out loud!), and for being solidly there for me in every way.

Special thanks to my brother John. He read my Covid dissident writings for a couple of years before saying, "I think you should submit your work to Brownstone." When my "latest article" kept growing until it became a book, John was my technical support and first editor. Thanks for always believing in me and the importance of the message.

In a February 27, 2020 journal entry I expressed concern about "the Coronavirus outbreak and how that is affecting the world, and may get closer to home." Within weeks, as the world shut down, it all felt wrong. I began to talk and write about the danger the official Covid-19 response was posing to our civil liberties, economies, and way of life. While the tide of the official narrative buffeted everyone and threatened to wash away so much we value, a few of you were especially supportive of my efforts to push back. Special thanks to Louise, Phil, Mark, Charlotte, Lori B., Joyce, Doug, and Ani. Because you listened and gave valuable feedback, I didn't feel so alone. To family and friends who have read my articles and expressed support, I thank you!

I wouldn't have been able to write this book without the brave people who have spoken out, often at great personal cost, as to how the official Covid response was unscientific, unprecedented, and a violation of human rights. Your courage has made all the difference.

To Logan and David, thank you both for your kindness and patience throughout the editing and publishing process. Of course, none of this would have happened without the remarkable vision and courage of Jeffrey Tucker. In the middle of darkness and censorship, he created a space for truth seeking and true collegiality. Thanks so much to him and all the team at Brownstone for recognizing the need, and working tirelessly to make it happen.

ABOUT THE AUTHOR

Lori Weintz has a bachelor's degree in Mass Communications from the University of Utah. She is a mother, grandmother, and independent journalist. Her work can be found at Brownstone Institute and OnTopicWithLori.com.

ENDNOTES

1 On Topic With Lori | https://www.ontopicwithlori.com/

2 Brownstone Institute | https://brownstone.org/about/

3 Dr. Aaron Kheriaty | https://aaronkheriaty.substack.com/

4 SARS in 2003 | https://www.ncbi.nlm.nih.gov/pmc/articles/PMC7106085/

5 2012 MERS | https://www.who.int/health-topics/middle-east-respiratory-syndrome-coronavirus-mers#tab=tab_1

6 influenza | https://ourworldindata.org/influenza-deaths

7 "Great Reset" | https://www.weforum.org/agenda/2020/06/now-is-the-time-for-a-great-reset/

8 Build Back Better (2 min)| https://www.youtube.com/watch?v=YkcaeaD45MY

9 March 2020 | https://www.ted.com/talks/bill_gates_how_we_must_respond_to_the_coronavirus_pandemic

10 April | https://www.cnbc.com/2020/04/30/bill-gates-what-you-need-to-know-about-a-coronavirus-vaccine.html

11 are an investment opportunity | https://www.cnbc.com/2019/01/23/bill-gates-turns-10-billion-into-200-billion-worth-of-economic-benefit.html

12 CEPI | https://cepi.net/

13 microbial planet | https://www.amazon.com/Fear-Microbial-Planet-Germophobic-Culture-ebook/dp/B0C29RC8ZB

14 pandemic wargames | https://brownstone.org/articles/what-is-crimson-contagion/

15 Event 201 | https://centerforhealthsecurity.org/our-work/tabletop-exercises/event-201-pandemic-tabletop-exercise#scenario

16 reversed much of that progress | https://phys.org/news/2021-06-large-extreme-poverty-neglected-covid.html

17 Dr. Jay Bhattacharya | https://brownstone.org/articles/pandemic-samizdat-in-the-us/

18 mRNA vaccine factories all over the world | https://www.bostonglobe.com/2023/06/23/business/moderna-after-covid-vaccines/

19 another pandemic soon | https://brownstone.org/articles/deadlier-pandemic-warns-new-york-times/

20 fund gain-of-function research | https://www.wsj.com/articles/doctor-anthony-faucis-parting-gift-nih-ecohealth-alliance-peter-daszak-coronavirus-research-11665002675?

21 here | https://www.bmj.com/content/374/bmj.n1656

22 here | https://www.usatoday.com/story/news/health/2023/02/27/covid-lab-leak-energy-department-theory-explained/11357354002/

23 here | https://brownstone.org/articles/timeline-the-proximal-origin-of-sars-cov-2/

24 allow the WHO to impose a response | https://brownstone.org/articles/the-who-has-changed-and-now-it-is-a-threat/

25 Declaration | https://brownstone.org/articles/the-uns-new-political-declaration-on-pandemics/

26 second largest donor | https://www.usnews.com/news/articles/2020-05-29/gates-foundation-donations-to-who-nearly-match-those-from-us-government

27 influence | https://www.swissinfo.ch/eng/politics/does-bill-gates-have-too-much-influence-in-the-who-/46570526

28 WHO's focus | https://reclaimthenet.org/gates-foundation-boosts-funding-for-digital-id-projects

29 Digital Vaccine Passport | https://ec.europa.eu/commission/presscorner/detail/en/ip_23_3043?

30 October 13, 2020 | https://cjhopkins.substack.com/p/the-covidian-cult

31 Facebook interview with Mark Zuckerberg | https://youtu.be/971QcDEha5I

32 2005 study | https://www.ncbi.nlm.nih.gov/pmc/articles/PMC1232869/

33 (a precursor to HCQ) | https://health-desk.org/articles/is-hydroxychloroquine-the-same-as-chloroquine

34 (see WHO list p. 55) | https://www.who.int/publications/i/item/WHO-MHP-HPS-EML-2021.02

35 December 13, 2023 | https://www.dossier.today/p/where-are-they-now-white-house-covid?utm_source=profile&utm_medium=reader2

36 Gilead spent $2.45 million lobbying Congress | https://www.npr.org/sections/health-shots/2020/05/02/849149873/gilead-lobbying-rose-as-interest-in-covid-19-treatment-climbed

37 2018 Ebola drug trials | https://www.nejm.org/doi/10.1056/NEJMoa1910993?

38 https://www.veklury.com/

39 put their name on patents | https://www.theepochtimes.com/non-profit-watchdog-uncovers-350-million-in-secret-payments-to-fauci-collins-others-at-nih_4454636.html?

40 Anthony Fauci's | https://www.foxnews.com/politics/fauci-wifes-net-worth-increased-5m-pandemic-analysis-finds

41 employees profiting | https://www.ncbi.nlm.nih.gov/pmc/articles/PMC545012/

42 provided by pharmaceutical companies | https://today.uconn.edu/2021/05/why-is-the-fda-funded-in-part-by-the-companies-it-regulates-2/

43 financial ties to Gilead | https://www.washingtontimes.com/news/2021/jun/9/follow-the-money-big-pharma-dr-fauci-and-the-death/

44 myriad relationships | https://www.covid19treatmentguidelines.nih.gov/about-the-guidelines/panel-financial-disclosure/

45 $3.5 billion | https://www.fiercepharma.com/pharma/gilead-buoyed-by-potential-remdesivir-covid-19-sales-elevates-2020-outlook-despite-weak-base

46 remdesivir trial https://classic.clinicaltrials.gov/ct2/show/NCT04280705

47 Protocol Details | https://clinicalstudies.info.nih.gov/ProtocolDetails.aspx?id=2020-I-0073

48 study conducted in China https://www.thelancet.com/journals/lancet/article/PIIS0140-6736(20)31022-9/fulltext

49 study | https://classic.clinicaltrials.gov/ct2/show/NCT04280705

50 $70 million in taxpayers' | https://www.politifact.com/factchecks/2020/jul/23/public-citizen/yes-taxpayers-have-sunk-least-70-million-developin/

51 significant number of adverse events in both arms of the study | https://clinicaltrials.gov/ct2/show/results/NCT04280705

52 Chinese study | https://www.thelancet.com/journals/lancet/article/PIIS0140-6736(20)31022-9/fulltext

53 News Release | https://www.nih.gov/news-events/news-releases/nih-clinical-trial-shows-remdesivir-accelerates-recovery-advanced-covid-19

54 FDA | https://www.fda.gov/news-events/press-announcements/coronavirus-covid-19-update-fda-issues-emergency-use-authorization-potential-covid-19-treatment

55 less than $6 for a course of treatment | https://theconversation.com/the-us-has-bought-most-of-the-worlds-remdesivir-heres-what-it-means-for-the-rest-of-us-141791

56 June 29, 2020 | https://stories.gilead.com/articles/an-open-letter-from-daniel-oday-june-29

57 fully approved in October 2020 | https://www.drugs.com/newdrugs/fda-approves-veklury-remdesivir-covid-19-5368.html

58 Remdesivir had not been tested for safety | https://www.medicalnewstoday.com/articles/veklury#breastfeeding

59 Nurse Gail Macrae | https://alphanews.org/former-nurse-describes-covid-19-protocols-as-medical-murder/?

60 serious adverse events | https://www.ncbi.nlm.nih.gov/pmc/articles/PMC7373689/

61 renamed remdesivir | https://brownstone.org/articles/why-are-hospitals-still-using-remdesivir/

62 advised against | https://www.nbcnews.com/health/health-news/massive-who-remdesivir-study-suggests-no-covid-19-benefit-doctors-n1243730

63 no evidence | https://www.who.int/news-room/feature-stories/detail/who-recommends-against-the-use-of-remdesivir-in-covid-19-patients

64 WHO's Solidarity trial | https://www.nejm.org/doi/10.1056/NEJMoa2023184

65 dismissed WHO's Solidarity trial | https://www.nbcnews.com/health/health-news/massive-who-remdesivir-study-suggests-no-covid-19-benefit-doctors-n1243730

66 the same trial that was ended by Fauci | https://www.nejm.org/doi/full/10.1056/NEJMoa2007764

67 October 2020 | https://www.fda.gov/news-events/press-announcements/fda-approves-first-treatment-covid-19

68 article | https://aapsonline.org/bidens-bounty-on-your-life-hospitals-incentive-payments-for-covid-19/

69 provides incentives | https://www.cms.gov/medicare/payment/covid-19/new-covid-19-treatments-add-payment-nctap

70 $10k funeral reimbursement | https://www.fema.gov/press-release/20210324/fema-help-pay-funeral-costs-covid-19-related-deaths

71 Thomas Renz and CMS (Centers for Medicaid Services) whistleblowers https://aapsonline.org/bidens-bounty-on-your-life-hospitals-incentive-payments-for-covid-19/

72 FDA approved | https://www.fda.gov/media/158036/download

73 here | https://www.nejm.org/doi/full/10.1056/NEJMoa2006100?query=featured_home

74 here | https://pubmed.ncbi.nlm.nih.gov/32179660/

75 substance abuse | https://health.ucsd.edu/news/press-releases/2021-08-24-how-adolescents-used-drugs-during-the-covid-19-pandemic/

76 depression, and anxiety | https://jamanetwork.com/journals/jamapediatrics/fullarticle/2782796

77 suicide attempts | https://www.cdc.gov/mmwr/volumes/70/wr/mm7024e1.htm

78 underlying health problems | https://www.bmj.com/content/377/bmj.o1431

79 reported | https://www.cdc.gov/nchs/data/health_policy/covid19-comorbidity-expanded-12092020-508.pdf

80 adverse events | https://brownstone.org/articles/why-are-hospitals-still-using-remdesivir/

81 known high mortality rate for those ventilated | https://www.michaelpsenger.com/p/the-great-covid-ventilator-death?utm_source=substack&utm_medium=email

82 reimbursement | https://www.hhs.gov/guidance/document/new-covid-19-treatments-add-payment-nctap

83 December 19, 2020 | https://ascpt.onlinelibrary.wiley.com/doi/10.1002/cpt.2145

84 approval | https://www.gilead.com/news-and-press/press-room/press-releases/2023/7/fda-approves-veklury-remdesivir-for-covid19-treatment-in-patients-with-severe-renal-impairment-including-those-on-dialysis

85 states | https://petermcculloughmd.substack.com/p/fda-approves-remdesivir-in-kidney?

86 Stories abound| https://brownstone.org/articles/whatever-happened-to-informed-consent/

87 killed | https://brownstone.org/articles/how-did-remdesivir-obtain-approval-for-kidney-disease/

88 class-action lawsuit | https://www.theepochtimes.com/article/class-action-lawsuit-filed-against-remdesivir-manufacturer-over-alleged-deceptive-practices-5500317

89 April 2020 the NIH did a short retrospective study | https://www.medrxiv.org/content/10.1101/2020.04.16.20065920v1.full

90 Associated Press| https://www.nbcnews.com/health/health-news/more-deaths-no-benefit-malaria-drug-va-virus-study-n1188981

91 June 17, 2023 | https://podcasts.apple.com/no/podcast/exposing-the-vaccine-military-machinery-behind-the/id1471411980?i=1000617761934

92 successful protocol | https://youtu.be/Y6gDletOCHc

93 study of 1,061 patients | https://www.ncbi.nlm.nih.gov/pmc/articles/PMC7199729/

94 political | https://www.defenddemocracy.press/why-france-is-hiding-a-cheap-and-tested-virus-cure/

95 poisonous substance | https://www.legifrance.gouv.fr/jorf/id/JORFTEXT000041400024

96 prescriptions | https://www.forbes.com/sites/alexledsom/2020/05/10/hydroxychloroquinenumber-of-prescriptions-explode-in-france/?sh=18c3fa7c180f

97 quickly acted | https://www.forbes.com/sites/alexledsom/2020/05/26/france-says-no-to-prescribing-hydroxychloroquine-after-lancet-study/?sh=2b438f3d24b5

98 study | https://www.medrxiv.org/content/10.1101/2023.04.03.23287649v1

99 French research bodies | https://www.france24.com/en/europe/20230528-french-researchers-slam-former-hospital-director-for-unauthorised-covid-trial

100 hundreds of studies | https://covid19criticalcare.com/studies/

101 two large HCQ studies | https://anthraxvaccine.blogspot.com/2020/06/who-trial-using-potentially-fatal.html

102 Recovery Trial | https://www.nejm.org/doi/full/10.1056/NEJMoa2022926

103 Solidarity Trial | https://www.who.int/emergencies/diseases/novel-coronavirus-2019/global-research-on-novel-coronavirus-2019-ncov/solidarity-clinical-trial-for-covid-19-treatments

104 Surgisphere study | https://www.thelancet.com/journals/lancet/article/PIIS0140-6736(20)31180-6/fulltext

105 scandal | https://www.the-scientist.com/features/the-surgisphere-scandal-what-went-wrong--67955

106 an analysis | https://pubmed.ncbi.nlm.nih.gov/37101974/

107 withdrawn | https://pubmed.ncbi.nlm.nih.gov/34601006/

108 did not support | https://petermcculloughmd.substack.com/p/retracted-covid-19-articles-significantly

109 https://zenodo.org/records/6564414

110 REMAP | https://vaccineimpact.com/2020/dr-meryl-nass-discovers-hydroxychloroquine-experiments-were-designed-to-kill-covid-patients-how-many-were-murdered/

111 given toxic doses | https://vaccineimpact.com/2020/dr-meryl-nass-discovers-hydroxychloroquine-experiments-were-designed-to-kill-covid-patients-how-many-were-murdered/

112 1979, H. Weniger | https://apps.who.int/iris/handle/10665/65773

113 patient-provider | https://childrenshealthdefense.org/defender/meryl-nass-maine-medical-board-lawsuit-first-amendment-rights/

114 probation | https://www.lifesitenews.com/news/maine-medical-board-puts-dr-meryl-nass-on-probation-imposes-draconian-sanctions/

115 countersuit | https://childrenshealthdefense.org/defender/meryl-nass-maine-medical-board-lawsuit-first-amendment-rights/

116 June 19, 2020 | https://vaccineimpact.com/2020/dr-meryl-nass-discovers-hydroxychloroquine-experiments-were-designed-to-kill-covid-patients-how-many-were-murdered/

117 FDA alert | https://www.fda.gov/news-events/press-announcements/coronavirus-covid-19-update-fda-revokes-emergency-use-authorization-chloroquine-and

118 Dr. Paul Marik | https://covid19criticalcare.com/experts/paul-e-marik/

119 U.S. Senate panel discussion | https://www.ronjohnson.senate.gov/2022/1/video-release-sen-ron-johnson-covid-19-a-second-opinion-panel-garners-over-800-000-views-in-24-hours

120 later hearing | https://rumble.com/v1ze4d0-covid-19-vaccines-what-they-are-how-they-work-and-possible-causes-of-injuri.html

121 explained | https://www.youtube.com/watch?v=y2BCOMTTxK4&t=962s

122 Northeastern (University) Global News | https://news.northeastern.edu/2021/12/13/virus-evolution/

123 later meta-analysis | https://www.medrxiv.org/content/10.1101/2021.07.08.21260210v2

124 over 80 percent of Covid deaths were in the 65 and older population | https://data.cdc.gov/NCHS/Provisional-COVID-19-Deaths-by-Sex-and-Age/9bhg-hcku

125 not one healthy child in the 0-12 age group died of Covid | https://www.medpagetoday.com/opinion/marty-makary/93029?

126 not one Covid death | https://www.nejm.org/doi/10.1056/NEJMc2026670

127 https://lymediseaseassociation.org/covid-19-and-lyme/peter-a-mccullough-md-mph-covid-19-treatment-protocols/

128 explains | https://archive.org/details/dr.-peter-mccullough-full-interview-on-covid-19-vaccines-the-great-reset

129 experience | https://covid19criticalcare.com/category/testimonials/

130 isolation | https://brownstone.org/articles/hallucinations-nightmares-despair-longing-for-human-contact-letter-to-editor/

131 ignored | https://brownstone.org/articles/why-are-hospitals-still-using-remdesivir/

132 ventilated | https://www.wsj.com/articles/hospitals-retreat-from-early-covid-treatment-and-return-to-basics-11608491436

133 mortality risk | https://brownstone.org/articles/how-many-people-did-ventilators-and-iatrogenesis-kill-in-april-2020/

134 elderly| https://www.thesun.co.uk/news/12100515/care-homes-accused-sedatives-coronavirus-die-quickly/

135 disabled| https://brownstone.org/articles/how-can-severe-mental-illness-be-the-deadliest-covid-comorbidity/

136 administered drugs | https://twitter.com/Ryan_L_Heath/status/1734651695207260160?s=20

137 rampant | https://brownstone.org/articles/sick-and-all-alone/

138 4:27:50 | https://rumble.com/vt62y6-covid-19-a-second-opinion.html

139 White Coat Summit | https://www.whitecoatsummit.com/

140 July 23, 2023 | https://ianmsc.substack.com/p/new-york-times-misleads-on-floridas

141 Big Pharma advertising dollars | https://childrenshealthdefense.org/defender/pfizer-vaccination-ads-news-sponsorships-research/

142 government pressure | https://www.wsj.com/articles/how-the-government-justifies-its-social-media-censorship-free-speech-supreme-court-doctrine-precedent-biden-laptop-twitter-fbi-facebook-af57b191

143 Twitter files | https://twitterfiles.substack.com/p/1-thread-the-twitter-files

144 https://thehill.com/opinion/judiciary/4198285-missouri-v-biden-and-the-crossroads-of-politics-censorship-and-free-speech/

145 October 3, 2023 | https://ktvz.com/money/cnn-business-consumer/2023/10/03/federal-appeals-court-extends-limits-on-biden-administration-communications-with-social-media-companies-to-top-us-cybersecurity-agency/

146 protocol | https://pubmed.ncbi.nlm.nih.gov/32771461/

147 World Council for Health | https://worldcouncilforhealth.org/

148 updated his protocol | https://lymediseaseassociation.org/covid-19-and-lyme/peter-a-mccullough-md-mph-covid-19-treatment-protocols/

149 gave testimony | https://www.hsgac.senate.gov/hearings/early-outpatient-treatment-an-essential-part-of-a-covid-19-solution/

150 testified at the Senate Committee | https://covid19criticalcare.com/dr-pierre-kory-flccc-alliance-testifies-to-u-s-senate-committee-about-i-mask/

151 FLCCC | https://covid19criticalcare.com/

152 (see WHO list p. 8) | https://www.who.int/publications/i/item/WHO-MHP-HPS-EML-2021.02

153 FDA's own rules | https://www.fda.gov/medical-devices/covid-19-emergency-use-authorizations-medical-devices/faqs-emergency-use-authorizations-euas-medical-devices-related-covid-19

154 advertising campaign | https://www.cbsnews.com/news/covid-vaccine-safety-250-million-dollar-marketing-campaign/

155 July 28, 2021 | https://www.wsj.com/articles/fda-ivermectin-covid-19-coronavirus-masks-anti-science-11627482393

156 safely taken by billions of people | https://www.ncbi.nlm.nih.gov/pmc/articles/PMC8383101/

157 article https://www.fda.gov/consumers/consumer-updates/why-you-should-not-use-ivermectin-treat-or-prevent-covid-19

158 articles | https://www.theepochtimes.com/health/growing-number-of-americans-believe-ivermectin-can-treat-covid-19-post-5532471

159 studies | https://covid19criticalcare.com/ivermectin/

160 here | https://www.nytimes.com/2022/08/29/technology/california-doctors-covid-misinformation.html

161 here | https://www.washingtonexaminer.com/news/california-scraps-2022-law-punishing-doctors-who-dissent-from-covid-consensus

162 filling ER's | https://www.insider.com/oklahomas-emergency-rooms-are-clogged-with-people-overdosing-on-ivermectin-2021-9

163 refuted by the ER departments | https://www.cnn.com/2021/09/07/politics/fact-check-oklahoma-ivermectin-story/index.html

164 Exaggerated claims | https://www.npr.org/sections/coronavirus-live-updates/2021/09/04/1034217306/ivermectin-overdose-exposure-cases-poison-control-centers

165 federal government | https://archive.cdc.gov/#/details?q=https://emergency.cdc.gov/han/2021/han00449.asp&start=0&rows=10&url=https://emergency.cdc.gov/han/2021/han00449.asp

166 here | https://news.ashp.org/news/ashp-news/2021/09/01/ama-apha-ashp-call-for-end-to-ivermectin-to-prevent-or-treat-covid-19

167 here | https://www.ama-assn.org/delivering-care/public-health/what-fda-wants-doctors-tell-patients-asking-ivermectin

168 Harvard Health | https://www.health.harvard.edu/medications/do-not-get-sold-on-drug-advertising

169 often compromised by Big Pharma | https://childrenshealthdefense.org/defender/pfizer-vaccination-ads-news-sponsorships-research/

170 Missouri v Biden | https://brownstone.org/articles/legal-updates-on-missouri-v-biden/

171 violations | https://brownstone.org/articles/the-white-houses-misinformation-pressure-campaign-was-unconstitutional/

172 Merck | https://www.merck.com/news/merck-statement-on-ivermectin-use-during-the-covid-19-pandemic/

173 EUA approval | https://www.fda.gov/news-events/press-announcements/coronavirus-covid-19-update-fda-authorizes-additional-oral-antiviral-treatment-covid-19-certain

174 molnupiravir | https://www.merck.com/news/merck-and-ridgebacks-molnupiravir-receives-u-s-fda-emergency-use-authorization-for-the-treatment-of-high-risk-adults-with-mild-to-moderate-covid-19/

175 Henderson | https://www.aier.org/people/david-r-henderson/

176 Hooper | https://www.aier.org/people/charles-l-hooper/

177 America Institute for Economic Research article | https://www.aier.org/article/the-fdas-war-against-the-truth-on-ivermectin/

178 TOGETHER trial | https://www.nejm.org/doi/full/10.1056/NEJMoa2115869

179 study of 223,128 people in Brazil | https://pubmed.ncbi.nlm.nih.gov/35070575/

180 Dr. Kory states | https://brownstone.org/articles/revamping-our-dysfunctional-drug-approval-process/

181 here | https://www.worldometers.info/coronavirus/#countries

182 Kerala | https://indianexpress.com/article/opinion/kerala-uttar-pradesh-covid-management-second-wave-7460292/

183 cases soared | https://www.bbc.com/news/world-asia-india-56799303

184 over-the-counter packet | https://newsrescue.com/the-undeniable-ivermectin-miracle-indias-240m-populated-largest-state-uttar-pradesh-horowitz/

185 dropped dramatically | https://www.hindustantimes.com/cities/lucknow-news/33-districts-in-uttar-pradesh-are-now-covid-free-state-govt-101631267966925.html

186 June 2022 | https://prsindia.org/covid-19/cases

187 dramatic drop | https://newsrescue.com/the-undeniable-ivermectin-miracle-indias-240m-populated-largest-state-uttar-pradesh-horowitz/

188 378 Covid deaths per million | https://www.worldometers.info/coronavirus/#countries

189 U.S.'s 3,492 | https://www.worldometers.info/coronavirus/#countries

190 peer-reviewed study | https://www.cureus.com/articles/172991-covid-19-excess-deaths-in-perus-25-states-in-2020-nationwide-trends-confounding-factors-and-correlations-with-the-extent-of-ivermectin-treatment-by-state#!/

191 Dr. Meryl Nass | https://substack.com/@merylnass

192 Dr. Harvey Risch | https://earlycovidcare.org/review-the-evidence/

193 Studies | https://c19ivm.org/meta.html

194 case | https://www.theepochtimes.com/health/doctors-can-prescribe-ivermectin-for-covid-19-fda-5456584

195 three doctors | https://www.theepochtimes.com/article/doctors-suing-food-and-drug-administration-over-ivermectin-4515103

196 21 United States Code Section 396 | https://uscode.house.gov/view.xhtml?req=granuleid:USC-prelim-title21-section396&num=0&edition=prelim

197 FDA didn't require anyone, or prohibit | https://covid19.onedaymd.com/2023/08/doctors-can-prescribe-ivermectin-for.html

198 "FDA is clearly acknowledging that doctors have the authority to prescribe human ivermectin to treat Covid | https://www.theepochtimes.com/article/doctors-can-prescribe-ivermectin-for-covid-19-fda-5456584

199 found false | https://brownstone.org/articles/how-to-create-a-fake-news-cycle/

200 immediately backtracked | https://www.theepochtimes.com/us/fda-says-ivermectin-remains-unapproved-for-covid-19-treatment-5476396

201 https://www.politifact.com/factchecks/2023/aug/17/maria-bartiromo/the-fda-didnt-reverse-course-ivermectin-is-still-n/

202 Snopes | https://www.snopes.com/fact-check/fda-admit-ivermectin/

203 consistently ignored | https://brownstone.org/articles/case-against-ivermectin-has-been-reversed-by-court/

204 emotional abusers | https://brownstone.org/articles/open-letter-to-an-emotionally-abused-world/

205 Paxlovid | https://www.paxlovid.com/?cmp=e5a7e3b5-6f21-4f9a-8560-89b4fc5cf66b&ttype=QRC&gclid=346893cd1f48152e967ff745edd4fdef&gclsrc=3p.ds

206 Dr. David Gortler | https://brownstone.org/articles/the-fdas-paxlovid-pandemonium/

207 September 8, 2020 | https://www.nature.com/articles/d41586-020-02544-6

208 speech | https://www.youtube.com/watch?v=IA2SCoYI8_U

209 decrease | https://www.flgov.com/2021/09/16/governor-ron-desantis-highlights-statewide-monoclonal-antibody-treatment-success/

210 Forbes | https://www.forbes.com/sites/jemimamcevoy/2021/09/10/nation-short-on-supply-of-key-covid-treatment-desperate-states-told-to-reduce-requests/

211 seven states | https://www.theguardian.com/us-news/2021/sep/21/monoclonal-antibodies-covid-treatment-vaccines

212 phased out | https://news.yahoo.com/most-monoclonal-antibody-treatments-dont-143000016.html

213 paxlovid | https://www.fda.gov/news-events/press-announcements/coronavirus-covid-19-update-fda-authorizes-first-oral-antiviral-treatment-covid-19

214 molnupiravir | https://www.fda.gov/news-events/press-announcements/coronavirus-covid-19-update-fda-authorizes-additional-oral-antiviral-treatment-covid-19-certain

215 rebound | https://www.goodrx.com/conditions/covid-19/paxlovid-rebound

216 President Joe Biden | https://www.npr.org/2022/07/21/1112740916/paxlovid-president-biden-covid-antiviral-treatment

217 Rochelle Wallensky | https://www.nbcnews.com/news/us-news/cdc-director-tests-positive-covid-paxlovid-rebound-case-rcna54870

218 Anthony Fauci | https://abcnews.go.com/US/fauci-taking-2nd-paxlovid-experiencing-rebound-antiviral-treatment/story?id=85922417

219 Dr. McCullough says of Paxlovid | https://www.theepochtimes.com/what-post-vaccination-autopsies-show-dr-peter-mccullough-on-new-analysis-removed-by-lancet-atlnow_5382545.html

220 April 2023 interview | https://www.youtube.com/watch?v=BFrjlPkwSaY

221 poor parameters | https://childrenshealthdefense.org/defender/fda-robert-califf-pfizer-covid-paxlovid/

222 Maryanne Demasi | https://maryannedemasi.substack.com/p/fda-chief-admits-to-cheerleading?utm_source=substack

223 interview | https://med.stanford.edu/cvi/events/2023-drug-discovery-conference/2023-day1-event-recordings.html

224 resigned | https://www.businessinsider.com/2-top-fda-officials-resigned-biden-booster-plan-reports-2021-9

225 pillars of medical ethics | https://brownstone.org/articles/medical-ethics-destroyed-in-covid-response/

226 states | https://www.theepochtimes.com/epochtv/dr-peter-mccullough-no-evidence-that-our-bodies-can-get-rid-of-vaccine-mrna-5550521?

227 ineffective PCR test | https://www.bbc.com/news/health-54000629

228 CDC.gov | https://www.cdc.gov/coronavirus/2019-ncov/your-health/treatments-for-severe-illness.html

229 $530 per treatment | https://arstechnica.com/health/2023/10/pfizer-more-than-doubles-price-of-life-saving-covid-antiviral-paxlovid/

230 $712 per treatment | https://www.businessinsider.com/how-merck-is-pricing-its-covid-19-pill-molnupiravir-globally-2021-10?op=1

231 $3,120 per treatment | https://www.cnn.com/2020/06/29/health/remdesivir-cost-coronavirus-treatment-bn/index.html

232 CDC article | https://scdhec.gov/sites/default/files/media/document/10454-CHU-03-28-2020-COVID-19.pdf

233 updated October 4, 2023 | https://www.cdc.gov/coronavirus/2019-ncov/hcp/clinical-care/outpatient-treatment-overview.html

234 Pfizer's revenues | https://www.cnbc.com/2023/01/31/the-covid-pandemic-drives-pfizers-2022-revenue-to-a-record-100-billion.html

235 https://investors.modernatx.com/news/news-details/2023/Moderna-Reports-Fourth-Quarter-and-Fiscal-Year-2022-Financial-Results-and-Provides-Business-Updates/default.aspx

236 Gilead Sciences | https://www.macrotrends.net/stocks/charts/GILD/gilead-sciences/gross-profit

237 Merck's | https://www.merck.com/stories/mercks-q4-and-full-year-2022-earnings-report/

238 CNN | https://www.cnn.com/2021/05/21/business/covid-vaccine-billionaires/index.html

239 Bernie Sanders | https://www.commondreams.org/news/2021/12/14/citing-multimillion-dollar-big-pharma-ties-sanders-vote-no-bidens-pick-fda-chief

240 December 14, 2021 | https://www.commondreams.org/news/2021/12/14/citing-multimillion-dollar-big-pharma-ties-sanders-vote-no-bidens-pick-fda-chief

241 lucrative positions | https://www.bmj.com/content/383/bmj.p2486

242 multi-billions of dollars | https://en.wikipedia.org/wiki/List_of_largest_pharmaceutical_settlements

243 December 13, 2023 | https://www.dossier.today/p/where-are-they-now-white-house-covid

244 $33.3 billion | https://nexus.od.nih.gov/all/2023/03/01/fy-2022-by-the-numbers-extramural-grant-investments-in-research/

245 Farewell Address | https://www.archives.gov/milestone-documents/president-dwight-d-eisenhowers-farewell-address

246 2006 *Harper*'s article | https://ahrp.org/out-of-control-aids-and-the-corruption-of-medical-science/

247 September 7, 2023 | https://econjwatch.org/File+download/1276/BhattacharyaHankeSept2023.pdf?mimetype=pdf

248 Great Barrington Declaration | https://gbdeclaration.org/

249 "quick and devastating takedown " | https://www.dailymail.co.uk/news/article-10324873/Emails-reveal-Fauci-head-NIH-colluded-try-smear-experts-called-end-lockdowns.html

250 Jay Bhattacharya | https://www.theepochtimes.com/opinion/american-pandemic-samizdat-5497566

251 September 13, 2023 | https://rumble.com/v3hwcgm-dr.-mcculloughs-speech-at-the-european-parliament.html

252 different platforms | https://www.nature.com/articles/s41586-020-2798-3

253 Dec 6, 2022 | https://podcasts.apple.com/us/podcast/dr-tess-lawrie-covid-19-vaccines-cause-inflammation/id1471411980?i=1000589790498

254 roundtable discussion | https://www.ronjohnson.senate.gov/2022/12/video-release-sen-johnson-please-share-expert-testimony-from-covid-19-vaccine-efficacy-and-safety-roundtable

255 (SEC) filing | https://www.theepochtimes.com/health/is-the-associated-press-lying-about-gene-therapy-shots_4980213.html

256 explained | https://www.theepochtimes.com/health/is-the-associated-press-lying-about-gene-therapy-shots-4980213

257 December 2022 U.S. Senate panel | https://rumble.com/v1ze4d0-covid-19-vaccines-what-they-are-how-they-work-and-possible-causes-of-injuri.html

258 article | https://www.nature.com/articles/s41578-021-00398-6

259 explain | https://rwmalonemd.substack.com/p/lipid-nanoparticles-and-mrna-shots

260 adenoviral vector| https://www.muhealth.org/our-stories/what-you-need-know-about-johnson-johnson-covid-19-vaccine

261 explained | https://healthtalk.unchealthcare.org/6-things-to-know-about-the-johnson-johnson-covid-19-vaccine/

262 article | https://cen.acs.org/pharmaceuticals/vaccines/Adenoviral-vectors-new-COVID-19/98/i19

263 rabies vaccine | https://cen.acs.org/pharmaceuticals/vaccines/Adenoviral-vectors-new-COVID-19/98/i19

264 FDA limited| https://www.cnn.com/2022/05/05/health/fda-johnson-johnson-vaccine-eua/index.html

265 dangerous clotting condition| https://heavy.com/news/jessica-berg-wilson/

266 Healthline| https://www.healthline.com/health/vaccinations/thrombosis-with-thrombocytopenia-syndrome

267 is no longer available | https://www.cnn.com/2023/05/15/health/johnson-johnson-covid-vaccine-end/index.html

268 Unherd interview | https://unherd.com/thepost/bret-weinstein-i-will-be/

269 study | https://onlinelibrary.wiley.com/doi/epdf/10.1002/prca.202300048

270 September 13, 2023 | https://rumble.com/v3hwcgm-dr.-mcculloughs-speech-at-the-european-parliament.html

271 29 proteins | https://cen.acs.org/biological-chemistry/infectious-disease/know-novel-coronaviruss-29-proteins/98/web/2020/04

272 April 9, 2020 | https://pubmed.ncbi.nlm.nih.gov/32292901/

273 Trojan horse | https://onlinelibrary.wiley.com/doi/10.1111/sji.12969

274 explains | https://www.youtube.com/watch?v=42uoERKuzo4

275 2017 *Journal of Autoimmunity* | https://pubmed.ncbi.nlm.nih.gov/28479213/

276 12,000 catalogued mutations | https://www.nature.com/articles/d41586-020-02544-6

277 original antigenic sin | https://en.wikipedia.org/wiki/Original_antigenic_sin

278 immune imprinting | https://rwmalonemd.substack.com/p/immune-imprinting-comirnaty-and-omicron

279 Q&A site | https://covidqanda.org/?p=1031

280 June 2022 | https://rwmalonemd.substack.com/p/immune-imprinting-comirnaty-and-omicron-520

281 May 18, 2020 | https://informedchoicewa.org/covid-19/covid-19-autoimmunity-via-pathogenic-priming/

282 animal trials | https://www.ncbi.nlm.nih.gov/pmc/articles/PMC4550498/

283 Dr. John Campbell | https://www.youtube.com/watch?v=HQgVV7DsZr0

284 mRNA does not stay in the arm muscle | https://brownstone.org/articles/cdc-lied-mrna-not-meant-to-stay-in-arm/

285 harmful | https://stateofthenation.co/?p=69654

286 study | https://onlinelibrary.wiley.com/doi/epdf/10.1002/prca.202300048

287 Respiratory Syncytial Virus | https://mvec.mcri.edu.au/references/vaccine-associated-enhanced-disease-vaed/

288 HIV vaccines | https://pubmed.ncbi.nlm.nih.gov/22811518/

289 https://www.reddit.com/r/DebateVaccines/comments/w6kdbk/anthony_fauci_speaks_out_against_aids_vaccine_a/?rdt=39910

290 points out | https://twitter.com/BretWeinstein/status/1687834523550302208

291 plans | https://www.technologyreview.com/2023/01/05/1066274/whats-next-mrna-vaccines/

292 August 5, 2023 | https://twitter.com/BretWeinstein/status/1687834523550302208

293 Dr. David Gortler | https://brownstone.org/articles/questioning-lipid-nanoparticles/

294 Joseph Ladapo | https://player.fm/series/american-thought-leaders/exclusive-dr-joseph-ladapo-on-why-hes-not-recommending-mrna-covid-vaccines-for-healthy-young-men-people-deserve-honesty

295 natural immunity | https://brownstone.org/?s=natural+immunity

296 insufficient | https://www.businessinsider.com/fauci-why-covid-vaccines-work-better-than-natural-infection-alone-2021-5?op=1

297 court order | https://www.reuters.com/legal/government/paramount-importance-judge-orders-fda-hasten-release-pfizer-vaccine-docs-2022-01-07/

298 June 17, 2023 | https://podcasts.apple.com/no/podcast/exposing-the-vaccine-military-machinery-behind-the/id1471411980?i=1000617761934

299 prevented transmission | https://www.youtube.com/watch?v=1_gdDUKh-U4

300 vaccine prevented transmission | https://www.pfizer.com/news/press-release/press-release-detail/pfizer-and-biontech-conclude-phase-3-study-covid-19-vaccine

301 largely ignored them | https://www.anecdotalsmovie.com/

302 "Vaccine Was '95% Effective How?" | https://brownstone.org/articles/vaccine-95-percent-effective-how/

303 https://www.canadiancovidcarealliance.org/wp-content/uploads/2021/12/The-COVID-19-Inoculations-More-Harm-Than-Good-REV-Dec-16-2021.pdf

304 clinically insignificant | https://en.wikipedia.org/wiki/Clinical_significance

305 advising doctors | https://www.ama-assn.org/delivering-care/public-health/what-fda-wants-doctors-tell-patients-asking-ivermectin

306 They knew | https://brownstone.org/articles/no-one-could-have-known/

307 vaccines hadn't been tested for stopping transmission | https://www.youtube.com/watch?v=1_gdDUKh-U4

308 seriously injured | https://celiafarber.substack.com/p/they-knew-foia-emails-sent-to-daily

309 Highly Effective (2:26 min) | https://www.youtube.com/watch?v=TSZMtSPX3iE

310 medical malfeasance | https://www.globalresearch.ca/dr-naomi-wolf-on-fdapfizer-malfeasance-in-pfizer-document-dump/5782684

311 Cause Unknown | https://www.amazon.com/Cause-Epidemic-Sudden-Childrens-Defense/dp/1510776397

312 explains | https://www.youtube.com/watch?v=PnJ5T1Enwq4&t=1042s

313 Cleveland Clinic | https://academic.oup.com/ofid/article/10/6/ofad209/7131292?login=false

314 Vaccines | https://www.mdpi.com/2076-393X/11/5/991

315 protection of the newborn | https://primaryimmune.org/about-primary-immunodeficiencies/specific-disease-types/transient-hypogammaglobulinemia-of-infancy/

316 June 13, 2022 | https://rumble.com/v18dhuj-the-debate-dr.-robert-malone-vs.-dr.-geert-vanden-bossche.html

317 $1.4 Billion dollars in taxpayer money | https://www.hhs.gov/about/news/2023/08/22/funding-1-billion-vaccine-clinical-trials-326-million-new-monoclonal-antibody-100-million-explore-novel-vaccine-therapeuti-technologies.html

318 floundering | https://medicalxpress.com/news/2023-10-million-americans-covid-shots.html

319 around 14% | https://abcnews.go.com/Health/36m-american-adults-received-updated-covid-vaccine-cdc/story?id=104874582

320 CDC Chart of Covid-19 Variants 20 Nov 2023 | https://covid.cdc.gov/covid-data-tracker/#variant-summary

321 government | https://www.fda.gov/news-events/press-announcements/fda-takes-action-updated-mrna-covid-19-vaccines-better-protect-against-currently-circulating

322 CDC | https://www.cdc.gov/respiratory-viruses/whats-new/covid-19-variant.html

323 broad immune response | https://www.youtube.com/watch?v=9RFMQEEmU8g&t=406s

324 ICWA article | https://informedchoicewa.org/education/researchers-concerned-about-sars-cov-2-vaccine-development/

325 interview | https://www.youtube.com/watch?v=pyPjAfNNA-U

326 open letter | https://doctors4covidethics.org/urgent-open-letter-from-doctors-and-scientists-to-the-european-medicines-agency-regarding-covid-19-vaccine-safety-concerns/

327 EMA responded | https://doctors4covidethics.org/reply-from-the-european-medicines-agency-to-doctors-for-covid-ethics-march-23-2021/

328 criminal charges | https://www.europereloaded.com/dr-sucharit-bhakdis-legal-case/

329 continuing to speak out | https://doctors4covidethics.org/letters/doctorsforcovidethics-letters/

330 became alarmed | https://www.theepochtimes.com/epochtv/pfizer-moderna-jj-vaccines-should-be-immediately-pulled-off-the-market-dr-peter-mccullough-and-john-leake-4516519

331 1976 during a swine flu | https://www.wired.com/2008/03/dayintech-0324/

332 https://www.midwesterndoctor.com/p/why-do-so-many-people-hate-the-vaccine-1d7

333 former employer | https://www.medscape.com/viewarticle/958916

334 various boards | https://petermcculloughmd.substack.com/p/dr-mccullough-fights-to-keep-his

335 open letter to the WHO | https://web.archive.org/web/20210317180828/https:/37b32f5a-6ed9-4d6d-b3e1-5ec648ad9ed9.filesusr.com/ugd/28d8fe_266039aeb27a4465988c37adec9cd1dc.pdf

336 speak out | https://www.voiceforscienceandsolidarity.org/scientific-blog/the-covid-19-mass-vaccination-program-violated-all-principles-of-science-and-the-hippocratic-oath

337 May 2021, Luc Montagnier | https://telanganatoday.com/mass-vaccination-during-pandemic-historical-blunder-nobel-laureate

338 Cause Unknown | https://www.amazon.com/Cause-Epidemic-Sudden-Childrens-Defense/dp/1510776397

339 last public appearances | https://rairfoundation.com/nobel-laureate-warns-doctors-vaccines-are-not-for-killing-they-are-for-shielding-video/

340 last paper | https://ijvtpr.com/index.php/IJVTPR/article/view/66

341 January 2022 | https://rairfoundation.com/nobel-laureate-warns-doctors-vaccines-are-not-for-killing-they-are-for-shielding-video/

342 biodistribution study | https://pandemictimeline.com/2021/05/japan-shares-biodistribution-study-of-pfizer-covid-19-vaccine/

343 harassed | https://www.guelphtoday.com/local-news/u-of-g-prof-says-he-is-receiving-workplace-harassment-after-sharing-vaccine-concerns-3888634

344 prevented from entering campus | https://www.jccf.ca/open-letter-to-the-president-of-the-university-of-guelph-from-dr-byram-bridle/

345 lost his job. | https://www.guelphmercury.com/news/professor-files-3m-lawsuit-against-university-of-guelph-researchers/article_77291677-3482-5540-b3a9-d53813a9dfc8.html

346 Wolf continued | https://naomiwolf.substack.com/

347 alarm about the Covid vaccines | https://dailyclout.io/our-story/

348 *The Bodies of Others* | https://www.amazon.com/Bodies-Others-Authoritarians-COVID-19-Against/dp/1737478560

349 court order | https://news.bloomberglaw.com/health-law-and-business/why-a-judge-ordered-fda-to-release-covid-19-vaccine-data-pronto

350 Daily Clout | https://dailyclout.io/category/pfizer-reports/

351 their work | https://dailyclout.io/the-war-room-dailyclout-pfizer-documents-analysis-volunteers-publish-e-book-available-on-dailyclout-ios-website/

352 Microbiolog | https://www.amazon.com/Fear-Microbial-Planet-Germophobic-Culture-ebook/dp/B0C29RC8ZB

353 four recirculating coronaviruses | https://pubmed.ncbi.nlm.nih.gov/31455974/

354 Covid shots could not prevent the spread of Covid-19 | https://www.usatoday.com/story/news/factcheck/2022/01/21/fact-check-vaccines-limit-serious-illness-and-death-covid-19/9185671002/

355 only diseases that have been eradicated | https://www.vox.com/2018/8/21/17588074/vaccines-diseases-wiped-out

356 push| https://www.weforum.org/press/2023/01/new-100-day-goal-set-for-faster-equitable-pandemic-response

357 mRNA vaccines within 100 days | https://100days.cepi.net/

358 June 13, 2022 | https://rumble.com/v18byhs-dr.-ryan-cole-covid-vaccine-side-effects-are-like-a-nuclear-bomb.html

359 disciplinary action | https://wmc.wa.gov/news/statement-charges-served-physician-license-ryan-cole

360 states | https://www.clarkcountytoday.com/news/health-nightmare-dr-robert-malone-spotlights-study-on-mrna-spike-protein/

361 May 2016 study| https://www.ncbi.nlm.nih.gov/pmc/articles/PMC5439223/

362 with Pfizer | https://www.npr.org/2021/01/31/960819083/vaccines-for-data-israels-pfizer-deal-drives-quick-rollout-and-privacy-worries

363 increase in myocarditis | https://www.gov.il/en/departments/news/01062021-03

364 states | https://twitter.com/venkmurthy/status/1409144713966465024

365 Covid hospitalizations occurring in the vaccinated | https://newsrescue.com/australia-israel-report-95-99-hospitalized-fully-vaccinated/

366 sharp rise in ER visits | https://www.nature.com/articles/s41598-022-10928-z

367 booster shots were required | https://globalnews.ca/news/8239533/israel-covid-green-pass-booster-shot/

368 Green Pass | https://allisrael.com/beach-chairs-for-vaccinated-only-in-tel-aviv-removed-but-not-before-photos-went-viral

369 Anecdotal word from Israel | https://rumble.com/vumoey-the-testimonies-project-vaccine-injuries-israeli-documentary-2022.html

370 Retsef Levi of MIT | https://rumble.com/v24mqk8-mit-professor-retsef-levi-talks-about-the-israeli-vaccine-safety-study.html

371 data was fiction | https://brownstone.org/articles/adverse-effects-of-the-pfizer-vaccine-covered-up-by-the-israeli-ministry-of-health/

372 Stat News reported | https://www.statnews.com/2021/06/10/officials-higher-than-expected-heart-inflammation-cases-covid-19-vaccination/

373 VAERS | https://openvaers.com/covid-data

374 established by the CDC | https://vaers.hhs.gov/about.html

375 Adverse Event | https://www.ncbi.nlm.nih.gov/books/NBK558963/

376 Serious Adverse Event | https://www.fda.gov/safety/reporting-serious-problems-fda/what-serious-adverse-event

377 VAERS | https://openvaers.com/covid-data

378 required by law| https://vaers.hhs.gov/faq.html

379 June 4, 2023| https://podcasts.apple.com/us/podcast/warner-mendenhall-medical-malpractice-unprecedented/id1471411980?i=1000618403100

380 Dr. Jessica Rose| https://podcasts.apple.com/gb/podcast/dna-contamination-in-vaccines-a-potential-cancer/id1471411980?i=1000634561053

381 comparing | https://vimeo.com/880654111

382 "Anecdotals" - Accounts of the Vaccine Injured https://www.anecdotalsmovie.com/

383 1976 flu pandemic | https://www.npr.org/templates/story/story.php?storyId=103582555

384 https://rumble.com/vt62y6-covid-19-a-second-opinion.html

385 JAMA Cardiology stated | https://jamanetwork.com/journals/jamacardiology/fullarticle/2781601?utm_source=silverchair&utm_medium=email&utm_campaign=article_alert-jamacardiology&utm_content=olf

386 https://rumble.com/v1ze4d0-covid-19-vaccines-what-they-are-how-they-work-and-possible-causes-of-injuri.html

387 5:09:50 | https://www.ronjohnson.senate.gov/2022/1/video-release-sen-ron-johnson-covid-19-a-second-opinion-panel-garners-over-800-000-views-in-24-hours

388 database had been altered | https://www.theepochtimes.com/us/us-military-confirms-myocarditis-spiked-after-covid-vaccine-introduction-5411759?

389 detailed timeline | https://www.theepochtimes.com/health/timeline-covid-19-vaccines-and-myocarditis-5317985

390 contractor hired | https://www.theepochtimes.com/article/vaccine-injury-reports-soared-above-red-line-after-covid-19-vaccine-authorizations-documents-show-5145186

391 multiple communications | https://celiafarber.substack.com/p/they-knew-foia-emails-sent-to-daily

392 link now reads| https://www.cdc.gov/coronavirus/2019-ncov/vaccines/safety/myocarditis.html

393 drowning | https://www.cdc.gov/drowning/facts/index.html

394 auto accidents | https://www.cdc.gov/transportationsafety/child_passenger_safety/cps-factsheet.html

395 lightning | https://www.cdc.gov/disasters/lightning/victimdata/infographic.html

396 death due to Covid | https://data.cdc.gov/NCHS/Provisional-COVID-19-Deaths-by-Sex-and-Age/9bhg-hcku

397 October 29, 2021 | https://www.fda.gov/news-events/press-announcements/fda-authorizes-pfizer-biontech-covid-19-vaccine-emergency-use-children-5-through-11-years-age

398 June 17, 2022 | https://www.prnewswire.com/news-releases/coronavirus-covid-19-update-fda-authorizes-moderna-and-pfizer-biontech-covid-19-vaccines-for-children-down-to-6-months-of-age-301570340.html

399 at risk | https://www.theepochtimes.com/cdc-removes-24-percent-of-child-covid-19-deaths-thousands-of-others_4345083.html

400 ACE2 receptors | https://jamanetwork.com/journals/jama/fullarticle/2766522

401 1:18:30 | https://rumble.com/v249a6s-bret-weinstein-the-joe-rogan-experience-1919-01.2023.html

402 here | https://adc.bmj.com/content/105/7/618

403 here | https://academic.oup.com/cid/article/71/15/825/5819060

404 here | https://www.eurosurveillance.org/content/10.2807/1560-7917.ES.2020.26.1.2002011

405 multiple European countries | https://www.forbes.com/sites/roberthart/2021/11/10/germany-france-restrict-modernas-covid-vaccine-for-under-30s-over-rare-heart-risk-despite-surging-cases

406 Sweden | https://www.theepochtimes.com/article/sweden-stops-recommending-covid-19-vaccines-for-children-4769450

407 June 13, 2022 | https://rumble.com/v18dhuj-the-debate-dr.-robert-malone-vs.-dr.-geert-vanden-bossche.html

408 microbiome | https://www.niehs.nih.gov/health/topics/science/microbiome/index.cfm

409 https://podcasts.apple.com/us/podcast/dr-sabine-hazan-the-gut-bacteria-thats-missing-in/id1471411980?i=1000606146013

410 transfer | https://www.ncbi.nlm.nih.gov/pmc/articles/PMC6200668/

411 breastfeeding| https://www.ncbi.nlm.nih.gov/pmc/articles/PMC5485682/

412 June 2022 study | https://pubmed.ncbi.nlm.nih.gov/35140064/

413 analysis | https://www.ncbi.nlm.nih.gov/pmc/articles/PMC9639653/#B17

414 encouraged | https://www.deseret.com/utah/2021/1/29/22256341/only-worst-cases-of-covid-19-put-pregnant-women-at-risk-study-finds

415 coerced | https://childrenshealthdefense.org/defender/pregnant-women-forced-covid-shots-cola/

416 video statement | https://globalcovidsummit.org/news/live-stream-event-physicians-alerting-parents

417 January 15, 2022 | https://rairfoundation.com/nobel-laureate-warns-doctors-vaccines-are-not-for-killing-they-are-for-shielding-video/

418 2:15:53 | https://rumble.com/v1ze4d0-covid-19-vaccines-what-they-are-how-they-work-and-possible-causes-of-injuri.html

419 June 13, 2022 | https://rumble.com/v18byhs-dr.-ryan-cole-covid-vaccine-side-effects-are-like-a-nuclear-bomb.html

420 2:51:49 | https://rumble.com/v1ze4d0-covid-19-vaccines-what-they-are-how-they-work-and-possible-causes-of-injuri.html

421 May 28, 2021 | https://torontosun.com/news/national/not-enough-data-on-kids-and-covid-vaccines-canadian-expert-cautions

422 2:29:34 | https://rumble.com/v1ze4d0-covid-19-vaccines-what-they-are-how-they-work-and-possible-causes-of-injuri.html

423 Sept. 23, 2021 | https://www.theepochtimes.com/epochtv/dr-scott-atlas-on-vaccine-mandates-for-children-natural-immunity-and-floridas-covid-19-surge-4013490

424 U.K. Doctors Express mRNA Vaccine Concerns (19 min) | https://rumble.com/embed/v1zyqyo/?pub=4

425 reports | https://www.express.co.uk/life-style/health/1840842/Chinese-pneumonia-Denmark-epidemic

426 Forbes | https://www.forbes.com/sites/joshuacohen/2023/11/25/mycoplasma-likely-main-culprit-of-outbreak-of-pediatric-cases-of-pneumonia-worldwide

427 here| https://www.unmc.edu/healthsecurity/transmission/2023/11/22/mystery-child-pneumonia-outbreak-reported-in-china-hospitals/

428 here | https://www.express.co.uk/life-style/health/1840842/Chinese-pneumonia-Denmark-epidemic

429 here | https://www.businessinsider.com/who-requests-data-china-mystery-illness-affecting-children-2023-11?op=1

430 article | https://themessenger.com/health/new-york-pneumonia-china-denmark-netherlands

431 calls| https://frontline.news/post/who-calls-for-masking-physical-distancing-over-mystery-illness

432 December 1, 2023 | https://www.coffeeandcovid.com/p/rats-friday-december-1-2023-c-and

433 3:53:04 | https://rumble.com/vt62y6-covid-19-a-second-opinion.html

434 National Childhood Vaccine Injury Act | https://www.hrsa.gov/vaccine-compensation/about

435 part of the routine vaccine schedule| https://www.hrsa.gov/vaccine-compensation/covered-vaccines

436 03:53:04 | https://www.ronjohnson.senate.gov/2022/1/video-release-sen-ron-johnson-covid-19-a-second-opinion-panel-garners-over-800-000-views-in-24-hours

437 3:45 to 4:47 | https://petermcculloughmd.substack.com/p/new-release-shot-dead

438 RSV shot | https://www.cdc.gov/respiratory-viruses/whats-new/rsv-update-2023-09-22.html

439 0:18 to 2:18 | https://petermcculloughmd.substack.com/p/new-release-shot-dead

440 childhood vaccination schedule | https://jeffereyjaxen.substack.com/p/the-fall-of-the-cdc-as-walensky-goes

441 seroprevalence study | https://www.cdc.gov/mmwr/volumes/71/wr/mm7117e3.htm

442 Ryan Cole | https://rumble.com/v18byhs-dr.-ryan-cole-covid-vaccine-side-effects-are-like-a-nuclear-bomb.html

443 Children's Health Defense Fund | https://childrenshealthdefense.org/press-release/chd-president-responds-to-cdc-officially-adding-covid-shots-to-the-recommended-schedule/

444 3:50:34 | https://rumble.com/vt62y6-covid-19-a-second-opinion.html

445 financial ties | https://www.lawfirms.com/resources/environment/environment-health/cdc-members-own-more-50-patents-connected-vaccinations

446 here | https://www.realclearpolitics.com/video/2022/01/24/full_hearing_covid-19_a_second_opinion_hosted_by_sen_ron_johnson.html#!

447 PREP Act | https://aspr.hhs.gov/legal/PREPact/Pages/PREP-Act-Question-and-Answers.aspx

448 introduced legislation | https://www.hydesmith.senate.gov/hyde-smith-backs-bill-help-those-hurt-covid-19-vaccines

449 report for fiscal years 2010-2023 | https://www.hrsa.gov/cicp/cicp-data

450 did not pass | https://www.govtrack.us/congress/bills/117/s3810

451 October 5, 2021 | https://thehill.com/policy/healthcare/public-global-health/575345-nba-star-andrew-wiggins-on-getting-vaccinated-i-guess/

452 Ramesh Thakur | https://brownstone.org/articles/they-profiled-citizens-according-to-their-degree-of-compliance/

453 evidence mounted | https://openvaers.com/covid-data

454 injured | https://www.anecdotalsmovie.com/

455 full approval | https://www.fda.gov/news-events/press-announcements/fda-approves-first-covid-19-vaccine

456 no plans | https://www.dossier.today/p/ghost-shot-pfizer-quietly-admits

457 addressed | https://www.whitehouse.gov/briefing-room/speeches-remarks/2021/09/09/remarks-by-president-biden-on-fighting-the-covid-19-pandemic-3/

458 Biden scolded and demeaned | https://www.cnn.com/us/live-news/coronavirus-pandemic-biden-speech-09-09-21/index.html

459 Austria | https://www.youtube.com/watch?v=y9io1MZz_7E

460 New Zealand | https://www.independent.co.uk/tv/news/jacinda-ardern-admits-new-zealand-will-become-a-twotier-society-between-vaccinated-and-unvaccinated-b2179915.html

461 here| https://nypost.com/2021/09/07/australias-covid-rules-are-a-warning-to-rest-of-the-world/

462 here | https://www.realclearpolitics.com/video/2021/12/02/inside_australias_howard_springs_covid_internment_camp_you_feel_like_youre_in_prison.html

463 here | https://www.ontopicwithlori.com/p/the-salt-lake-tribune-shouts-fire-in-a-crowded-theater

464 The Unvaccinated are the Problem (3:43 min)| https://twitter.com/tomselliott/status/1657021799652024324

465 HIPAA | https://www.hhs.gov/hipaa/for-professionals/privacy/laws-regulations/index.html

466 Nuremberg Code | https://encyclopedia.ushmm.org/content/en/article/the-nuremberg-code

467 No One is Safe Until We're All Safe (11:24 min) | https://www.youtube.com/watch?v=zl3yU5Z2adI

468 https://twitter.com/realDailyWire/status/1436157799852032002?

469 did | https://thehill.com/homenews/media/580119-ben-shapiros-media-company-sues-biden-administration-over-vaccine-mandate/

470 won | https://www.bbc.com/news/world-us-canada-59989476

471 Supreme Court | https://thehill.com/policy/healthcare/4353550-supreme-court-wipes-rulings-federal-employee-military-vaccine-mandates/

472 Cause Unknown | https://www.amazon.com/Cause-Epidemic-Sudden-Childrens-Defense/dp/1510776397

473 here | https://jamanetwork.com/journals/jamacardiology/fullarticle/2775372

474 here | https://www.nfl.com/news/nfl-major-sports-leagues-announce-study-showing-low-rate-of-heart-disease-from-a

475 athletes began dropping | https://rairfoundation.com/alert-politician-questions-soaring-number-of-top-athletes-with-heart-problems-video/

476 states | https://brownstone.org/articles/questioning-lipid-nanoparticles/

477 Damar Hamlin | https://www.cnn.com/videos/sports/2023/01/03/damar-hamlin-collapse-buffalo-bills-nfl-vpx.cnn

478 Jamie Foxx | https://www.theguardian.com/film/2023/jul/22/jamie-foxx-speaks-publicly-after-health-problem

479 Grant Wahl | https://www.reuters.com/world/us/us-sportswriter-grant-wahl-dies-qatar-during-world-cup-npr-us-soccer-2022-12-10/

480 Heather McDonald | https://www.youtube.com/watch?v=ctPViK3LK-o

481 Nick Nemeroff | https://twitter.com/alexstein99/status/1542174031406923778

482 Jessica Sutta| https://podcasts.apple.com/us/podcast/jessica-sutta-former-pussycat-dolls-member-i-was-severely/id1471411980?i=1000604798362

483 Sergio Aguero | https://www.bbc.com/sport/football/59660727

484 Oscar Cabrera Adames | https://www.totalprosports.com/nba/oscar-cabrera-adames-dies-heart-attack/

485 David Renne | https://petermcculloughmd.substack.com/p/italian-fashion-designer-david-renne?utm_campaign=email-post&r=15zdhf&utm_source=substack&utm_medium=email

486 Bronny James | https://www.cbssports.com/college-basketball/news/bronny-james-suffers-cardiac-arrest-collapses-on-usc-court-lebron-james-son-now-in-stable-condition/

487 Brooke Shields | https://www.yahoo.com/news/brooke-shields-reveals-she-had-213208782.html

488 India | https://www.thenationalnews.com/world/asia/2023/10/23/garba-gujarat-heart-attack/

489 replies | https://www.coffeeandcovid.com/p/low-tech-thursday-october-26-2023

490 asks | https://dailysceptic.org/2023/04/04/unacceptable-calculating-the-covid-vaccine-fatality-rate

491 posts | https://markcrispinmiller.substack.com/

492 Substack | https://substack.com/@coffeeandcovid

493 Until Proven Otherwise (4 min) | https://rumble.com/embed/v1oyvsc/?pub=4

494 highest ever seen | https://www.thecentersquare.com/indiana/indiana-life-insurance-ceo-says-deaths-are-up-40-among-people-ages-18-64/article_71473b12-6b1e-11ec-8641-5b2c06725e2c.html

495 disability payouts | https://crossroadsreport.substack.com/p/breaking-fifth-largest-life-insurance?utm_source=substack

496 Cause Unknown | https://www.amazon.com/Cause-Epidemic-Sudden-Childrens-Defense/dp/1510776397

497 OneAmerica's CEO | https://www.thecentersquare.com/indiana/indiana-life-insurance-ceo-says-deaths-are-up-40-among-people-ages-18-64/article_71473b12-6b1e-11ec-8641-5b2c06725e2c.html

498 Thomas Buckley | https://brownstone.org/author/thomas-buckley/

499 compromised industry | https://disentanglement.substack.com/p/james-c-smith-pfizers-propaganda

500 Thomson Reuters' Chairman of the Board | https://www.trust.org/about-us/

501 Director at Pfizer Inc | https://www.pfizer.com/people/leadership/board-of-directors/james_smith

502 funded | https://www.generation95.com/is-factcheck-org-owned-by-bill-gates-the-controversy-of-the-annenburg-public-policy-center/

503 December 20, 2023 | https://www.theepochtimes.com/epochtv/dr-peter-mccullough-no-evidence-that-our-bodies-can-get-rid-of-vaccine-mrna-5550521?

504 analysis | https://rumble.com/vwjmjm-bombshell-naomi-wolf-interviews-edward-dowd-about-pfizer-fraud-and-criminal.html

505 states | https://www.cdc.gov/coronavirus/2019-ncov/long-term-effects/index.html

506 called out the fallacy | https://rumble.com/v3hwcgm-dr.-mcculloughs-speech-at-the-european-parliament.html l

507 6:50 | https://wethepatriotsusa.org/shot-dead-movie/

508 12:54 | https://wethepatriotsusa.org/shot-dead-movie/

509 denying a connection | https://www.2ndsmartestguyintheworld.com/p/insurance-industry-execs-alarmed

510 Group Life Covid-19 Mortality Survey Report | https://www.soa.org/resources/experience-studies/2023/group-life-covid-mort-06-23/

511 warning| https://www.youtube.com/watch?v=pyPjAfNNA-U

512 shows | https://vigilance.pervaers.com/p/usa-the-mass-vaccine-casualty-event

513 regressions | https://www.investopedia.com/terms/r/regression.asp

514 scatterplot | https://www.thoughtco.com/frequently-used-statistics-graphs-4158380

515 examples | https://www.khanacademy.org/math/statistics-probability/describing-relationships-quantitative-data/introduction-to-scatterplots/a/scatterplots-and-correlation-review

516 Substack | https://vigilance.pervaers.com/p/us-summer-deaths-of-2021

517 Robert W. Chandler | https://robertchandler.substack.com/?nthPub=851

518 statement | https://dailyclout.io/report-52-nine-months-post-covid-mrna-vaccine-rollout-substantial-birth-rate-drops/

519 extensive research | https://dailyclout.io/report-52-nine-months-post-covid-mrna-vaccine-rollout-substantial-birth-rate-drops/

520 36:08 | https://wethepatriotsusa.org/shot-dead-movie/

521 38:34 | https://wethepatriotsusa.org/shot-dead-movie/

522 44:35 | https://wethepatriotsusa.org/shot-dead-movie/

523 https://margaretannaalice.substack.com/p/dissident-dialogues-dr-naomi-wolf

524 45:45 | https://wethepatriotsusa.org/shot-dead-movie/

525 Pfizer report| https://archive.org/details/cumulative-analysis-of-post-authorization-adverse-event-reports-of-pf-07302048-bnt162b2

526 46:40 | https://wethepatriotsusa.org/shot-dead-movie/

527 December 20, 2023 | https://www.theepochtimes.com/epochtv/dr-peter-mccullough-no-evidence-that-our-bodies-can-get-rid-of-vaccine-mrna-5550521

528 June 17, 2023 | https://podcasts.apple.com/no/podcast/exposing-the-vaccine-military-machinery-behind-the/id1471411980?i=1000617761934

529 How Bad Is My Batch | https://howbadismybatch.com/index.html

530 Dr. Marik | https://odysee.com/@FrontlineCovid19CriticalCareAlliance:c/Weekly_Webinar_June15:d

531 [2:25:45] | https://www.ronjohnson.senate.gov/2022/12/video-release-sen-johnson-please-share-expert-testimony-from-covid-19-vaccine-efficacy-and-safety-roundtable

532 fired from her position | https://wa.childrenshealthdefense.org/news/pediatricians-contract-terminated-by-wsu-after-reporting-to-senate-roundtable-on-covid-shot-harms/

533 5:03:53 | https://www.ronjohnson.senate.gov/2022/2/a-second-opinion-on-covid

534 October 5, 2023 | https://boriquagato.substack.com/p/did-the-pfizer-vaccine-even-really?utm_source=profile&utm_medium=reader2

535 Guetzkow explains | https://boriquagato.substack.com/p/did-the-pfizer-vaccine-even-really?utm_source=profile&utm_medium=reader2

536 article | https://brownstone.org/articles/vax-gene-files-accidental-discovery/

537 'complete recipe' | https://brownstone.org/articles/vax-gene-files-accidental-discovery/

538 explains | https://rwmalonemd.substack.com/p/what-is-adulteration-of-pseudo-mrna

539 fragments of the DNA virus SV-40 | https://osf.io/mjc97/

540 article | https://www.nature.com/articles/3302074

541 McKernan | https://brownstone.org/articles/vax-gene-files-accidental-discovery/

542 interviewed Dr. Vibeke Manniche | https://www.youtube.com/watch?v=KgIdG9r-i9M

543 study | https://onlinelibrary.wiley.com/doi/10.1111/eci.13998

544 expressed concern | https://dailysceptic.org/2023/04/04/unacceptable-calculating-the-covid-vaccine-fatality-rate

545 printed | https://onlinelibrary.wiley.com/doi/full/10.1111/eci.13998

546 https://www.anecdotalsmovie.com/

547 September 13, 2023 | https://rwmalonemd.substack.com/p/lipid-nanoparticles-and-mrna-shots

548 proclaimed | https://www.cdc.gov/coronavirus/2019-ncov/vaccines/reporting-systems.html

549 Published studies | https://aaronsiri.substack.com/p/v-safe-part-2-what-is-v-safe-what

550 July 2020 | https://www.ncbi.nlm.nih.gov/pmc/articles/PMC7377258/#ap2

551 October 2020 | https://jamanetwork.com/journals/jama/fullarticle/2772137

552 https://www.ncbi.nlm.nih.gov/pmc/articles/PMC6760227/

553 explains | https://aaronsiri.substack.com/p/v-safe-part-2-what-is-v-safe-what

554 464 days | https://aaronsiri.substack.com/p/v-safe-part-1-after-464-days-cdc

555 deceitful | https://aaronsiri.substack.com/p/v-safe-part-7-cdc-deceived-the-public

556 states| https://aaronsiri.substack.com/p/v-safe-part-4-cdc-designs-v-safe?utm_source=profile&utm_medium=reader2

557 archived zip file | https://data.cdc.gov/Public-Health-Surveillance/v-safe-COVID-19/dqgu-gg5d

558 screenshot | https://icandecide.org/v-safe-data/

559 David Gortler | https://brownstone.org/articles/cdc-refusing-new-covid-vaccine-adverse-event-reports/

560 V-safe | https://www.cdc.gov/vaccinesafety/ensuringsafety/monitoring/v-safe/index.html

561 obscure | https://crsreports.congress.gov/product/pdf/LSB/LSB10589

562Section 264a | https://uscode.house.gov/view.xhtml?req=(title:42%20section:264%20edition:prelim)%20OR%20(granuleid:USC-prelim-title42-section264)&f=treesort&edition=prelim&num=0&jumpTo=true

563 order| https://apnews.com/article/race-and-ethnicity-homeless-shelters-coronavirus-pandemic-699b0f31a20cb83b039c2e97d627e88a

564 Executive authority| https://abcnews.go.com/Politics/biden-order-federal-workers-vaccinated-strategy-combat-delta/story?id=79905030

565 mask mandate| https://law.justia.com/cases/federal/district-courts/florida/flmdce/8:2021cv01693/391798/53/

566 vaccine mandates | https://supreme.justia.com/cases/federal/us/595/21a244/

567 no-eviction rule | https://www.npr.org/2021/08/06/1025212834/cdc-new-eviction-ban-moratorium-renters-landlords

568 overreach | https://www.npr.org/2021/08/26/1024668578/court-blocks-biden-cdc-evictions-moratorium

569 contested | https://popularrationalism.substack.com/p/ninth-circuit-court-judges-shocked

570 appealed | https://www.rochesterfirst.com/coronavirus/arguments-end-in-appeal-case-over-covid-19-quarantine-regulations/

571 state's power grab | https://coxlawyers.com/borrello-fellow-plaintiffs-win-case-against-ny-on-quarantine-isolation/

572 Bobbie Ann Cox | https://substack.com/@attorneycox

573 eviction ban| https://www.npr.org/2021/06/29/1003268497/the-supreme-court-leaves-the-cdcs-moratorium-on-evictions-in-place

574 upheld | https://www.thebalancemoney.com/federal-appeals-court-lets-eviction-moratorium-stand-5198349

575 new 60-day eviction moratorium| https://www.npr.org/2021/08/06/1025212834/cdc-new-eviction-ban-moratorium-renters-landlords

576 struck down| https://www.scotusblog.com/2021/08/court-lifts-federal-ban-on-evictions/

577 stated | https://www.supremecourt.gov/opinions/20pdf/21a23_ap6c.pdf

578 ongoing court case | https://www.theepochtimes.com/opinion/american-pandemic-samizdat-5497566

579 The Federalist| https://thefederalist.com/2021/03/18/one-of-the-lockdowns-greatest-casualties-could-be-science/

580 agreed| https://www.washingtonexaminer.com/policy/courts/supreme-court-agrees-to-take-up-major-big-tech-free-speech-case

581 require Covid shots | https://substack.com/@nocollegemandates

582 here | https://www.mercurynews.com/2023/10/31/bay-area-masking-up-again-as-mandates-go-into-effect-nov-1-in-health-care-settings/

583 here| https://www.atlantanewsfirst.com/2023/08/22/morris-brown-college-reinstates-mask-mandate-due-rise-on-campus-covid-cases/

584 here | https://globalnews.ca/news/10029147/more-alberta-hospitals-masking-requirements/

585 here | https://ianmsc.substack.com/p/the-definition-of-insanity-exhibit

586 face masks | https://www.cochrane.org/news/featured-review-physical-interventions-interrupt-or-reduce-spread-respiratory-viruses

587 harmful | https://www.city-journal.org/article/the-harm-caused-by-masks

588 complete failure | https://health.wusf.usf.edu/health-news-florida/2022-02-02/a-johns-hopkins-study-says-ill-founded-lockdowns-did-little-to-limit-covid-deaths

589 charts | https://www.resistbiden.org/2022/01/im-charts-see-ianmsc-on-twitter-archive.html

590 Ian Miller | https://ianmsc.substack.com/

591 coerced/forced | https://www.independent.co.uk/tv/news/jacinda-ardern-admits-new-zealand-will-become-a-twotier-society-between-vaccinated-and-unvaccinated-b2179915.html

592 locked down the unvaccinated | https://www.youtube.com/watch?v=y9io1MZz_7E

593 October 11, 2023 | https://www.youtube.com/watch?v=42uoERKuzo4

594 *The Death of Science* | https://www.amazon.com/Death-Science-Retreat-Reason-Post-Modern-ebook/dp/B0CKS6X6HS

595 Terrorism Advisory | https://www.dhs.gov/ntas/advisory/national-terrorism-advisory-system-bulletin-february-07-2022

596 Preliminary Injunction | https://www.wsj.com/articles/a-key-ruling-against-social-media-censorship-missouri-v-biden-government-covid-9b457364

597 upheld in part | https://www.telegraph.co.uk/world-news/2023/09/12/biden-social-media-meta-first-amendment-violation-covid/

598 Supreme Court | https://www.washingtonexaminer.com/policy/courts/supreme-court-agrees-to-take-up-major-big-tech-free-speech-case

599 Twitter Files | https://twitterfiles.substack.com/p/1-thread-the-twitter-files

600 Linda Yaccarino | https://www.forbes.com/sites/roberthart/2023/05/12/what-to-know-about-linda-yaccarino-musks-pick-for-twitter-ceo/?sh=480da3521c42

601 World Economic Forum's | https://www.thenationalnews.com/business/technology/2023/05/12/linda-yaccarino-twitter-ceo-who/

602 assures | https://twitter.com/ReclaimTheNetHQ/status/1689710983571181568

603 June 29, 2023 | https://www.youtube.com/watch?v=EXUpMMde51E

604 January 17, 1961 | https://www.archives.gov/milestone-documents/president-dwight-d-eisenhowers-farewell-address

605 January 18, 2023 | https://brownstone.org/articles/lockdowns-counterterrorism-not-public-health/

606 documented | https://brownstone.org/articles/governments-national-security-arm-led-the-covid-response/

607 White House | https://www.whitehouse.gov/nsc/

608 article | https://brownstone.org/articles/lockdowns-counterterrorism-not-public-health/

609 Sasha Latypova | https://brownstone.org/articles/proof-vaccines-were-military-backed-countermeasure/

610 U.S. Department of Defense | https://www.militaryfactory.com/dictionary/military-terms-defined.php?term_id=1353

611 Pandemonium | https://reportfromplanetearth.substack.com/p/pandemonium

612 May 2023 interview | https://podcasts.apple.com/gb/podcast/dr-pierre-kory-deadly-conflicts-of-interest-and/id1471411980?i=1000613740456

613 August 2020| https://www.cidrap.umn.edu/covid-19/commentary-public-healths-share-blame-us-covid-19-risk-communication-failures

614 1:18:25 | https://www.youtube.com/watch?v=ry3Xo-EJZ00

615 Oxford | https://www.oxfordlearnersdictionaries.com/us/definition/english/lockdown?q=lockdown

616 Cambridge | https://dictionary.cambridge.org/dictionary/english/lockdown

617 states | https://brownstone.org/articles/lockdowns-counterterrorism-not-public-health/

618 April of 2020 | https://rorate-caeli.blogspot.com/2020/04/guest-op-ed-immorality-of-indefinite.html

619 https://brownstone.org/articles/at-the-white-house-march-10-2020-reconstructed/

620 https://brownstone.org/articles/ecohealth-alliances-wuhan-virus-dalliances/

621 here| https://www.washingtonexaminer.com/news/fauci-worked-behind-scenes-cast-doubt-wuhan-lab-leak-hypothesis

622 here | https://www.dailymail.co.uk/news/article-9702125/Dr-Fauci-says-scientists-secret-February-2020-call-said-COVID-possibly-engineered-virus.html

623 here| https://www.nature.com/articles/S41591-020-0820-9

624 here| https://brownstone.org/author/randall-s-bock/

625 conspiracy theorist | https://nypost.com/2021/01/16/doctor-who-denied-covid-was-leaked-from-a-lab-had-this-major-bias/

626 submitted a letter | https://www.thelancet.com/journals/lancet/article/PIIS0140-6736(20)30418-9/fulltext

627 coordinated by | https://www.usatoday.com/story/opinion/2021/06/17/covid-19-fauci-lab-leaks-wuhan-china-origins/7737494002/

628 vested interest | https://dailycaller.com/2021/02/13/world-health-organization-peter-daszak-conflict-of-interest-china-coronavirus/

629 provided funding | https://nypost.com/2021/10/21/nih-admits-us-funded-gain-of-function-in-wuhan-despite-faucis-repeated-denials/

630 https://www.the-scientist.com/the-nutshell/moratorium-on-gain-of-function-research-36564

631 summarizes | https://brownstone.org/articles/covid-mrna-vaccines-required-no-safety-oversight/

632 March 15, 2020 | https://www.ft.com/content/d6c96612-67c0-11ea-800d-da70cff6e4d3

633 March 17, 2020| https://www.statnews.com/2020/03/17/a-fiasco-in-the-making-as-the-coronavirus-pandemic-takes-hold-we-are-making-decisions-without-reliable-data/

634 March 19, 2020| https://time.com/5806657/donald-trump-coronavirus-war-china/

635 March 26, 2020| https://www.who.int/director-general/speeches/detail/who-director-general-s-remarks-at-the-g20-extraordinary-leaders-summit-on-covid-19---26-march-2020

636 Prof. John Ioannidis | https://www.vox.com/2015/2/16/8034143/john-ioannidis-interview

637 integrity | https://en.wikipedia.org/wiki/Why_Most_Published_Research_Findings_Are_False

638 A State of Fear | https://www.amazon.com/State-Fear-government-weaponised-Covid-19-ebook/dp/B08ZSYN14J

639 Dodsworth, Laura. *A State of Fear: How the UK government weaponised fear during the Covid-19 pandemic* (p. 243). Pinter & Martin. Kindle Edition.

640 March 16, 2020 | https://www.imperial.ac.uk/media/imperial-college/medicine/sph/ide/gida-fellowships/Imperial-College-COVID19-NPI-modelling-16-03-2020.pdf

641 noted | https://www.cato.org/blog/how-one-model-simulated-22-million-us-deaths-covid-19

642 Dodsworth, *Laura. A State of Fear: How the UK government weaponised fear during the Covid-19 pandemic* (p. 275). Pinter & Martin. Kindle Edition.

643 explained | https://www.youtube.com/watch?v=37VgawZN6ZE

644 Sweden never locked down | https://brownstone.org/articles/sweden-did-exceptionally-well/

645 Phillip W. Magness | https://www.aier.org/article/the-failure-of-imperial-college-modeling-is-far-worse-than-we-knew/

646 clarified | https://www.aier.org/article/the-failure-of-imperial-college-modeling-is-far-worse-than-we-knew/

647 study published | https://www.imperial.ac.uk/news/237591/vaccinations-have-prevented-almost-20-million/

648 prevented 20 million deaths | https://www.thelancet.com/journals/laninf/article/PIIS1473-3099(22)00320-6/fulltext#%20

649 76.4 years | https://www.cdc.gov/nchs/fastats/life-expectancy.htm

650 start | https://www.cdc.gov/nchs/pressroom/nchs_press_releases/2020/202012.htm

651 reasons | https://brownstone.org/articles/a-closer-look-at-the-us-2020-mortality-data/

652 states | https://brownstone.org/articles/a-common-sense-look-at-20-million-saved-lives/

653 Brownstone Institute | https://brownstone.org/

654 Deborah Birx | https://brownstone.org/articles/deborah-birxs-guide-to-destroying-a-country-from-within/

655 dubious credentials | https://brownstone.org/articles/how-did-deborah-birx-get-the-job/

656 March 19, 2020 | https://www.youtube.com/watch?v=971QcDEha5I

657 Bill Rice Jr.'s | https://substack.com/@billricejr

658 here | https://www.ncbi.nlm.nih.gov/pmc/articles/PMC7864798/

659 here | https://pubmed.ncbi.nlm.nih.gov/33714813/

660 2019 Military World Games | https://www.dailymail.co.uk/news/article-8327047/More-competitors-reveal-ill-World-Military-Games.html

661 U.S. Navy | https://brownstone.org/articles/first-confirmed-cases-in-america-were-on-us-aircraft-carrier/

662 fateful day | https://trumpwhitehouse.archives.gov/articles/15-days-slow-spread/

663 May 7, 2020 | https://news.yahoo.com/dr-atlas-coronavirus-too-widespread-011311026.html

664 traveled the country | https://www.tabletmag.com/sections/news/articles/deborah-birx-guide-destroying-america

665 resigned | https://www.bbc.com/news/world-us-canada-55419954

666 states | https://brownstone.org/articles/deborah-birxs-guide-to-destroying-a-country-from-within/

667 writes | https://brownstone.org/articles/was-the-covid-response-a-coup-by-the-intelligence-community/

668 warned | https://www.dailymail.co.uk/news/article-11668157/Bill-Gates-warns-Australia-pandemic-just-corner-man-made.html

669 Catastrophic Contagion | https://catastrophiccontagion.centerforhealthsecurity.org/

670 commented | https://www.foxnews.com/media/podcaster-sam-harris-covid-killed-children-f-ing-patience-vaccine-skeptics

671 said | https://www.oxfordreference.com/display/10.1093/acref/9780191826719.001.0001/q-oro-ed4-00002041

672 climate change| https://frontline.news/post/uk-developing-vaccines-for-next-pandemic

673 Nuremberg Code | https://research.unc.edu/human-research-ethics/resources/ccm3_019064/

674 Nazi doctors | https://encyclopedia.ushmm.org/content/en/article/the-doctors-trial-the-medical-case-of-the-subsequent-nuremberg-proceedings

675 JAMA in 2017 | https://jamanetwork.com/journals/jama/fullarticle/2649074

676 Universal Declaration of Human Rights | https://www.un.org/en/about-us/universal-declaration-of-human-rights

677 October 26, 2022 | https://brownstone.org/articles/duplicitous-fauci-backpedaling/

678 no-mask | https://www.newsweek.com/fauci-said-masks-not-really-effective-keeping-out-virus-email-reveals-1596703

679 yes-mask | https://www.pbs.org/newshour/show/what-dr-fauci-wants-you-to-know-about-face-masks-and-staying-home-as-virus-spreads

680 two masks | https://www.today.com/health/dr-fauci-shows-how-wear-2-masks-correctly-today-t208765

681 Harvard Institute | https://iop.harvard.edu/events/conversation-dr-anthony-fauci

682 mask mandate| https://www.deseret.com/utah/2020/6/23/21300523/coronavirus-covid19-governor-no-plans-to-shut-down-economy-case-counts-rise-restrictions-masks

683 Matt Hancock| https://www.bbc.com/news/uk-64848106

684 violating | https://www.bbc.co.uk/news/57611369

685 cited | https://www.standard.co.uk/news/uk/covid-britons-refuse-wear-mask-fines-b968782.html

686 testify | https://www.facebook.com/watch/?v=1025558331227117

687 aerosols| https://www.nap.edu/read/25769/chapter/1

688 no difference | https://www.ncbi.nlm.nih.gov/pmc/articles/PMC2493952/pdf/annrcse01509-0009.pdf

689 CDC study | https://wwwnc.cdc.gov/eid/article/26/5/19-0994_article

690 guilt | https://brownstone.org/articles/open-letter-to-an-emotionally-abused-world/

691 Dr. Harvey Risch | https://drdrew.com/2022/epidemiologist-masks-like-a-chain-link-fence-to-block-mosquitoes-dr-harvey-risch-professor-emeritus-at-yale-dr-kelly-victory-on-ask-dr-drew/

692 February 5, 2020 | https://www.newsweek.com/fauci-said-masks-not-really-effective-keeping-out-virus-email-reveals-1596703

693 November 26, 2023 | https://youtu.be/AFA5fTeAJno

694 OSHA | https://www.osha.gov/respiratory-protection

695 respirators | https://www.military.com/equipment/m50-m51-joint-service-general-purpose-mask

696 pamphlet | https://apps.state.or.us/forms/served/le8626.pdf

697 CDC Director | https://www.cbsnews.com/newyork/news/cdc-director-dr-mandy-cohen-offers-covid-caution-for-the-holidays/

698 All About the Air - Oct. 2023 (1:52 min) | https://twitter.com/justin_hart/status/1715115591386697745

699 books | https://www.amazon.com/Unmasked-Global-Failure-COVID-Mandates/dp/1637583761/ref=sr_1_1

700 hundreds of articles | https://brownstone.org/?s=face+masks

701 ineffectiveness | https://brownstone.org/articles/the-mask-studies-you-should-know/

702 historic symbolism | https://brownstone.org/articles/mask-enslavement-history-meaning-escrava-anastacia/

703 psychological | https://frontline.news/post/mcdonald-masks-mirror-mental-illness

704 physical | https://www.bmj.com/content/369/bmj.m1435/rr-40

705 damage to young children | https://www.medrxiv.org/content/10.1101/2021.08.10.21261846v1.full.pdf

706 Political satirist | https://cjhopkins.substack.com/p/the-covidian-cult

707 respected | https://www.cochranelibrary.com/cdsr/doi/10.1002/14651858.CD006207.pub5/full

708 up-to-date | https://wwwnc.cdc.gov/eid/article/26/5/19-0994_article

709 Denmark | https://www.acpjournals.org/doi/10.7326/M20-6817

710 clarified | https://jamanetwork.com/journals/jama/fullarticle/2762694

711 here| https://thefederalist.com/2022/03/15/mitt-romney-just-voted-to-keep-masking-two-year-olds/

712 here| https://nypost.com/2022/06/02/parents-blast-mayor-eric-adams-for-keeping-toddler-mask-mandate/

713 here | https://www.nytimes.com/2022/09/07/us/head-start-masks-toddlers.html

714 Toddler cries as he is forced into a mask (1:19 min) | https://www.youtube.com/watch?v=wpoMEgUd_hE

715 explained | https://brownstone.org/articles/what-they-did-to-the-children/

716 April 2022 | https://www.theepochtimes.com/epochtv/dr-mark-mcdonald-how-we-sacrificed-our-children-in-the-name-of-covid-protection-4403533

717 Dodsworth, Laura. *A State of Fear: How the UK government weaponised fear during the Covid-19 pandemic* (p. 239). Pinter & Martin. Kindle Edition.

718 Cochrane | https://www.cochrane.org/about-us

719 latest review | https://www.cochranelibrary.com/cdsr/doi/10.1002/14651858.CD006207.pub6/full

720 Tom Jefferson | https://jeffereyjaxen.substack.com/p/the-fall-of-the-cdc-as-walensky-goes

721 under fire | https://maryannedemasi.substack.com/p/exclusive-lead-author-of-cochrane

722 said | https://www.independent.co.uk/news/health/dr-fauci-covid-masks-cochrane-review-b2404761.html

723 Cochrane | https://www.cochranelibrary.com/cdsr/doi/10.1002/14651858.CD006207.pub6/full

724 Mayo Clinic | https://www.mayoclinic.org/diseases-conditions/coronavirus/in-depth/coronavirus-mask/art-20485449

725 walk it back | https://www.statnews.com/2020/06/09/who-comments-asymptomatic-spread-covid-19/

726 China | https://www.nature.com/articles/s41467-020-19802-w

727 Germany | https://wwwnc.cdc.gov/eid/article/27/4/20-4576_article

728 Inventor of the PCR Test| https://www.youtube.com/watch?v=0zEoRmJgt00

729 explains | https://brownstone.org/articles/mass-testing-the-fatal-conceit/

730 process | https://www.genome.gov/about-genomics/fact-sheets/Polymerase-Chain-Reaction-Fact-Sheet

731 Prof. Carl Henegen | https://www.bbc.com/news/health-54000629

732 study| https://www.cebm.net/study/covid-19-testing-and-correlation-with-infectious-virus-cycle-thresholds-and-analytical-sensitivity/

733 exhaustively highlighted | https://coronavirus.jhu.edu/map.html

734 article | https://www.wsj.com/articles/the-college-covid-scare-11600990425

735 lateral flow | https://www.itv.com/news/2021-04-10/covid-holidays-what-are-the-differences-between-lateral-flow-and-pcr-coronavirus-tests-and-how-do-they-work

736 multiple companies | https://www.nbcnews.com/business/companies-produce-tens-millions-rapid-covid-tests-meet-surge-rcna9717

737 article | https://www.cdc.gov/coronavirus/2019-ncov/testing/self-testing.html

738 article | https://pharmaceutical-journal.com/article/feature/how-reliable-are-lateral-flow-covid-19-tests

739 said | https://youtu.be/CVTIH5KwK_c

740 misconstrued | https://brownstone.org/articles/does-new-york-city-2020-make-any-sense/

741 propagandized | https://brownstone.org/articles/the-freezer-truck-canard/

742 50 percent drop | https://brownstone.org/articles/more-questions-about-spring-2020-covid-in-new-york-city-hospitals/

743 research | https://pandata.org/what-the-diamond-princess-tells-us-about-nyc-in-spring-2020/

744 study | https://www.ncbi.nlm.nih.gov/pmc/articles/PMC3840149/

745 2018 Time article | https://time.com/5107984/hospitals-handling-burden-flu-patients/

746 field hospitals | https://www.deseret.com/utah/2020/4/6/21210159/utah-prepares-alternative-site-where-patients-can-be-treated-during-covid-19-outbreak

747 never used | https://kslnewsradio.com/1935012/utah-ready-to-set-up-temporary-covid-19-hospital-but-health-officials-hope-they-wont-need-it/

748 New York City| https://www.cbsnews.com/news/field-hospital-that-treated-coronavirus-patients-in-central-park-to-close/

749 treated less than 200 patients | https://www.navytimes.com/news/your-navy/2020/04/30/hospital-ship-comfort-departs-nyc-having-treated-fewer-than-200-patients/

750 Cuomo | https://apnews.com/article/new-york-andrew-cuomo-us-news-coronavirus-pandemic-nursing-homes-512cae0abb55a55f375b3192f2cdd6b5

751 Covid-19 Map of the World | https://coronavirus.jhu.edu/map.html

752 1,600 people a day | https://www.cancer.org/research/cancer-facts-statistics/all-cancer-facts-figures/cancer-facts-figures-2020.html

753 CDC's Covid-19 death data | https://townhall.com/tipsheet/bronsonstocking/2020/08/30/heres-the-shockingly-small-number-of-people-who-died-from-just-the-coronavirus-n2575306

754 financial motives | https://www.usatoday.com/story/news/factcheck/2020/04/24/fact-check-medicare-hospitals-paid-more-covid-19-patients-coronavirus/3000638001/

755 perspective | https://www.facebook.com/watch/?v=616887239159067

756 target | https://www.facebook.com/watch/?v=272572927153273

757 actions were eventually dropped | https://www.kttc.com/2023/03/24/minnesota-board-medical-practice-drops-charges-against-former-minnesota-governor-candidate-scott-jenson/

758 speech | https://www.youtube.com/watch?v=IA2SCoYI8_U

759 Nurse Gail Macrae | https://alphanews.org/former-nurse-describes-covid-19-protocols-as-medical-murder

760 closely examined the data | https://www.theblaze.com/news/denominatorgate-how-public-health-agencies-are-skewing-the-statistics-on-vaccine-effectiveness

761 CDC definition | https://data.cdc.gov/Public-Health-Surveillance/Rates-of-COVID-19-Cases-or-Deaths-by-Age-Group-and/54ys-qyzm/about_data

762 Dr. Pierre Kory| https://pierrekorymedicalmusings.com/p/proof-the-cdc-knowingly-lied-when?utm_source=profile&utm_medium=reader2

763 Data from multiple countries| https://brownstone.org/articles/this-is-not-a-pandemic-of-the-unvaccinated/

764 New South Wales | https://brownstone.org/articles/a-pandemic-of-the-triple-vaccinated/

765 study | https://academic.oup.com/jamiaopen/article/6/2/ooad026/7117831?login=false\

766 56:56 to 58:07 | https://www.cbsnews.com/news/transcript-dr-anthony-fauci-on-face-the-nation-november-28-2021/

767 metaphorical whiplash | https://pacificlegal.org/most-ridiculous-arbitrary-covid-restrictions/

768 *The Death of Science* | https://www.amazon.co.uk/Death-Science-retreat-reason-post-modern/dp/1854571133

769 *The Treason of the Experts* | https://www.amazon.com/Treason-Experts-Covid-Credentialed-Class-ebook/dp/B0C4G4785Y

770 Professor Steve Templeton | https://brownstone.org/articles/ten-examples-where-experts-were-wrong/

771 De Kai | https://dek.ai/

772 interview | https://www.cidrap.umn.edu/covid-19/special-episode-masks-and-science

773 serious flaws | see page 18 | https://www.cidrap.umn.edu/sites/default/files/downloads/special_episode_masks_6.2.20_0.pdf

774 article | https://www.vanityfair.com/news/2020/05/masks-covid-19-infections-would-plummet-new-study-says

775 article | https://theconversation.com/masks-help-stop-the-spread-of-coronavirus-the-science-is-simple-and-im-one-of-100-experts-urging-governors-to-require-public-mask-wearing-138507

776 explained | https://www.cidrap.umn.edu/covid-19/podcasts-webinars/special-ep-masks.

777 article | https://brownstone.org/articles/how-zeynep-tufekci-and-jeremy-howard-masked-america/

778 article | https://www.telegraph.co.uk/news/2022/08/15/peddlers-environmental-doom-have-shown-true-totalitarian-colours/

779 video | https://www.youtube.com/watch?v=--QS_UyW2SY

780 Cleveland Clinic | https://health.clevelandclinic.org/how-long-will-coronavirus-survive-on-surfaces

781 aerosols| https://www.nap.edu/read/25769/chapter/1

782 households| https://www.medrxiv.org/content/10.1101/2020.04.04.20053058v1

783 asymptomatic spread| https://www.nature.com/articles/s41467-020-19802-w

784 Adrian Monk | https://en.wikipedia.org/wiki/Adrian_Monk

785 Bob Wiley | https://www.imdb.com/title/tt0103241/

786 unintended consequences | https://www.sas.upenn.edu/~haroldfs/540/handouts/french/unintconseq.html

787 cost | https://www.wsj.com/articles/how-much-covid-19-cost-those-businesses-that-stayed-open-11592910575

788 customers| https://www.businessinsider.com/companies-pass-on-costs-ppe-cleaning-supplies-customers-2021-2?op=1

789 Small businesses | https://www.chamberofcommerce.org/coronavirus-small-business-guide

790 article| https://www.nature.com/articles/d41586-021-00251-4

791 finally acknowledged | https://www.cnn.com/2021/05/07/health/cdc-coronavirus-transmission/index.html

792 compulsion | https://www.cnet.com/health/new-cdc-cleaning-guidance-how-to-disinfect-and-sanitize-your-home-of-covid-19/

793 protocol | https://www.aapa-ports.org/files/FileDownloads/COVID-19%20Protocols%20-%20Updated%20April%2029.pdf

794 64 cargo ships waiting | https://calmatters.org/newsletters/whatmatters/2021/10/california-ports-backlog-labor/

795 open letter | https://www.ics-shipping.org/press-release/joint-open-letter-transport-heads-call-on-world-leaders-to-secure-global-supply-chains/

796 article | https://health.clevelandclinic.org/how-long-will-coronavirus-survive-on-surfaces

797 time warp| https://www.cdc.gov/coronavirus/2019-ncov/prevent-getting-sick/how-covid-spreads.html

798 church choir | https://www.livescience.com/covid-19-superspreader-singing.html

799 study| https://docs.iza.org/dp13670.pdf

800 Johns Hopkins | https://www.cbsnews.com/news/sturgis-motorcycle-rally-superspreader-johns-hopkins-research-doubt/

801 reality| https://www.usatoday.com/story/news/factcheck/2020/09/17/fact-check-sturgis-rallys-covid-19-cases-misstated-online-post/3458606001/

802 dire warnings| https://www.nytimes.com/2020/12/02/health/coronavirus-quarantine-period.html

803 solemn statements | https://www.washingtonpost.com/health/2021/02/05/covid-spread-super-bowl/

804 just 57 cases | https://www.usatoday.com/story/sports/nfl/super-bowl/2021/03/03/super-bowl-55-not-covid-19-super-spreader-tampa/6905658002/

805 accused| https://www.youtube.com/watch?v=r0tbYuuA1j0

806 no outbreaks | https://frontofficesports.com/rangers-averaging-largest-crowds-no-outbreaks-linked/

807 feared | https://www.nbcdfw.com/news/coronavirus/70000-boxing-fans-converge-on-att-stadium-may-be-biggest-indoor-event-since-pandemic/2627154/

808 "experts"| https://source.wustl.edu/2021/06/washu-expert-without-requiring-vaccines-filled-stadiums-are-unsafe/

809 danger | https://www.indystar.com/story/opinion/columnists/james-briggs/2020/07/01/indy-500-recipe-disaster/5352485002/

810 August 18, 2022 | https://www.weforum.org/agenda/2022/08/one-health-disease-climate-future-pandemics/

811 Bill Gates | https://www.today.com/health/health/bill-gates-now-time-prepare-next-pandemic-rcna27146

812 here | https://news.un.org/en/story/2021/02/1084982

813 here | https://news.un.org/en/story/2021/08/1098472

814 here | https://www.usnews.com/news/world/articles/2023-05-22/dont-delay-reforms-to-prepare-for-next-pandemic-who-chief

815 World Economic Forum | https://www.weforum.org/agenda/2021/03/new-treaty-aims-to-unite-nations-in-preparation-for-future-pandemics/

816 7 million | https://coronavirus.jhu.edu/map.html

817 Spanish flu | https://archive.cdc.gov/#/details?url=https://www.cdc.gov/flu/pandemic-resources/1918-pandemic-h1n1.html

818 aspirin | https://academic.oup.com/cid/article/49/9/1405/301441

819 *Fear of a Microbial Planet* | https://www.amazon.com/Fear-Microbial-Planet-Germophobic-Culture-ebook/dp/B0C29RC8ZB

820 Jan 25, 2021 | https://www.youtube.com/watch?v=WK4M9iJrgto

821 *Covidian Cult* | https://www.bitchute.com/video/V978p3lgvvQM

822 Scott Gottlieb | https://www.forbes.com/sites/graisondangor/2021/09/19/cdcs-six-foot-social-distancing-rule-was-arbitrary-says-former-fda-commissioner/?sh=4ac4d3fa33e4

823 testified| https://oversight.house.gov/release/wenstrup-releases-statement-following-dr-faucis-two-day-testimony/

824 3-feet apart | https://www.pbs.org/newshour/health/cdc-changes-school-guidance-allowing-desks-to-be-closer

825 activities | https://www.edweek.org/policy-politics/cdc-eases-social-distancing-guidelines-seen-as-a-hurdle-to-school-reopening/2021/03

826 said | https://www.cdc.gov/media/releases/2022/t0225-covid-19-update.html

827 January 2022 | https://nationalpost.com/news/world/johns-hopkins-university-study-covid-19-lockdowns

828 economists released | https://health.wusf.usf.edu/health-news-florida/2022-02-02/a-johns-hopkins-study-says-ill-founded-lockdowns-did-little-to-limit-covid-deaths

829 Neil Ferguson | https://www.dailymail.co.uk/news/article-10473937/Now-Prof-Lockdown-slams-shock-study-draconian-curbs-reduced-Covid-deaths-0-2.html

830 vitamin D | https://www.sciencedirect.com/science/article/pii/S0899900720303890?via%3Dihub.

831 https://bjsm.bmj.com/content/55/19/1099

832 weight loss | https://www.ncbi.nlm.nih.gov/pmc/articles/PMC8073853/

833 Pandemonium | https://reportfromplanetearth.substack.com/p/pandemonium

834 winners | https://www.science.org/content/article/mrna-discovery-paved-way-covid-19-vaccines-wins-nobel-prize-physiology-medicine

835 study | https://pubmed.ncbi.nlm.nih.gov/37650258/

836 study published | https://www.nature.com/articles/s41586-023-06800-3 l

837 December 18, 2023 | https://www.theepochtimes.com/epochtv/bombshell-study-on-vaccine-ribosomal-frameshifting-dr-paul-marik-atlnow-5548921

838 glowing report | https://www.pennmedicine.org/news/news-releases/2023/october/katalin-kariko-and-drew-weissman-win-2023-nobel-prize-in-medicine

839 tumbling | https://247wallst.com/investing/2023/10/23/pfizer-and-moderna-stock-wrecked-as-covid-vaccine-sales-dry-up/

840 coauthored a paper | https://www.nature.com/articles/nrd.2017.243

841 notes | https://petermcculloughmd.substack.com/p/nobel-prize-awarded-for-modification

842 petition | https://nobelprizeprotest.com

843 Inamori Center | https://case.edu/inamori/mission

844 increased | https://www.theepochtimes.com/article/faucis-net-worth-nearly-doubled-during-the-pandemic-financial-disclosure-forms-show-4763456

845 announced | https://thedaily.case.edu/dr-anthony-fauci-to-be-awarded-2024-inamori-ethics-prize-by-inamori-international-center-for-ethics-and-excellence/

846 President | https://www.dossier.today/p/anthony-fauci-to-be-awarded-with

847 attacks on science | https://www.youtube.com/watch?v=z-tfZr8lv0s

848 https://www.publicnow.com/view/F3D733D4F6FF926E2F49B1DD7D13509D9EC6F348

849 Award | https://www.idsociety.org/about-idsa/awards/peter-hotez---anthony-fauci-courage-award/

850 states | https://johnanderson.net.au/conversations-konstantin-kisin-ii/

851 April 29, 2022 | https://johnanderson.net.au/conversations-konstantin-kisin-ii/

852 writes | https://lists.youmaker.com/archive/hCdBJF04V/5kdG8QBtO/1RIZS9AgY1U

853 David Bell | https://brownstone.org/articles/we-must-save-ourselves-from-the-public-health-professionals/

854 resigned | https://www.businessinsider.com/2-top-fda-officials-resigned-biden-booster-plan-reports-2021-9

855 opinion piece | https://nypost.com/2023/09/14/the-real-data-behind-the-new-covid-vaccines-the-white-house-is-pushing/

856 latest Covid shots | https://www.npr.org/sections/health-shots/2023/09/11/1198719166/new-covid-vaccines-get-fda-approval

857 CBS News | https://www.cbsnews.com/news/new-covid-vaccine-shots-booster-2023/

858 September 2023 | https://www.cbsnews.com/news/new-covid-vaccine-shots-booster-2023/

859 taboo | https://brownstone.org/articles/most-important-meeting-in-history-of-world-that-never-happened/

860 John Ioannidis | https://www.washingtonpost.com/dc-md-va/2020/12/16/john-ioannidis-coronavirus-lockdowns-fox-news/

861 here | https://www.whitehouse.gov/briefing-room/statements-releases/2023/05/09/fact-sheet-actions-taken-by-the-biden-harris-administration-to-ensure-continued-covid-19-protections-and-surge-preparedness-after-public-health-emergency-transition/

862 here | https://www.cdc.gov/coronavirus/2019-ncov/your-health/end-of-phe.html

863 here | https://www.hhs.gov/about/news/2023/05/09/fact-sheet-end-of-the-covid-19-public-health-emergency.html

864 November 10, 2023 | https://alexberenson.substack.com/p/yep-theyre-still-trying-to-make-fetch

865 May 9, 2023 | https://www.whitehouse.gov/briefing-room/statements-releases/2023/05/09/fact-sheet-actions-taken-by-the-biden-harris-administration-to-ensure-continued-covid-19-protections-and-surge-preparedness-after-public-health-e-mergency-transition/

866 September 3, 2023 | https://www.theepochtimes.com/article/dr-david-bell-on-100-days-vaccines-and-new-lockdown-policies-5485004

867 September 9, 2023 | https://www.foxnews.com/health/covid-booster-warning-florida-surgeon-general-advises-people-not-get-new-vaccine

868 September 10, 2023 | https://www.deseret.com/2023/9/11/23868413/fauci-covid-mask-mandate-vaccine-fda-cdc

869 September 11, 2023 | https://www.drugs.com/news/fda-takes-action-updated-mrna-covid-19-vaccines-better-protect-against-currently-circulating-114915.html

870 September 12, 2023 | https://covid19criticalcare.com/flccc-statement-on-fdas-rollout-of-covid-19-vaccine-boosters/

871 September 12, 2023 | https://covid19criticalcare.com/flccc-statement-on-fdas-rollout-of-covid-19-vaccine-boosters/

872 September 12, 2023 | https://www.whitehouse.gov/briefing-room/statements-releases/2023/09/12/statement-from-president-biden-on-fda-and-cdc-actions-on-updated-covid-19-vaccines/

873 September 12, 2023 | https://www.cdc.gov/media/releases/2023/p0912-COVID-19-Vaccine.html

874 September 13, 2023 | https://rumble.com/v3hwcgm-dr.-mcculloughs-speech-at-the-european-parliament.html

875 September 15, 2023 | https://brownstone.org/articles/the-fda-has-gone-rogue/

876 September 19, 2023 | https://www.wbur.org/news/2023/09/19/ashish-jha-covid-vaccine-flu-and-white-house

877 September 19, 2023 | https://enigmachronicle.com/2023/09/20/naomi-wolf-interviews-ed-dowd/

878 November 4, 2023 | https://www.mayoclinic.org/diseases-conditions/coronavirus/in-depth/different-types-of-covid-19-vaccines/art-20506465

879 https://ourworldindata.org/covid-hospitalizations | https://ourworldindata.org/covid-hospitalizations

880 https://ourworldindata.org/covid-deaths | https://ourworldindata.org/covid-deaths

881 Jan 17, 2023| https://youtu.be/z3EsX-mjfHE

882 18:15 | https://www.youtube.com/watch?v=Cz5DNVbkzml

883 U.N., noted | https://news.un.org/en/story/2023/05/1136367

884 counters | https://brownstone.org/articles/the-uns-new-political-declaration-on-pandemics/

885 notes | https://unherd.com/thepost/matthew-b-crawford-the-ongoing-state-of-emergency/

886 Fifth National Climate Assessment | https://nca2023.globalchange.gov/chapter/focus-on-3/

887 scientific evidence | https://brownstone.org/articles/timeline-the-proximal-origin-of-sars-cov-2/

888 5:30 | https://www.projectveritas.com/news/part-1-cnn-director-admits-network-engaged-in-propaganda-to-remove-trump

889 (ICMRA) | https://www.icmra.info/drupal/en/covid-19

890 stop complying | https://www.theepochtimes.com/article/dr-david-bell-on-100-days-vaccines-and-new-lockdown-policies-5485004

891 May 2023 interview | https://podcasts.apple.com/gb/podcast/dr-pierre-kory-deadly-conflicts-of-interest-and/id1471411980?i=1000613740456

892 August 23, 2022 | https://www.foxnews.com/media/fauci-slammed-denying-school-lockdowns-irreparably-damaged-kids-no-remorse

893 interview | https://www.youtube.com/watch?v=D6l_XYCA4es

894 Excelsior Pass | https://www.governor.ny.gov/news/governor-cuomo-announces-launch-excelsior-pass-help-fast-track-reopening-businesses-and

895 required | https://www.marketwatch.com/story/heres-how-to-get-the-excelsior-pass-now-that-nyc-will-soon-require-proof-of-vaccination-11628104401

896 mask mandates | https://www.governor.ny.gov/news/amid-ongoing-covid-19-pandemic-governor-cuomo-issues-executive-order-requiring-all-people-new

897 business closures | https://www.governor.ny.gov/news/governor-cuomo-issues-guidance-essential-services-under-new-york-state-pause-executive-order

898 loss of jobs | https://www.cbsnews.com/newyork/news/nyc-vaccine-mandate-workers-fired/

899 loss of life | https://www.cnn.com/2020/10/01/politics/andrew-cuomo-nursing-homes-fact-check/index.html

900 overreaching | https://www.nytimes.com/2020/11/18/nyregion/nyc-schools-covid.html

901 unconstitutional | https://www.cnbc.com/2020/07/16/coronavirus-new-york-gov-cuomo-cites-significant-evidence-nyc-restaurants-are-violating-social-distancing-rules-threatens-closures.html

902 outright creepy | https://www.youtube.com/watch?v=NXaP76musWM

903 But we did know | https://brownstone.org/articles/the-great-overreaction/

904 interview | https://www.politico.com/news/2023/09/10/newsom-covid-california-00114888

905 French Laundry| https://calmatters.org/politics/2020/11/newsom-dinner-california-medical-lobby-french-laundry-pandemic/

906 strict masking | https://www.newsweek.com/coronavirus-california-gavin-newsom-new-rules-gatherings-thanksgiving-1541402

907 social distancing | https://www.gov.ca.gov/2020/03/12/governor-newsom-issues-new-executive-order-further-enhancing-state-and-local-governments-ability-to-respond-to-covid-19-pandemic/

908 "everything differently" | https://www.floridapoliticalreview.com/the-fantastical-tale-of-deathsantis/

909 exception| https://www.npr.org/sections/coronavirus-live-updates/2020/04/01/825383186/florida-governor-orders-statewide-lockdown

910mask mandates| https://www.forbes.com/sites/nicholasreimann/2020/11/25/florida-gov-extends-ban-on-cities-imposing-their-own-masks-mandates-critic-calls-move-a-killing-spree/?sh=14b1accd1b27

911 September 2020 | https://www.washingtonexaminer.com/news/desantis-we-will-never-do-any-of-these-lockdowns-again

912 passed laws | https://www.cbsnews.com/miami/news/governor-ron-desantis-to-sign-legislation-banning-vaccine-mandates/

913 CDC mortality data | https://data.cdc.gov/NCHS/Provisional-COVID-19-Deaths-by-Sex-and-Age/9bhg-hcku/data_preview

914 here | https://nypost.com/2021/12/31/aoc-pictured-maskless-in-miami-beach-as-omicron-cases-soar/

915 here| https://apnews.com/article/general-news-health-government-and-politics-florida-pandemics-2c103ee2532fc1339075472ab1a925c2

916here| https://www.insider.com/celebrity-influencer-party-partying-coronavirus-covid-19-pandemic-2021-3

917 exodus of residents | https://www.snopes.com/fact-check/u-haul-california/

918 children | https://www.latimes.com/california/story/2023-02-09/more-than-150-000-california-public-school-students-may-be-missing

919 small businesses | https://www.nytimes.com/2021/02/19/business/newsom-coronavirus-california.html

920 held accountable | https://www.youtube.com/watch?v=EhvRi4LEOZc

921 significant influence | https://www.dailysignal.com/2023/04/26/4-highlights-from-teachers-union-head-randi-weingartens-testimony-on-school-lockdowns/

922 failure | https://www.vice.com/en/article/y3zxp5/teachers-are-making-their-own-gravestones-and-coffins-to-protest-going-back-to-school

923 investigate parents | https://thepostmillennial.com/national-teachers-union-boss-celebrates-the-fbi-targeting-concerned-and-angry-parents

924 testified | https://nypost.com/2023/05/09/randi-weingarten-fact-checked-by-twitter-after-claiming-she-advocated-to-reopen-schools-during-covid/

925 reported | https://nypost.com/2023/05/09/randi-weingarten-fact-checked-by-twitter-after-claiming-she-advocated-to-reopen-schools-during-covid/

926 string of texts | https://nypost.com/2023/06/02/texts-reveal-exchange-between-cdc-director-teachers-union-boss-before-school-reopening-memo/

927 tweeted | https://twitter.com/rweingarten/status/1651959322388094977?s=20

928 sidestepped | https://nypost.com/2023/04/26/biden-transition-team-asked-randi-weingartens-advice-on-school-reopening/

929 donated | https://nypost.com/2023/04/25/teacher-unions-had-bigger-hand-in-cdc-school-reopening-plan/

930 article | https://nypost.com/2023/04/26/biden-transition-team-asked-randi-weingartens-advice-on-school-reopening/

931 Levi's | https://www.amazon.com/Levis-Unbuttoned-Woke-Took-Voice/dp/1958682241

932 explains | https://brownstone.org/articles/truth-about-randi-weingarten-school-closures/

933 spokeswoman said | https://nypost.com/2023/06/02/texts-reveal-exchange-between-cdc-director-teachers-union-boss-before-school-reopening-memo/

934 amnesty | https://www.theatlantic.com/ideas/archive/2022/10/covid-response-forgiveness/671879/

935 Prof. Scott Galloway calls for "grace" Oct. 29, 2023 (41 seconds) | https://twitter.com/RobSchneider/status/1718526841911480451?s=20

936 Prof. Galloway dumps on unvaccinated Oct. 2021 (51 seconds) | https://twitter.com/TheChiefNerd/status/1718641685251379514

937 Alex Washburne | https://brownstone.org/articles/covid-amnesty-is-mercy-the-answer/

938 here | https://margaretannaalice.substack.com/p/mistakes-were-not-made-an-anthem

939 Mistakes Were NOT Made (4 min.) | https://youtu.be/ueUXNL-A3Zg

940 Westminster Declaration | https://westminsterdeclaration.org/

941 Student Declaration 2023 | https://studentdeclaration2023.org/

942 No College Mandates | https://nocollegemandates.com/

943 Alliance for Responsible Citizenship | https://www.arcforum.com/

Index

A
Adams, Jerome, 244
Adams, Jessica, 60
Adenoviral vector, 305
Adenovirus, 72, 305
Adjuvant, 305
Adulteration, 178
Adverse event (AE), 180, 182
Covid shots, 90
definition, 121
neurological, 48
remdesivir, 18
reproductive system, 170
VAERS, 123–126
V-safe, 185–190
Adverse Events of Special Interest (AESI), 128–129
Advisory Committee on Immunization Practices (ACIP), 142
Aerosol borne diseases, 229–231, 233
Agamben, Giorgio, 284
Airborne respiratory virus, 223
Alice, Margaret Anna, 300
Alliance for Responsible Citizenship (ARC), 303
Ambitious activists, 253
American Association of Port Authorities, 256
American College of Obstetrics and Gynecology, 174
American Medical Association, 232, 233
Anderson, Christine, 283
Anergy, 305
Angiotensin-converting enzyme 2 (ACE2) receptor, 131, 133, 305
Antibiotic, 305–306
Antibiotic resistant gene fragments, 178
Antibody, 306
Antibody dependent enhancement (ADE), 77–78, 83, 95, 306
Antigen, 306
Antigen escape, 79–80, 306
Antigenic imprinting, 306
Antigenic priming, 311
Antigen-specific antibodies, 106, 306
Anti-public health, 213
Antiviral drugs, 61, 306
Apocalyptic disaster, 303
Asymptomatic spread, 237–242, 254, 256
Atlas, Scott, 216–217, 223–224, 253
Attenuated vaccine, 307
Attkisson, Sharyl, 13
Autism, 141
Autoimmune diseases, 267, 307
Autoimmunity, 77

B
Bacteria, 307
Bait and switch, 176–178
Baker, Clayton J., 60
Baker, Norman, 217
Bannon, Steve, 110
B cell, 307
Behavioural Insights Team, 235
Belfer, Isaac, 55
Bell, David, 234, 284, 290, 292
Bhakdi, Sucharit, 73, 103–105, 160, 260
Bhattacharya, Jay, 6, 67
Biden, Joe, 58, 146–148, 192, 245, 257–258
Biden-Harris Administration, 278
Bifidobacteria, 134, 307
Big Pharma, 62, 66, 85, 142, 157–158, 192, 244
Bill & Melinda Gates Foundation, 5, 27
Biodefense strategy, 208
Biological medical products, 177
Biologics, 177, 307, 313
Biomedical research ethics, 227
Biondi, Yuri, 238
Biopharmaceuticals, 307
Biosecurity threat

CISA, 213–214
lockdowns, definition, 212–213
medical lockdowns, 211–212
military countermeasure, 206–210
military-industrial complex, 205–206
VAERS, 210–211
Birth rates, 169–170
Birx, Deborah, 222
Blumen, Robert, 91
B lymphocyte, 307
BNT162b2 (Pfizer) vaccine, 180
The Bodies of Others (Wolf), 109
Boreing, Jeremy, 149
Bossche, Geert Vanden, 168
Brandeis, Louis, 225
Bridle, Byram, 69, 108
Buckley, Thomas, 158

C

Califf, Robert, 59–60
Campbell, John, 81, 179–181, 183, 198
Canadian Covid Care Alliance analysis, 89
Cardiac events
CDC data, 151
excess mortality, 155, 156
fact-checkers, 152–153, 157–158
fatalities after Covid shots, 154
lipid nanoparticles (LNPs), 152
sudden illness/medical event, 152, 153
Cardiovascular adverse conditions, 118
Cause Unknown (Dowd), 156
Cavazzoni, Patricia, 19
CDC policy, 164
Censorship of science, 182–184
Centers for Disease Control and Prevention, 191
Chandler, Robert W., 169–170
Character assassination doctors, 40
Chester, Charlie, 285
Chickenpox, 86
Child and adolescent immunization schedule, 140
Childers, Jeff, 154
Childhood Vaccine Injury Compensation Program, 143
Chloroquine, 9
Chromosomal DNA, 179
Cleveland Clinic study, 96
Climate change, 226, 285–286
Clinical trials, Covid-19 shots
absolute risk, 89–91
adverse events, 88
hypothesis, 87
"95% effective," 88, 93
placebo group, Pfizer, 91
profits, 92
relative risk, 89
Clot formation, brain veins, 104, 160
Clotting, 246
Coalition for Epidemic Preparedness Innovations (CEPI), 5
Cobb, Clayton, 246, 248
Cochrane review, 235, 236
Cohen, Mandy, 230
Cole, Ryan, 113, 114, 142
Collins, Francis, 67
Conspiracy theories, 4, 200, 202
Coronavirus Aid, Relief and Economic Security (CARES) Act, 11, 19
Coronaviruses, 308
See also Covid-19 infection
Corruption
Big Pharma, Government, and Media, 143
medical journals, 27–28
Countermeasures Injury Compensation Program (CICP), 143, 144
Counterterrorism, 206
The Courage to Face Covid-19 (Mccullough), 268
Covid-19 infection, 33–34, 308
children, 132

chronic illnesses, 34
Covid treatment protocol, 37–38
cytokine storms, 34
death counts, 243–245
diabetes and obesity, 34–35
Hippocratic Oath, 38
Long Covid, 158–159
SARS-CoV-2 virus, 34
standard protocol, 36
vaccination, 167–168
Wuhan and Delta variants, 35
Covid-19 shots
ACE2 receptor, 131
bifidobacteria, 134
for children, 135–137, 142
early warning, 103–110
European countries, 132
immune systems, 137–138
infants and toddlers, 132–134
Covid authoritarians, 195–198
Covid boosters, protest of, 275–277
Covid dissidents, 1
Covid hospitalizations, 117–118
Crawford, Matthew B., 284, 286
Creutzfeldt-Jakob disease (CJD), 107, 308
Cuomo, Andrew, 243, 293, 294
Current Good Manufacturing Practices (CGMP), 183, 184
Cybersecurity and Infrastructure Security Agency (CISA), 213–214
Cycle threshold (Ct), 239
Cytokine, 308
Cytokine storms, 34, 114, 308

D

Daily Wire, 148–149
Dalgleish, Angus, 77–78, 95, 113, 198, 249
Daszak, Peter, 214
Davison, Scott, 156
The Deadly Rise of Anti-Science: A Scientist's Warning (Hotez), 269, 272
The Death of Science (Dagleish), 198, 249
Demasi, Maryanne, 59–60
Denison, Leigh, 128
Deoxyribonucleic acid (DNA), 308
Department of Defense (DoD), 209–210
Department of Homeland Security (DHS), 199, 207, 210
DeSantis, Ron, 294
Digital health partnership, 8
Digital Vaccine Passport system, 8
Dingwall, Robert, 263
Dirty placebo, 14, 308
D-Med, military medical record system, 127–128
DNA contaminations, 177, 178
DNA vaccine, 308–309
DNA virus, 309
Dodsworth, Laura, 216, 219, 241
Domestic terrorism, 199–204
Doughty, Terry, 193, 201, 203–204
Dowd, Edward, 124, 155, 156, 158, 159, 281

E

Ecological imprinting, 95
Effector cell, 309
Eisenhower, Dwight D., 205–206
Electronic medical record (EMR), 247
Emergency Support Function Leadership Group (ESFLG), 207
Emergency Use Authorized (EUA), 58, 146, 188, 277, 309
Endemic, 309
Endothelium, 309
Epidemic, 1, 218, 309
Epistemology, 309
Epitope, 309
Eron, Joseph J., 72
European Journal of Clinical Investigation, 182

European Medical Association (EMA), 105
Event 201, 6
Evidence-based studies, 236
Excess deaths, 309
Excess mortality, 155, 156

F

Face masks, 227–237
Fact-checkers, 152–153, 157–158
Farber, Celia, 66
Farley, John, 91, 92
Fauci, Anthony, 8, 12, 14, 15, 23, 59, 66, 83, 112, 140–141, 214, 228, 229, 236, 257–258, 262, 266, 268, 269
Fear of a Microbial Planet (Templeton), 260
Federal Emergency Management Agency (FEMA), 207
Federal government
  public transportation, 191
  social media, 192–193
Feigl-Ding, Eric, 251
Ferguson, Neil, 218, 263
Financial incentives, remdesivir, 19
Focused protection, 67
Food and Drug Administration (FDA), 13
Frameshifting, 266
Freedom of Information Act of 1967, 86, 174
Freedom of speech, 202–203, 286
Front Line Covid Critical Care (FLCCC), 44

G

Gain-of-function, 7, 309
Galloway, Scott, 297–298
Garland, Merrick, 295
Gates, Bill, 5, 8
Gene therapy, 179, 310
Genetically modified organisms (GMOs), 249
Genome, 310
Genuine scientists, 66
Gershman, Michelle, 171
Glycans, 133, 310
Goddard, Paul R., 249
Gortler, David, 57, 59, 72, 85, 152, 190
Gottlieb, Scott, 262
Government interference, doctor/patient relationship, 21
Great Barrington Declaration (GBD), 67
Great Reset, 4
Gruber, Marion, 275
Guetzkow, Joshua, 177
Guillain-Barré syndrome, 246
Gupta, Sunetra, 242
Gut microbiome, 134

H

Halpern, David, 235
Hancock, Matt, 228
Hansen, Peter Riis, 180
Harries, Dame Jenny, 226
Harrington, Thomas S., 250
Harris, Sam, 225
Hazan, Sabine, 134
HCQ. *See* Hydroxychloroquine (HCQ)
Health and Human Services (HHS), 65, 143, 158, 206, 207, 214
Health-care associated infection, 313
Heart attack, 246
Heart/neurological issues, 246
Heneghan, Carl, 239
Heterogeneity, 310
Hippocratic Oath, 38, 138
Histopathology, 310
HIV vaccines, 83
Hochul, Kathy, 191
Hoeg, Tracy Beth, 276
Homologous/homology, 310
Honold, Ashley Cheung, 54
Hopkins, Johns, 243, 257

Horst, Manfred, 220
Hospital-acquired infection, 313
Hospital charting system, 246–247
Hotez, Peter J., 269–270, 272
Howard, Jeremy, 252, 253
Human immune system, 6
Humanitarian catastrophes, 291–292
Humanity, 300
Hyde-Smith, Cindy, 144
Hydroxychloroquine (HCQ), 8, 9
  benefits, 31–32
  corruption, medical journals, 27–28
  doctors and off-label drugs, 39–40
  EUA approval, 30
  FDA's withdrawal of EUA, 31
  and ivermectin, 26
  NIH study, 23–24
  official narrative, 26
  poisonous substance, France, 25–26
  Raoult, 25
  Recovery and Solidarity trials, 27
  side effects, 24
  toxic doses, 28–30
  war on ivermectin and, 47–55
  Zelenko, 24
Hygiene theater, 255

I

Iatrogenic, 310
IgG4 antibodies, 97–98, 310
Illusion of Consensus, 277–281
Immune, 310
Immune escape, 79–80, 311
Immune imprinting, 80, 311
Immune suppression, 179
Immune systems, 137–138
Immunization, 311
Immunization Information Systems (IIS), 248
Immunogen, 311
Immunogenic, 311
Immunological escape, 311
Immunological imprinting, 306, 313
Imperfect knowledge, 299
Inactivated vaccine, 311
Inamori Center, 268
Infectious Diseases Society of America (IDSA), 269
Inflammatory cardiomyopathy. *See* Myocarditis
Information Environment, 286–287
Informed consent, 95, 179–181
Innate immune system, 167
Inoculate, 311
Insanity, 283
Intelligence community, 214, 224–225
International Coalition of Medicines Regulatory Authorities (ICMRA), 289
Intravenous remdesivir, 13
In vitro, 312
In vivo, 312
Ioannidis, John, 34, 216, 276
Israel Ministry of Health (IMOH), 119
Ivermectin, 43–45
  See also War on ivermectin and HCQ

J

Jefferson, Tom, 236
Jensen, Scott, 244–245
Johnson, Boris, 228
Johnson & Johnson Covid-19 vaccine, 312

K

Kai, De, 251–252
Kaler, Eric W., 268
Kalil, Andre, 18
Kampe, Olle, 266
Karikó, Katalin, 265–268
Kennedy, Robert F., Jr., 12, 14, 32, 65, 66, 140, 141, 269, 290
Kheriaty, Aaron, 211–212

Kidney disease, 22
Kisin, Konstantin, 270–271
Koops, Roger W., 223
Kory, Pierre, 44, 45, 47, 51, 143, 210–211, 247–248, 290
Krause, Philip, 276

L
Lab-leak theory, 202, 214
Ladapo, Joseph, 86
Lateral flow antigen tests, 240–241
Latypova, Sasha, 175, 176, 209
Lawrie, Tess, 302
Lee, Mike, 144
Left-libertarianism, 290–292
Lerman, Debbie, 206, 208, 213
Lewis, C. S., 226
Lipid nanoparticles (LNPs), 71–72, 81, 82, 114, 152, 170, 267, 312
Lockdowns
  definition, 212–213
  medical brand in 2020, 211–212
  propaganda, 221–227
Long Covid, 158–159
Lymphocyte, 312
Lyons-Weiler, James, 77, 113

M
Macrae, Gail, 17, 18, 245–247
Macrophage, 312
Magness, Phillip W., 219
Mainstream media, 50–51, 138
Makary, Marty, 276
Malone, Robert, 71, 72, 79–80, 114, 132, 135–136, 177–178
Manniche, Vibeke, 179–183
Manufacturing costs, remdesivir, 16
Marik, Paul, 31, 175, 266
Marks, David, 263
MASKS4ALL, 252
Mask wearing, 262
Mass media, 291
Mass spectrometry, 266
Mass Vaccine Casualty Event of 2021, 161
McCarthy, Hugh, 234
McCormick, Rich, 21
McCullough, Peter, 22, 28, 35, 36, 40–41, 43–44, 47, 52, 59, 60, 76, 105–106, 140, 152, 157, 159, 267
McDonald, Mark, 235
McKernan, Kevin, 177–179
Memory B and T cells, 312
Mendenhall, Warner, 124
Merck, 51
Mercola, Joseph, 71
Messenger RNA (mRNA), 312
Microbiome, 312
Miller, Ian, 195
Miller, Mark Crispin, 154
Ministry of Truth, 203–204
Miscarriages, 169
Missouri vs. Biden, 192–193
Moderna, 7
Molnupiravir, 51
Monoclonal antibodies, 57–58
Montagnier, Luc, 107–108, 168
Moon, Renata, 176
mRNA shots/vaccines, 71–72, 98, 112, 132, 176, 178, 265–270
  vs. adenoviral vector shots, 72
  antibody dependent enhancement (ADE), 77–78
  autoimmunity, 77
  definition, 313
  European countries, 132
  immune escape, 79–80
  infants and toddlers, 132–134
  spike protein, 75–76, 84
Mulli, Kary, 238
Murthy, Venkatesh, 117
Myocarditis, 117, 121, 313
  Covid shots, 130

D-Med, military medical record system, 127–128
FDA Emergency Use Authorization, 128–130
and pericarditis, 129
VAERS data, 122–126

N
N95 respirators, 230, 232, 233
Nanotechnology, 71–72, 313
Nass, Meryl, 27–30, 53
National Center for Biotechnology Information, 243
National Childhood Vaccine Injury Act (NCVIA) of 1986, 139–140
National Institutes of Health (NIH), 11
Covid-19 Treatment Guidelines panel, 13
employees profit, drug patents, 12–13
remdesivir, 18, 19
research and studies, 65–68
vaccine-induced spike protein, 76
National Security Council (NSC), 206, 207, 208, 213, 224
National Vaccine Injury Compensation Program (NVICP), 139
Naugle, John F., 213
Newsom, Gavin, 294, 295
New-technology vaccines, 70
New York Post (Makary and Hoeg), 276
Noem, Kristi, 257
Nosocomial spread, 313
Novavax, 73
Nudge Unit, 235
Nuremberg Code, 227

O
Occupational and Safety Health Administration (OSHA), 230
Off-label prescribing, 31, 51
Off-label use, 313
Omicron, 80, 99, 111–112, 133
Operation Warp Speed (OWS), 209
Original antigenic sin, 313
Orwell, George, 203–204
Osterholm, Michael, 251, 252

P
PANDA research, 242
Pandemic
definition, 314
wargames, 5–6
Pandemic response
climate change, 285–286
goal, 283
state of exception, 284
Parks, Christina, 140
Pathogenic, 314
Pathogenic priming, 314
See also Antibody dependent enhancement (ADE)
Paxlovid
antivirals, 61
Biden's speech, 58
Big Pharma, 62
Califf, promoting, 59–60
Emergency Use Authorization, 57
ineffective and expensive, 59
monoclonal antibodies, 57–58
Pericarditis, 314
Peterson, Jordan, 253
Pfizer, 59–61
Pharmacokinetics, 314
Pharmacovigilance, 314
Phony placebo, 14
A Plague Upon Our House (Atlas), 216
Plasmid, 315
Plasmid DNA fragments, 177, 178
Policy decisions, 249
Polyethylene glycol (PEG), 314
Polymerase chain reaction (PCR), 238–239
Pornography, 202

Post-Covid vaccination
  maternal and fetal complications, 171–172
  reproductive system adverse events, 170
Poverty and hunger, 6
Powered air-purifying respirator, 232
Precautionary principle, 236
Preterm births, 169
Primary immune response, 315
Prion disorders., 308
Propaganda techniques
  asymptomatic spread, 237–242
  censorship, 291
  Covid-19 death counts, 243–245
  experts, 250–253
  face masks, 227–237
  foreign enemies, 216
  hospitals, 242–243
  lockdowns and 15 days, 221–227
  messaging, master class, 260–263
  meta-analysis protocols, 216
  policy decisions, 218–221
  prizes and awards, 265–274
  sanitizing surfaces, 254–256
  science, 248–250
  scientific and medical data, 216–217
  Spanish flu, 259–260
  super-spreader events, 257–259
  unvaccination, 245 248
Prophylactic, 315
Pseudouridine, 265, 266
Public Health Emergency (PHE), 277–278
Public health professionals, 223, 257
Public health safety, 261
Public Health Service Act of 1944, 191
Public opinion, 249
Public policy, 253
Public Readiness and Emergency Preparedness Act (PREP Act) of 2005, 143, 144
Public transportation, 191
Public trust, 294

Q
Quality control, mRNA shots, 175
Quarter 3/2021 vaccine deaths, 165–166

R
Randomized controlled trials (RCTs), 235
Raoult, Didier, 25–26, 31
Reactogenicity, 186, 315
The Real Anthony Fauci (Kennedy), 12, 269
Recombinant DNA, 315
Redfield, Robert, 228–229
Remdesivir, 8
  adverse events, 18
  causes and side effects, 16
  class-action lawsuit claims, 22
  costs, manufacture/treatment, 16
  Covid-19 infection, 20
  dirty placebo, 14
  Ebola trial, 11
  emergency use authorization, 15–16
  FDA's approval, 15
  financial incentives, 19
  government interference, doctor/patient relationship, 21
  for infants and children, 19–20
  kidney disease, 22
  nickname (run-death-is-near), 17
  NIH employees profit, drug patents, 12
  rise of, 13–14
  standard of care in U.S., 15
  WHO's Solidarity trial, 18
Renz, Thomas, 19, 128
Reproductive system adverse events, 170
Repurposed drug, 315
Repurposed use. *See* Off-label use
Respiratory illnesses, 252
Respiratory syncytial virus (RSV), 190
Respiratory virus, vaccine development, 111
Retroviruses, 315

Revenues, drug companies, 61–62
Ribonucleic acid (RNA), 315
Risch, Harvey, 53, 229
Risk communication consultant, 211
RNA vaccine, 315
RNA viruses, 315
Rose, Jessica, 28, 124, 178, 179

S

Sandberg, Rickard, 265
Sanders, Bernie, 62
Sandman, Peter M., 211, 223
SARS-CoV-2 RNA/antigen, 246–247
Sayers, Freddie, 284
Scatterplot charts, Covid deaths, 161–164
Schmeling, Max, 180
Schwab, Klaus, 4, 8
Scientific consensus, 249
Secondary immune response, 312, 316
Self-justification, 297–300
Senger, Michael, 224, 253
Serious Adverse Event (SAE), 121–122
Seropositivity, 316
Seroprevalence, 142, 316
Severe Acute Respiratory Syndrome Coronavirus 2 (SARS-CoV-2), 34, 76–78, 101, 106, 208, 214, 229–230, 316
Sewage disposal systems, 259
Sey, Jennifer, 296
Shahar, Eyal, 154, 182
Sharav, Vera, 15
Shir-Raz, Yaffa, 119–120
Signal masks, 235
Sikora, Karol, 249
Social distancing, 224
Social media platforms, 50, 192–193, 199
Society of Actuaries Group Life Covid-19 Mortality Survey Report, 159–160
Sowell, Thomas, 260–261
Spanish flu, 259–260
Spieker, Fabian, 161, 163–165, 167, 168
Spiked placebo, 14
Staley, Peter, 228
A State of Fear (Dodsworth), 219
Statistical tools, correlation and regression, 162
Stieber, Zachary, 128
Stillbirths, 169
Stroke, 246
Subunit vaccine, 316
Sudden Adult Death Syndrome (SADS), 154
Super-spreader events, 257–259
Supply chain, 256
Surgical masks. *See* Face masks
Surgisphere, 28
Suspected adverse effects (SAEs), 179

T

T cell, 316
Team reality *vs.* narrative, 277–278
Templeton, Steve, 251, 260
Test-tube experiments, 312
Therapeutic, 316
Thermocycler machine, 239
Thorp, James, 171
Threat environment, 200
Thrombosis, 316
Thrombosis with thrombocytopenia syndrome (TTS), 73
T lymphocyte. *See* T cell
Top-down control, 7–8
Totalitarianism, 294–295
Traditional vaccines, 69–70
Transfection, 316
Triple threat, 195
Trump, Donald, 216
Tucker, Jeffrey, 272–274, 299
Tufekci, Zeynep, 253
Turbo cancer, 113, 316

U

Underlying Cause of Death (UCOD), 163
United States Government (USG), 209
Universal face masking, 252
Unvaccinated Americans, 146–148
Uridine, 265
Urso, Richard, 38, 40, 142
U.S. Fifth National Climate Assessment (NCA 5), 285
U.S. Food and Drug Administration, 279

V

Vaccine Adverse Event Reporting System (VAERS), 121, 170, 210–211
  adverse event reports, 123–126
  myocarditis/pericarditis reports (2021), 122
  preliminary list of AESI, 128–129
  safety signals, 130
  vaccine related deaths, 122
  Vaccines/vaccination
  adverse events, 119–120
  Biden administration, 146–148
  CDC's Covid Data Tracker, 101
  chickenpox, 86
  childhood schedule, 139–140
  Covid-19 variants, CDC chart, 100
  definition, 71, 247, 316–317
  development, 85–86
  gene therapy, 70
  injury compensations, 144
  mRNA shots, 71–72
  mRNA *vs.* adenoviral vector shots, 72
  new-technology, 70
  Novavax, 73
  rollouts, 172–174
  susceptibility, Covid-19 infection, 167–168
  traditional, 69–70
  updated, 99
  XBB.1.5, 99
Vaccinology, 267
VAERS. *See* Vaccine Adverse Event Reporting System (VAERS)
Vanden Bossche, Geert, 106–107, 132, 133
Vanity Fair, 252
Van Kerkhove, Maria, 237
Vector, 317
Viral epidemiology, 252
Viral escape, 79–80, 317
Virus, 317
Virus management, 213
V-safe
  CDC enrollment, 188–190
  defined, 185
  health impacts, 186
  legal action, 188
  mRNA shots, 186–188

W

Walensky, Rochelle, 111, 167, 257–258, 262, 296
War on ivermectin and HCQ
  Covid treatments, 48
  Emergency Use Authorization, 47–48
  emotional abuse, FDA, 55
  false rumors of hospitals, 50
  FDA's communication, 53–54
  Honold's statement, 54
  mainstream media, 50–51
  McCullough's statement, 52
  Merck, 51
  Nobel Prize-winning medicine, 49
  social media, 50
  suppression, 51
Washburne, Alex, 298
Water and sanitation systems, 259
Weingarten, Randi, 295–297
Weissman, Drew, 265–268
Wen, Leana, 293, 294
White House Coronavirus Task Force,

207, 216
White *vs.* black hats, 272–274
Whole-of-government approach, 287
Wiseman, David, 70, 176
Wolf, Naomi, 109–110
World Economic Forum (WEF), 4, 259
World Health Organization (WHO), 238, 259
Solidarity trial, 18
top-down control, 7–8
Wrong protein, 114
Wrong shot, 113–114
Wrong virus, 114

X
XBB.1.5 vaccines, 99

Y
Yaccarino, Linda, 202–203

Z
Zelenko, Vladimir, 24, 31
Zoonotic disease, 285
Zuckerberg, Mark, 8, 11, 83

www.ingramcontent.com/pod-product-compliance
Lightning Source LLC
Chambersburg PA
CBHW022101210326
41518CB00039B/349

* 9 7 8 1 6 3 0 6 9 2 6 2 9 *